Pork Quality:
Genetic and Metabolic Factors

Pork Quality:
Genetic and Metabolic Factors

Papers presented at an OECD workshop in Helsinki, Finland, June 8-10, 1992

Edited by

E. Puolanne and D.I. Demeyer

with

M. Ruusunen and S. Ellis

C·A·B International

C·A·B International
Wallingford
Oxon OX10 8DE
UK

Tel: Wallingford (0491) 832111
Telex: 847964 (COMAGG G)
Telecom Gold / Dialcom: 84: CAU001
Fax: (0491) 833508

A catalogue record for this book is available from the British library.

ISBN 0 85198 836 9

Printed and bound in the UK by Redwood Books, Trowbridge

Contents

Introduction to the Project and the Workshop

The Organization for Economic Co-operation and Development (OECD), part of the system of Western international institutions developed after World War II, is a multilateral and intergovernmental organization that today comprises 24 democratic nations with advanced market economies. It is the main forum for monitoring economic trends in its member countries of North America, western Europe and the Pacific.

It is in the OECD that government representatives meet to compare and co-ordinate their policies, both domestic and international. The OECD is the largest source of comparative data on the industrial economies in the world. It produces a wide range of publications: economic surveys of each of the member countries, statistics, analyses, and policy recommendations in a vast variety of subjects - trade, banking and financial markets, employment, social policies, the environment, agriculture, energy, industry, development aid, science and technology, R and D, taxation, education, transport and much more.

Within the activities of the Directorate for Food, Agriculture and Fisheries, a new kind of co-operation in agricultural research was proposed for support to the OECD by the USA in 1975! The objective was to advance agricultural technology by identifying and filling gaps and establishing a program of co-operative research among OECD member countries.

From the beginning, a long-term fundamental research approach was taken. The ambitious program worked out by experts was divided into four themes: 1) Photosynthesis; 2) Nitrogen Fixation; 3) Mycotoxins; and 4) Cellulose Utilization (towards an improved utilization of lignocellulose materials).

In 1990, a new co-operative Research Project was started on "Management of Biological Resources", covering three themes: 1) Modification of soil/plant/microbial interactions to reduce inputs in farming systems; 2) Use of biotechnology to improve animal health status and meat quality in domestic animals; and 3) Assessment of benefits and risks of introducing new organisms in agricultural practice.

Two types of activities are promoted by the Project :

- Fellowships to encourage exchange between scientists interested in the research themes: the OECD pays transportation costs and subsistence to post-doctoral level research scientists wishing to work in a laboratory in another participating country. These fellowships may be from 2 to 26, in exceptional cases, up to 52 weeks.
- organization of workshops and publication of proceedings : three or four workshops are organized, or jointly organized by the OECD each year in different countries, in order to study in depth an aspect of each one of the themes.

Currently, 22 countries are participating members of the Project: Australia, Austria, Belgium, Denmark, Finland, France, Germany, Greece, Ireland, Italy, Japan, Netherlands, Norway, Portugal, Spain, Sweden, Switzerland, Turkey, United Kingdom, United States, and Yugoslavia.

Only scientists from these countries are eligible for Project support. The Project theme on "Animal Biotechnology" recognizes that production and consumption of meat from farm animals within OECD countries has considerable economical, as well as nutritional, ecological, and social impact. Imminent possibilities of the newer biotechnologies emphasize even more the need to evaluate "quality" of the end product "meat" in a fundamental, integrated manner, involving *inter alia* animal health and welfare, eating quality of meat, and the problem of chemical residues. Fundamental research on several points of focus can be considered under this theme:

- Transgenic animals, cloning and animal uniformity.
- Embryo storage (freezing) and transplantation, germ plasm preservation.
- Identification of key enzyme complexes involved in resistance to certain diseases.
- Relation of animal physiology to animal welfare and quality of animal products.
- Nutritional engineering of animal products.
- Food safety, reduced chemical residues (antibiotics, hormones, medicines), reduced disease and improved hygiene (salmonella).

Since 1990, 78 fellowships were awarded under this theme and two workshops took place:

- 7-9 November 1990: Melle (Belgium) - Animal Biotechnology and the Quality of Meat Production.
- 9-11 April 1991: Nottingham (UK) - The Role of Muscle Proteinases in Muscle Growth and Meat Quality.

The Expert Group convened by the OECD in Bristol (UK) in March 1990 to define the main lines of work under the theme "Animal Biotechnology" concluded that attention should be focussed on Animal Biotechnology in its widest sense, as related to meat production and meat quality, in view of the considerable economical, ecological and ethical aspects involved. In recent years, considerable progress has been made into the molecular genetics of stress susceptibility in pigs. Application of that knowledge in practical breeding and in optimization of (sensory) meat quality is of the utmost importance both for the fresh and processed meat industries. The modern pig is different from the "original" wild pig from which it has been bred. There are large differences in growth rate, muscle/bone ratio, the oxidative capacity of muscles etc. Selective breeding, which has produced these differences, may have also influenced the overall physiology of the modern pig. This stresses the importance of studying the quantitative physiology and metabolism of the modern pig. Research must be focused on the physiology of the living animal, as well as on porcine muscle physiology, in order to relate these research areas to meat quality and animal welfare.

The present workshop aims at an integration of fundamental knowledge in genetics, breeding and physiology of pigs with progress in muscle physiology and bio-physics, as related to measurement and optimization of (sensory) meat quality.

C. Raichon
D. Demeyer
E. Puolanne

Acknowledgements

The workshop was organized by Dr. Eero Puolanne and Dr. Daniel Demeyer (Belgium). The Workshop secretary was Ms. Marita Ruusunen, Lic.Sc., who was in charge of all the technical arrangements of the Workshop. The staff of the Department of Food Technology, Meat Section, was of valuable help in executing the Workshop. Mr. Steven Ellis, B.Sc., (Connecticut, USA), and in the final stage also Mr. Petri Papinaho, M.Sc. and Mr. Jukka Partanen, M.Sc., did the editing and typesetting of the Proceedings. The chairpersons and the secretaries of the meetings and the authors of the Proceedings as well as the participants were responsible for the scientific quality of the meeting. The Organizers would like to thank all those who have contributed to this Workshop.

The Workshop was financially supported by the OECD and the Ministry of Agriculture and Forestry in Finland, and Scandinavian Association of Agricultural Scientists (NJF).

An Overview of Production, Slaughter and Processing Factors that Affect Pork Quality - General Review

P.V. Tarrant

The National Food Centre, Teagasc, Dunsinea, Castleknock,
Dublin 15

Introduction

In the production of pig meat (and other foods of animal origin) the general aim is to provide a product that is safe, of good quality, i.e. meets the customer's needs and does so consistently, is good value for money, and is grown and slaughtered under acceptably humane conditions.

The purpose of pig meat research is to enable the industry to better meet these market requirements. The research activities may be divided into two general areas, quality and development. The former is concerned with improving the safety and quality of products, while the latter concerns the development of new processes and new products.

The aim of this introductory paper is to provide an overview of the main research areas, including reference to commercial applications, and to introduce topics that are considered in more detail elsewhere in the proceedings. Choice of material had to be selective in view of the wide scope of this seminar, and many important aspects of pig meat quality have necessarily been excluded from this introduction.

The subject matter has been divided into three general areas as follows: first, production factors affecting meat quality; second, transport and handling aspects; and third, the effects of processing on meat quality.

Production factors affecting meat quality

Carcass leanness and meat quality

Current trends in healthy eating have created a demand for leaner meat. The industry has responded by using leaner breeds, diets that reduce fat deposition, techniques for monitoring fatness, and growth promoters. For example, over the past 15 years, the average British pig carcass (weight 63kg) has undergone a 33% reduction in P_2 (the thickness of the backfat measured at a point 65mm from the dorsal mid-line at the level of the last rib), a 35% reduction in separable fat and a 27% reduction in intramuscular, or "marbling" fat (Wood and Warriss, 1991). Marbling fat is located in the perimysial connective tissue which surrounds bundles (or fasciculi) of muscle fibers. As all the indices of fat content have declined, there is concern that the eating quality of pork has also declined, and in particular, that the meat from the leaner carcasses lacks tenderness and juiciness (Wood *et al.*, 1986; Kempster *et al.*, 1986; Bejerholm and Barton-Gade 1986) probably as a result of cold-toughening or the absence of marbling fat.

Marbling

Attention has focused on the content of marbling as the aspect of fat most closely related to eating quality. The correlations between the concentration of marbling and taste panel scores are generally low, but positive, indicating that both tenderness and juiciness are improved by marbling. The optimum level of intramuscular fat predicted by different authors varies from 2.5% upwards (Smith and Carpenter, 1976; Bejerholm, 1984; Cameron *et al.*, 1990). Nowadays, in most European countries, carcasses contain less than 2% intramuscular fat (Wood and Warriss, 1991). For example, O'Mahony (1991) found that intramuscular fat content of pork chops at retail outlets in Dublin averaged only 1.4%, and ranged from 0.6% to 1.9%. When buying pork chops, leanness was found to be the first preference among Irish consumers, and they preferred the look of chops with lower levels of intramuscular fat. Furthermore, they found no significant difference in the eating quality of chops with different amounts of marbling fat. This finding contrasts with much of the earlier work done in the US, UK, and Denmark. However, the amount of marbling in the earlier studies was far in excess of that presently found in pork from Irish or English herds (O'Mahony, 1991; Wood *et al.*, 1991).

The recent results suggest that the absolute levels of marbling may now be below the threshold at which consumers can detect differences in eating quality and that the concentration of intramuscular fat may have to be in excess of 2% for consumers to detect benefits in eating quality. On the other hand, the presence of obvious marbling is likely to be a deterrent, in view of the strong preference for leanness. More research is needed to see if it is possible to reconcile these apparently conflicting demands.

In the meantime, commercial breeding companies are seeking techniques for measuring marbling fat in live animals and in carcasses. Although marbling fat increases with total carcass fat, there is considerable breed variation, and the dark-haired breeds have generally high values in relation to P_2 backfat thickness, and among these, the Duroc appears to have the highest concentration of marbling fat (Wood *et al.*, 1991).

Tenderness

Arguably the most important component of eating quality, there is good evidence that production of lean pigs is adversely affecting tenderness through additional mechanisms over and above the effect on marbling. These include the increased risk of cold toughening in leaner carcasses with less fat cover - this aspect is considered later in the section on chilling - and also the increased toughness reported in pigs with slow growth rates, and in pigs given β-adrenergic agonists to increase leanness. These effects may be explained by changes in the activities of muscle proteolytic enzymes responsible for the increased tenderness of conditioned (aged) meat. The main enzymes concerned are thought to be the calcium-dependent neutral proteases and the cathepsins (notably B, D, probably also L and H). When growth rate is slowed through underfeeding, there is a reduced rate of protein breakdown in muscle, which, in the post-mortem period, may lead to reduced tenderization. Also, there may be a higher proportion of older, i.e. more cross-linked, connective tissue proteins in muscle that has turned over less rapidly, and this would also reduce tenderness (Wood and Warriss, 1991).

Growth promoters

Despite the potential for better economic returns, and a reduction in unwanted animal fats, the use of the so-called lipomodulators - the β-agonists and porcine somatotropin - in meat production is severely limited or blocked in many countries by opposition from consumers and regulatory agencies on the basis of safety or welfare considerations. The 1986 ban on hormone implants in the European Community has contributed to producer interest in β-agonists, although these feed additives are not licenced for use except for veterinary purposes. Meisinger (1989) estimated that lipomodulators could increase net profits to individual pork producers by as much as US$8 to US$15 per pig slaughtered.

The repartitioning effect of the β-agonists is somewhat lower in pigs than in ruminants, but they improve the efficiency of production of carcass lean, and this is commercially important, given the relatively high cost of feed and housing for pigs. Likewise, the toughening effect of the β-agonists on meat is greater in ruminants than in pigs.

The suggested mechanism is a decrease in proteolytic activity both in the live animal and in the meat post-mortem. Perhaps the greater catheptic activity in pig muscle (Etherington *et al.*, 1987) accounts for the fact that pig meat is affected less by β-agonists. There is also some evidence that pigs treated with β-agonists had less ether extractable fat (marbling) and less cholesterol in *M. longissimus dorsi* (Hedrick *et al.*, 1991).

A critical factor in the usefulness of β-agonists is likely to be the degree to which their effects are retained after withdrawal, since a recommended (if not statutory) period is likely. The growth rate declines, perhaps drastically, immediately following withdrawal. The carcass gains are not lost so rapidly but the advantages are lost after a prolonged period of withdrawal. Only when the withdrawal period necessary to eliminate residues has been determined can the true potential impact of β-agonists on the efficiency of carcass lean production be determined (Moloney *et al.*, 1991).

Recombinant porcine somatotropin (PST), administered intramuscularly, appears to have a more dramatic effect on growth rate and carcass composition than the β-agonists. Exogenous somatotropin increases growth rate (5-15%), efficiency (10-35%), and lean (5-20%), and decreases intake (5-20%), and fat (10-40%) in pigs (Enright, 1991). Loss of marbling fat and tenderness may be less significant with somatotropin than with β-agonists.

Stress susceptibility

Over several decades, the improvement in the physical performance of the growing animals, including growth rate, feed conversion efficiency and, in particular, selection for leanness or muscularity, has lead to an increased incidence of a porcine stress syndrome (PSS).

An early breakthrough in the search for the basic biochemical lesion in PSS came with the discovery in 1966 that PSS was very similar to the inherited muscle disorder malignant hyperthermia (MH) (Hall *et al.*, 1966). There is now ample evidence that the two syndromes are similar, and may in fact be identical, and are the expression of the same myopathy but are triggered by different stimuli. For example, carazolol, a β-adrenergic receptor antagonist, could not prevent halothane-induced lactacidaemia (Gregory and Wilkins, 1984), but reduced the generation of lactate in pigs exposed to normal management stress, and thus prevented PSE muscle. It appears therefore that the activation of β-adrenergic receptors is not important in halothane-induced stress, whereas β-adrenergic receptor linked mechanisms are associated with management-induced stress.

MH is triggered by anesthetics such as halothane and by depolarizing muscle relaxants such as succinylcholine. The classic presentation is of a profoundly accelerated muscle metabolism with intense production of heat, carbon dioxide and lactic acid, with contraction of the skeletal muscles and tachycardia. Once initiated, a futile metabolic cycle is established and a fulminant syndrome evolves in which the body temperature may exceed 43°C (Gronert, 1986). The hyperthermia seems to depend on muscle contraction.

There is evidence that the intracellular free calcium in intact, untriggered, MH muscle fibers is elevated above the normal resting level (Lopez *et al.*, 1985; Lopez, 1986). The syndrome appears to follow from a sudden and sustained rise in the concentration of calcium in the sarcoplasm, and can be reversed by drugs that lower myoplasmic calcium, notably dantrolene sodium.

The defect in MH appears to be hypersensitive gating of the calcium-release channel of the skeletal muscle sarcoplasmic reticulum (SR); channel opening is facilitated and closing is inhibited (Iaizzo *et al.*, 1988; Nelson, 1983; Ohnishi *et al.*, 1983; Mickelson *et al.*, 1988).

The calcium channel of the SR is a protein called the ryanodine receptor (because it binds the plant alkaloid ryanodine with very high affinity). It is probable that a polymer of the ryanodine receptor protein constitutes the junctions, known as "feet", between the transverse tubule and the terminal vesicle of the SR. Ryanodine causes contracture of skeletal muscle, both *in vivo* and *in vitro*, and it appears to act by fixing the channel in the open configuration (Heffron, 1991). This results in an uncontrolled elevation of cytoplasmic calcium. Mickelson *et al.* (1988) presented evidence that MH is due to a mutation in the ryanodine receptor protein. This causes it to bind ryanodine with greater affinity in the presence of micromolar concentrations of calcium than does the receptor in normal pig muscle. In pigs that are heterozygous for the halothane gene, and therefore not susceptible to MH, the properties of the calcium release channel are intermediate to those of the MH-susceptible and those of the normal pigs, suggesting that both normal and abnormal genes for this protein are expressed in the heterozygous pigs (Mickelson *et al.*, 1989).

Fujii *et al.* (1991) identified a single point mutation in the pig muscle ryanodine receptor gene that was found to be correlated with MH in five major breeds of lean, heavily muscled swine (Duroc, Poland China, Yorkshire, Pietrain and Landrace). A single nucleotide polymorphism led to an alteration in amino-acid sequence from an arginine at position 615 in normal pigs, to a cystine in the MH animals. Haplotyping suggested that the mutation in all five breeds has a common origin. Assuming that this is the causal mutation for MH, the development of a non-invasive diagnostic test will provide the basis for elimination of the MH gene, or its controlled inclusion in pig breeding programs.

The results of breeding experiments in pigs suggest that MH is inherited as the effect of a single recessive gene (HAL^n, the normal allele is HAL^N). During the past 15 years it was considered that the HAL^n gene was recessive for both PSS and PSE, but additive for muscularity. This gave rise to a crossbreeding policy in order to produce heterozygotes combining the advantages of the two homozygotes, for muscularity and meat quality. For example, the expected frequency of the Nn genotype in the Canadian population is now approximately 23% (Aalhus *et al.*, 1991). Nowadays, this strategy is questionable because there are more and more reports in favor of an additive inheritance of the HAL^n allele for meat quality (e.g. Murray *et al.*, 1989) and the meat quality of the heterozygote may have a considerable impact on the overall incidence of PSE pork. If this turns out to be the case, then it will be more urgent than ever to use the newly available molecular genetic techniques to identify carriers of the halothane gene to control, or even, if necessary, to eradicate this gene.

Exercise, fitness and muscle metabolism

Muscle fiber size and capillary density influence the rate of lactate accumulation in muscle at slaughter and, consequently, the degree of protein denaturation, paleness and drip loss in the meat (Essén-Gustavsson *et al.*, 1992). In comparison with other mammalian muscle, pig muscle in general has a lower percentage of type 1 fibers (slow-twitch, oxidative) and stress-susceptible strains in particular tend to have larger muscle fibers than non stress-susceptible pigs. The larger fibers and therefore lower capillary density in stress-susceptible pigs are considered to be important factors in determining anaerobic metabolic rate at slaughter.

Intensive production systems are often criticized because of the lower level of fitness in the finished pigs. Observations on the effect of genotype on muscle physiology and meat quality have given rise to speculation that increased physical activity during growth and development might increase the level of physical fitness and capacity for aerobic activity in pigs, and that this might be achieved, for example, through enforced exercise, or the use of free-range production systems, and might result in a lower rate of acid production in the musculature at slaughter, and hence reduce PSE.

Although the oxidative capacity of pig muscle was found to increase with moderate exercise (see Hansson *et al.*, 1991 for references), the experimental results show little evidence of improvement in meat color or water-holding capacity. Hansson and colleagues concluded that much heavier exercise may be necessary to produce desirable effects on carcass lean/fat composition and meat quality and that this was not a commercially realistic approach.

Similarly, neither Warriss *et al.* (1983b), nor van der Wal (1991) found any improvement in pork color or water-holding capacity in pigs from free-range versus intensive production systems. Indeed, consumers' preference for free-range meat appears to be quite subjective, te Nijenhuis and Bon (1988) found that sensory ratings increased markedly when the free-range product was labelled as such. In contrast to the powerful genetic influences on muscle metabolism, the above findings suggest that manipulating the environment to increase the level of physical activity during rearing of pigs has had relatively little effect.

Transport and handling

Timing of the last feed

Several aspects of meat quality and yield are affected by increasing the time of feed withdrawal before slaughter. Advantages include lower mortality rate in transport (Lendfers, 1974; von Mickwitz, 1982; Guise, 1987), better meat color and water-holding capacity (Warriss, 1982; Jones *et al.*, 1985; Eikelenboom *et al.*, 1991), lower carcass weight losses during chilling and cutting (Warriss and Brown, 1983; Lopez-Bote and Warriss, 1988), and lower risk of bacterial contamination from gut contents at evisceration; the main disadvantages are a lower yield of dressed carcass (Jones *et al.*, 1985; Fischer *et al.*, 1988; Eikelenboom *et al.*, 1991), and more DFD meat (Fischer *et al.*, 1988; Eikelenboom *et al.*, 1991).

The main incentive for feed withdrawal is to reduce the risk of deaths in transit, particularly in hot weather and in stress-susceptible pig strains. There is also interest in its potential to increase the final meat pH to control the yield of cooked hams and as an alternative to the use of polyphosphates. This is important because PSE presents a problem even in halothane-negative pigs (Murray *et al.*, 1989) and there are reports that meat water-binding capacity is actually worse in PSE meat from halothane-negative pigs compared with halothane reactors of the same breed (Eikelenboom and Nanni Costa, 1988). Fasting for 24h followed by 4h of transport and lairage brought the mean ultimate pH of the large muscles above pH 5.8; reducing the fasting period to 16h had a smaller effect (Eikelenboom *et al.*, 1991). In contrast, Fernandez *et al.* (1992) found that losses of muscle glycogen after lairage for 2h or 24h were similar, and were independent of the duration of lairage. These unexpected results indicate that the behavior and physiology of pigs during lairage at the abattoir deserves further study, particularly the factors that influence muscle carbohydrate metabolism.

Loss of carcass yield due to fasting before slaughter amounts to about 1% in the first 24h without food, and can increase to about 2%, or more, if dehydration occurs during transportation. Longer periods of fasting give correspondingly higher losses in carcass yield (for references see Tarrant, 1989).

The evidence would suggest that a period of 12 to 18h without feed before slaughter is sufficient to reap most of the advantages outlined above, without incurring a loss of carcass yield. The balance of advantages for some processors may require a longer period of fasting, where loss of carcass yield is more than compensated for by gains in processing yield. Commercial practice is highly variable (Warriss and Bevis, 1987) and the available evidence indicates that there is much scope for improvement by getting more pigs into the optimal period of feed withdrawal before slaughter.

Stocking density

In practice, recommended stocking densities are rarely adhered to, and space allowances as little as 0.3m^2 per 100kg pig are frequently encountered (Guise and Penney, 1989). Such tight packing is unacceptable, as pigs have barely enough room to avoid standing or lying on each other. More fights were observed in densely packed pigs, and there was more skin damage and a higher incidence of rectal prolapse. Low density is less of a problem than overloading because pigs tend to lie down during transport and are therefore less likely to be thrown about as a result of careless driving.

The recommended stocking density for slaughter weight pigs varies considerably, for example, the minimum recommended space allowance for 100kg pigs varies from 0.35m^2 in Australia and Denmark to 0.5m^2 in Germany. Differences probably reflect weather conditions, and the proportion of the local pig population that is stress-susceptible. In the case of long-distance transport, e.g. 24h journeys or longer, the results of Lambooy and Engel (1991) indicate that the minimum space allowance should be increased to 0.43m^2 per 100kg pig. Draft European Community guidelines suggest that a space allowance of 0.42m^2 for 100kg pigs may be agreeable, decreased by 5% for short journeys (< 3h) and increased by 5% for ambient temperatures over 25°C.

Body temperature

The variation in temperature encountered by pigs during transport may be up to 20°C, or even higher. Temperature in the stock crate is determined by the air temperature outside, the heat produced by the animals, and the ventilation rate. It is probable that in many transport conditions pigs are at their upper level of thermal tolerance (Lambooy *et al.*, 1987). Therefore, cooling is beneficial and the stocking density is very important if cooling during transport, by natural or forced ventilation, is to be effective.

Factors affecting body temperature at slaughter are of great interest because of the role of muscle temperature in PSE development and the possibility of control measures. Researchers have looked for a linkage between ambient temperature, body temperature at slaughter, and meat quality. Using pigs from a single source, Long and Tarrant (1990) found no differences in body temperature or *M. longissimus dorsi* (LD) muscle temperature between pigs slaughtered in summer or winter (mean temperatures 19.5°C and 11°C respectively). Over a wider range of ambient temperatures (12° to 24°C), Warriss (1991) found a linkage between ambient temperature, LD muscle temperature at 1h post-mortem and drip loss. It was not clear whether the effect on muscle was due to the influence of the ambient temperature on the live animal, or whether the main effect was on the rate of cooling of the carcass on the killing line.

The use of cold water sprays is effective in cooling pigs before slaughter. Long and Tarrant (1990) found that showering with cold water brought about a drop in the temperature of the loin muscle that was sufficient to reduce paleness and drip in pork chops.

These experiments show that muscle temperature at slaughter affects meat quality, and that it responds to treatment. It would appear that sizeable ambient differences are necessary to produce measurable improvements in meat quality. However, the beneficial effects of cooling before slaughter would obviously be more important in commercial shipments, where overheating is likely to occur, compared with experimental pigs handled under good conditions. Lambooy *et al.* (1987) found that activity-related heat production was decreased by showering pigs with water at 15°C, and concluded that both the heat load of stressed pigs and the activity could be reduced by showering during transport and lairage.

Deaths during transport

Typical mortality rates are reported to be well below 0.1%, although the rate may rise tenfold in stress susceptible strains. Length of journey is not a major factor, but losses are greater where pigs are fed on the day of transport, regardless of distance travelled. Losses often double on hot humid days, and Williams (1985) found that the plot of death rate versus temperature was hyperbolic, rising steeply above 18°C and almost vertically above 25°C.

Effective mechanical ventilation in Denmark was credited with reducing deaths in trucks by 50% and reducing deaths in lairage at the abattoir by one-third (Nielsen, 1982).

Hyperthermia and heart failure are probably the major causes of transport deaths. The highest heart rates occur at loading and unloading (Augustini and Fischer, 1982) and body temperature also increased rapidly at loading. At high stocking densities ($0.35m^2/100kg$) pigs were unable to dissipate heat effectively, even though they panted intensively (von Mickwitz, 1982; Augustini and Fischer, 1982). Transportation at very low ambient temperatures was found to be associated with increased stress in pigs in Finland (Honkavaara, 1991). This was due to closed ventilation slots in the stockcrate that were not opened until external air temperature rose above 0°C.

While good information is available on the main causes of deaths during transport, implementation of good practices is a problem, and here again good stockmanship, effective training, and well-designed equipment are essential for good results.

Dehydration

When fasting pigs are also denied access to water, additional weight loss occurs due to dehydration. Muscle mass shrinks because of its high water content (Becker *et al.*, 1989). Loss of weight through dehydration commences earlier than in the case of feed withdrawal, but is greatly influenced by ambient conditions. The data of Fischer *et al.* (1988), suggests that the additional reduction in killing-out percentage caused by the withdrawal of water as well as feed can be as much as 1% per day. However, the rate of dehydration is greatly accelerated by road transportation, and Warriss *et al.* (1983a) found a 2.1% reduction in killing-out percentage after only 6h of transportation. Factors contributing to dehydration of pigs during transport are increased ambient temperature and decreased humidity in the stockcrate, increased air flow, and increased body temperature. Lambooy *et al.* (1987) found that both heat production and weight loss were lower at an ambient temperature of 16°C then at 24°C or 8°C, indicating that there was an optimal temperature for minimal metabolic rate and weight loss.

Provision of water in trucks did not materially alter carcass weight loss on long road journeys because water intake by the pigs was too low (Lambooy *et al.*, 1985). However, wetting down pigs with water sprays in the truck significantly lowered carcass weight loss (Lambooy and Engel, 1989).

More information is needed about the role and extent of dehydration during transport, and also what preventative measures are possible. Ways of facilitating rehydration after transport are needed, and in this respect, the effectiveness of isotonic solutions should be investigated. Provision of an electrolyte drink to cattle after transportation reduced both liveweight and carcass weight losses (Schaefer *et al.*, 1990).

Pre-slaughter handling

In commercial abattoirs, animals are inevitably subjected to unaccustomed social, environmental and physical stresses. Situations that increase the amount of unaccustomed physical activity and psychological or emotional stress trigger responses through the voluntary and autonomic nervous systems that lead to a build-up of heat, lactic acid and carbon dioxide in the muscles and bloodstream.

Rapid glycogen breakdown in living muscle may be triggered by adrenalin release or by vigorous muscle contraction, or by both mechanisms acting in concert. Simultaneous adrenergic and contractile activation of muscle results in the maximum rate of glycogen breakdown. This situation may arise during pre-slaughter handling and loading of pigs, where unaccustomed, or heavy physical activity may coincide with sympathetic arousal and adrenalin release. A drop in pH is caused by a transient increase in the lactic acid content within the muscle fibers, which may be sufficient to reduce the tissue pH from just above 7.0 to about 6.4 during heavy exercise (Sahlin *et al.*, 1978). Clearly, if an animal is slaughtered in this condition, it is more likely to develop pale, soft, exudative (PSE) meat than if rested. On the other hand, if the stress is allowed to persist for some hours before the animal is slaughtered, the capacity of the muscles to produce lactic acid will be depleted and dark, firm, dry (DFD) meat will result.

Effects of processing on meat quality

Humane considerations

The purpose of stunning is to render the animal instantaneously insensible to pain, and to ensure that it remains insensible until it is dead. The law in the EC and elsewhere permits several methods of stunning, of which electrical and carbon dioxide are the most commonly used. There is concern about the humaneness of both methods, and also about damage to meat quality during stunning.

The argument that carbon dioxide stunning is an inhumane procedure is based on the assumption that the animals may experience distress during exposure to the gas, for a period of perhaps 20s before loss of consciousness. The increased physical activity or convulsive behavior, commonly observed during carbon dioxide stunning, is seen as a manifestation of distress. Recent work at the Swedish Meat Research Institute found that the electroencephalogram pattern typical of anaesthesia had developed in all swine a few seconds before the appearance of increased motor activity, indicating that the swine were unconscious then (Forslid, 1991).

Electrical stunning is considered more humane because it is capable of rendering pigs insensible within 1s of application of the electrodes, provided there is a sufficient flow of current across the brain. Data from electroencephalograms showed that the minimum current necessary to achieve an effective stun is 1.3amps. To obtain this amperage in all pigs, under practical conditions in the slaughterhouse, at least 240volts is necessary (Gregory, 1988).

Low-voltage stunning (c. 75-90v) is still permitted in many countries, but is criticized on the grounds that it fails to guarantee stunning in all animals, and in particular, the time to loss of consciousness is longer at low voltages, and there is also a possibility that some animals may appear unconscious but are only immobilized while the hand-operated tongs are in position. Recently, Anil (1991) found that stunning with 100v for 3s produced unsatisfactory results, as 38% of the test pigs were not stunned for more than 10s.

Cardiac arrest stunning is claimed to give a relaxed carcass that shows less kicking and is easier to handle. It does not lead to poor bleeding in pigs, provided they are stuck within 2min (Gregory, 1987). A minimum current of 1.3amps will effectively stun and cause a cardiac arrest in 99% of pigs if it is applied using head-to-back electrodes positioned so they span both the brain and heart (Gregory, 1988).

Bone fractures and blood splash

The major drawback to electrical stunning is the higher incidence of damage compared with carbon dioxide stunning. Broken shoulder blades may result from electrical stunning of pigs while their feet are in contact with the floor, and broken vertebrae may occur in pigs stunned with head-to-back tongs if the voltage is too high. The hemorrhage associated with broken bones requires trimming and is costly; bleeding can be reduced if cardiac arrest is induced at stunning (Gregory, 1988).

The causes of blood splash (petechial or diffuse hemorrhages in soft tissues, particularly skeletal muscle) are not well established, but it is likely that any muscular contractions which cause aneurysms in the blood vessels while systolic blood pressure is high enough to force blood through the perforations are of importance (Gregory, 1987). High-voltage stunning in a restraining conveyor, probably the most widespread method of stunning pigs today, leads to intense muscular contractions and to a high incidence of blood splash, particularly in the shoulder muscles. It is not known whether the rise in blood pressure that occurs during and after electrical stunning actually causes muscle hemorrhages, or merely increases bleeding from existing hemorrhages. Either way, reducing the interval between stunning and sticking reduces the expression of blood splash. If possible, pigs should be stuck within 10s of electrical stunning, well before the kicking (clonic) phase of muscle activity sets in.

Developments in stunning

As none of the existing methods of stunning pigs are entirely satisfactory from both the humane and meat quality viewpoints, there is a need for better stunning methods. Preliminary experiments using a high-pressure water jet, as is used for cutting and drilling in solid materials, show that a 1s application to a pig's head was sufficient to mechanically destroy the brain to such an extent that it caused immediate loss of consciousness (Lambooy, 1991). There was little evidence of blood splash. This method warrants further investigation.

In summary, there is a need for more information in at least three key areas: firstly, confirmation on whether the onset of unconsciousness precedes the kicking phase seen during carbon dioxide stunning; secondly, better designs for electrical stunning equipment to reduce the problems of bone fractures and blood splash; thirdly, alternative stunning techniques that satisfy the requirements for humaneness, cost effectiveness and meat quality more effectively than the present commercial methods.

Chilling and drip loss

There is good evidence that the high drip loss and softness of PSE meat is caused by myosin denaturation, while denaturation of sarcoplasmic proteins may be responsible for the paleness and high light-scattering properties (Staborsvik *et al.*, 1984; Honikel and Kim, 1986; Offer, 1991). Myofibrils occupy over 80% of the volume of muscle fibers and *in vivo* they hold most of the cellular water in the spaces between the thick and thin filaments of the myofibrillar lattice. Lateral shrinkage of the myofibrillar lattice at the onset of rigor causes some drip in normal meat. Offer (1991) suggested that the shrinkage results from two events associated with rigor in normal meat - the fall in pH, which reduces the negative charge on the thick and thin filaments, thereby lessening the repulsive forces between them and allowing some shrinkage, and secondly, the irreversible attachment of myosin heads to actin at rigor, which also diminishes the filament lattice spacing - at the onset of rigor there is a 4.4% decrease in spacing corresponding to a 9% volume decrease.

In PSE meat, the denaturation of the doublet heads of the myosin molecules may cause additional shrinkage and more fluid to be expelled from the myofibrillar lattice, increasing the amount of drip formation. The heads of native myosin are 19nm long and they shrink to 17nm when subjected to PSE-type conditions, e.g. pH 6.0 and 35°C. Allowing for a 50° angle of attachment of myosin heads to thin filaments, the loss of 2nm in length could reduce the center-to-center spacing between thick and thin filaments by 1.4nm or 6% - equivalent to a further 12% reduction in myofibrillar volume. It is not known what proportion of myosin molecules would have to be denatured to allow this additional shrinkage of the myofibrillar lattice to occur - between 20 and 50% has been observed in PSE muscle using differential scanning calorimetry (Staborsvik *et al*., 1984; Honikel and Kim, 1986).

To minimize drip, carcass chilling should be quick and efficient, but within the limits necessary to avoid cold toughening (Honikel, 1987). While rapid chilling is expected to reduce the expression of PSE in affected carcasses, it may cause cold-toughening of normal meat. Using hot-boned pork muscles, Honikel (1987) noted that fast chilling reduced drip loss and paleness of PSE muscles, but not sufficiently to give the quality attributes of normal pork. He concluded that fast chilling can diminish, but not solve the PSE problem.

With ultra-rapid chilling, the benefits of reducing myosin denaturation may be outweighed by the increase in drip caused by cold-induced shortening of the sarcomeres and a possible further increase in drip caused by partial or "crust" freezing of the meat surface. Honikel (1987) reported a linear relationship between muscle sarcomere length and drip loss in pork pieces, with drip doubling as sarcomeres shortened to half the resting length.

Offer (1991) has pointed out that the benefits of fast chilling are greatest for carcasses with slow or intermediate rates of pH fall, where appreciable cooling can take place before the carcasses set in rigor, thereby reducing the amount of myosin denatured. With very fast rates of pH fall, very rapid chilling is much less effective in limiting myosin denaturation because this is already largely precluded by the formation of the actomyosin complex (which is relatively resistant to denaturation) before chilling can take effect. To be effective in lowering drip loss, cooling must occur pre-rigor, when myosin heads are unattached and susceptible to denaturation.

Particular attention needs to be paid to the carcass temperature immediately post-slaughter. A rise of 10°C increases the rate of myosin denaturation about 12-fold. Offer (1991) estimated that a decrease of 2°C in the starting temperature would reduce the initial rate of denaturation of myosin by 37%, with a corresponding reduction in drip loss. Factors that increase carcass temperature or delay chilling need attention, including handling situations that raise body temperature ante-mortem and the effect of scalding the carcass post-mortem. Likewise, air temperature on the killing line may make an appreciable difference in the rate of cooling.

Paleness

Rapid chilling reduces paleness in pork, as was shown by lower probe light-scattering values in pork sides pre-chilled at -20°C for 2 or 3h (Long and Tarrant, 1990). Although the beneficial effect is small, it is nonetheless important in view of the strong discrimination against pale pork in some markets.

Toughness

The main risk in ultra-rapid chilling systems is that of increased toughness. This may be caused by sarcomere shortening (Honikel, 1987) together with a slow-down of the natural enzymatic conditioning process in meat at lower temperatures (Troy and Tarrant, 1987). Toughening has been associated with ultra-rapid chilling employing blasts at -30°C or lower and is greater in carcasses with a slow development of rigor mortis (for references see Tarrant, 1989). Post-mortem electrical stimulation of pig carcasses has been shown to prevent cold toughening (Taylor and Tantikov, 1989), and more research is needed to establish its benefits in ultra rapid pig-chilling systems and in particular how to eliminate cold shortening without adverse effects on drip and paleness.

Conclusions

To meet industry needs for pig welfare, meat quality and yield, new information is needed in several parts of the chain from farm to factory. These include:

- Ways to improve eating quality without compromising leanness.
- Concerns about ethics and safety must be resolved before the powerful potential of PST for increasing the efficiency of production of lean meat can be realized.
- A diagnostic test-kit for the ryanodine receptor gene should facilitate the elimination of the negative aspects of the halothane gene.

- Major improvement is needed, and is increasingly being demanded, in methods and equipment for handling, transport and stunning. For this we need a better understanding of pig behavior and tolerance, to stress.
- Control over dehydration during transport and ways to facilitate rehydration before slaughter are needed.
- Ways to keep pigs calm and cool before slaughter and to bring the carcass quickly below 30°C are urgently needed.

References

Aalhus, J.L., Gariepy, C. Murray, C.A., Jones, S.D.M. and Tong, A.K.W. 1991. Stunning and shackling influences on quality of porcine *Longissimus dorsi* and *Semimembranosus* muscles. Meat Sci. 29, 323-34.

Anil, M.H. 1991. Studies on the return of physical reflexes in pig following electrical stunning. Meat Sci. 30, 13-21.

Augustini, C. and Fischer, K. 1982. In: Moss, R. (ed.) Transport of Animals Intended for Breeding Production and Slaughter. Martinus Nijhoff, The Hague, pp. 125-35.

Becker, B.A., Mayes, H.F., Hahn, G.L., Nienaber, J.A., Jesse, G.W., Andersen, M.E. Heymann, H. and Hedrick, H.B. 1989. Effect of fasting and transportation on various physiological parameters and meat quality of slaughter hogs. J. Anim. Sci. 67, 334-41.

Bejerholm, C. 1984. Experience in taste panel testing fresh pork at the Danish Meat Research Institute. Proc. 30th Meeting of Meat Research Workers, Bristol, pp. 196-7.

Bejerholm, C. and Barton-Gade, P.A. 1986. Effect of intramuscular fat level on eating quality of pig meat. Proc. 32nd European Meeting of Meat Research Workers, pp 389-91.

Cameron, N.D., Warriss, P.D., Porter, S.J. and Enser, M.B. 1990. Comparison of Duroc and British Landrace pigs for meat and eating quality. Meat Science, 27, 227-47.

Eikelenboom, G. and Nanni Costa, L. 1988. Fibre optic probe measurements in Landrace pigs of different halothane phenotypes. Meat Sci. 23, 9-19.

Eikelenboom, G., Bolink, A.H. and Sybesma, W. 1991. Effects of feed withdrawal before delivery on pork quality and carcass yield. Meat Sci. 29, 25-30.

Enright, W.J. 1991. Biotechnology in meat production: Meeting report. Livest. Prod. Sci. 27(4), 353-57.

Essén-Gustavsson, B., Karlström, K. and Lundström, K. 1992. Muscle fibre characteristics and metabolic response at slaughter in pigs of different halothane genotypes and their relation to meat quality. Meat Sci. 31, 1-11.

Etherington, D.J., Taylor, M.A.J. and Dransfield, E. 1987. Conditioning of meat from different species. Relationship between tenderizing and the levels of cathepsin B, cathepsin L, calpain I, calpain II and β-glucuronidase. Meat Sci. 20, 1-18.

Fernandez, X., Magard, M. and Tornberg, E. 1992. The variation in pig muscle glycolytic potential during lairage - an *in vivo* study. Meat Sci. 32, 81-91.

Fischer, K., Augustini, C. and McCormick, R. 1988. Effect of fasting time before slaughter on the quality of pigmeat. Fleischwirtschaft, 68, 485-88.

Forslid, A. 1991. Pre-slaughter CO_2-anaesthesia in swine - Neurophysiological and ethical aspects. Proc. 37th International Congress of Meat Science and Technology, Kulmbach, pp. 402-4.

Fujii, J., Otsu, K., Zorzato F., De Leon, S., Khanna, V.K., Weiler, J.E., O'Brien, P.J. and MacLennan, D.H. 1991. Identification of a mutation in porcine ryanodine receptor associated with malignant hyperthermia. Science, 253, 448-51.

Gregory, N.G. 1987. In: Tarrant, P.V., Eikelenboom, G. and Monin, G. (eds) Evaluation and Control of Meat Quality in Pigs. Martinus Nijhoff, Dordrecht, pp. 265-72.

Gregory, N.G. 1988. In: Devine, C.E. and Shaw, F.D. (eds) Workshop on Stunning of Livestock. Proc. 34th International Congress of Meat Science and Technology, Brisbane., pp. 55-58 and 59-62.

Gregory, N.G. and Wilkins, L.J. 1984. The intravenous halothane test as an experimental method for quantifying stress sensitivity in pigs. J. Sci. Food Agric. 35, 147-53.

Gronert, G.A. 1986. Malignant hyperthermia. In: Eagle, A. and Banker, B. (eds) Myology. McGraw Hill, New York pp. 1763-83.

Guise, J. 1987. Moving pigs from farm to factory. Pig International, December, 8-12.

Guise, H.J. and Penny, R.H.C. 1989. Factors influencing the welfare and carcass and meat quality of pigs. 2. Mixing unfamiliar pigs. Anim. Prod. 49, 517-21.

Hall, L.W., Woolf, N., Bradley, J.W.P. and Jolly, D.W. 1966. Unusual reaction to suxamethonium chloride. Brit. Med. J. 2, 1305.

Hansson I., Lundström, K., Enfalt, A-C., Karlsson, A., Essén-Gustavsson, B. and Håkansson, J. 1991. Effect of moderate indoor exercise on carcass composition, meat quality and muscle enzyme activities in pigs. In: Proc. 37th International Congress of Meat Science and Technology, Kulmbach 1, 103-6.

Hedrick, H.B., Fennewald, K.J., Bailey, M.E., Heymann, H., Mawhinney, T.P., Chang, C.H. and Wallace, D.H. 1991. Effect of β-agonists on composition of pork *longissimus* muscle. Proc. International Congress Meat Science and Technology, Kulmbach, pp. 377-80.

Heffron, J.J.A. 1991. Human malignant hyperthermia: is it a calcium channel disorder? Paper read to the Royal Irish Academy, Dublin 2, on 11 October 1991.

Honikel, K.O. 1987. Influence of chilling on meat quality attributes of fast glycolysing pork muscles. In: Tarrant, P.V., Eikelenboom, G. and Monin, G. (eds) Evaluation and Control of Meat Quality in Pigs. Martinus Nijhoff Publishers, Dordrecht, pp. 273-83.

Honikel, K.O. and Kim, C. 1986. Causes of the development of PSE pork. Fleischwirtschaft, 66, 349-53.

Honkavaara, M. 1991. Effect of transport on porcine stress. In: Proc. 37th International Congress of Meat Science and Technology, Kulmbach, 2, 257-60.

Iaizzo, P.A., Klein, W. and Lehmann-Horn, F. 1988. Fura-2 detected myoplasmic calcium and its correlation with contracture force in skeletal muscle from normal and malignant hyperthermia susceptible pigs. Pfluegers Arch. 411, 648-53.

Jones, S.D.M., Rompala, R.E. and Haworth, C.R. 1985. Effects of fasting and water restriction on carcass shrink and pork quality. Can. J. Anim. Sci. 65, 613-18.

Kempster, A.J., Dilworth, A.W., Evans, D.G. and Fischer, K.D. 1986. The effects of fat thickness and sex on pig meat quality with special reference to problems associated with overleanness. 1: Butcher and consumer panel results. Anim. Prod. 43, 517-33.

Lambooy, E., Garssen, G.J., Walstra, P., Mateman, F. and Merkus, G.S.M. 1985. Transport of pigs by car for two days: Some aspects of watering and loading density. Livest. Prod. Sci. 13, 289-99.

Lambooy, E., van der Hel, W., Hulsegge, B. and Brandsma, H.A. 1987. Effect of environmental temperature and air velocity two days pre-slaughter on heat production, weight loss and meat quality in non-fed pigs. In: Verstegen, M.W.A. and Henren, A.M. (eds) Energy Metabolism in Farm Animals: Effects of Housing, Stress and Disease. Nijhoff Publishers, Dordrecht. pp. 164-79.

Lambooy, E. and Engel, B. 1989. Transport of slaughter pigs by truck over a long distance. Proc. 40th Annual Meeting of the European Association for Animal Production, Dublin.

Lambooy, E. and Engel, B. 1991. Transport of slaughter pigs by road over a long distance: some aspects of loading density and ventilation. Livest. Prod. Sci. 28, 163-74.

Lendfers, L.H.H.M. 1974. Gevoeligheid van het Nederlands slachtvarken voor transportinvloeden. Ph.D. Thesis, University of Utrecht, The Netherlands.

Long, V.P. and Tarrant, P.V. 1990. The effect of pre-slaughter showering and post-slaughter rapid chilling on meat quality in intact pork sides. Meat Sci. 27, 181-95.

Lopez, J.R. 1986. Use of intercostal muscle biopsies and calcium-sensitive microelectrodes as an approach to malignant hyperthermia. Muscle Nerve, 9, Suppl. 5S, 32.

Lopez, J.R., Alamo, L., Caputo, C., Wininski, J. and Ledezma, D. 1985. Intracellular ionized calcium concentration in muscles from humans with malignant hyperthermia. Muscle Nerve, 8, 355-8.

Lopez-Bote, C. and Warriss, P.D. 1988. A note on the relationships between measures of water-holding capacity in the *M. longissimus dorsi* and total drip loss from butchered pig carcasses during storage. Meat Sci. 23, 227-34.

Meisinger, D.J. 1989. Potential economic impact of carcass modifiers. J. Anim. Sci. 67, 2150-4.

Mickelson, J.R., Gallant, E.M., Litterer, L.A., Johnson, K.M., Rempel, W.E. and Louis, C.F. 1988. Abnormal sarcoplasmic reticulum ryanodine receptor in malignant hyperthermia. J. Biol. Chem. 263, 9310-15 .

Mickelson, J.R., Gallant, E.M., Rempel, W.E., Johnson, K.M., Litterer, L.A., Jacobson, B.A. and Louis, C.F. 1989. Halothane-sensitivity gene and sarcoplasmic reticulum function in malignant hyperthermia. Am. J. Physiol. 257, C787-94.

Moloney, A., Allen, P., Joseph, R. and Tarrant, V. 1991. Influence of β-adrenergic agonists and similar compounds on growth. In: Growth Regulation in Farm Animals. Adv. Meat Res. Vol. 7, Elsevier, Amsterdam pp. 455-513.

Murray, A.C., Jones, S.D.M. and Sather, A.P. 1989. The effect of pre-slaughter feed restriction and genotype for stress susceptibility on pork lean quality and composition. Can. J. Anim. Sci. 69, 83-91.

Nelson, T.E. 1983. Abnormality in calcium release from skeletal sarcoplasmic reticulum of pigs susceptible to malignant hyperthermia. J. Clin. Invest. 72, 862-70.

Nielsen, N.J. 1982. In: Moss, R. (ed.) Transport of Animals Intended for Breeding Production and Slaughter. Martinus Nijhoff, The Hague, pp. 115-24.

Offer, G. 1991. Modelling of the formation of pale, soft and exudative meat: effects of chilling regime and rate and extent of glycolysis. Meat Sci. 30, 157-84.

Ohnishi, S.T., Taylor, S.R. and Gronert, G.A. 1983. Calcium-induced Ca^{2+} release from sarcoplasmic reticulum of pigs susceptible to malignant hyperthermia. FEBS Lett 161, 103-7.

O'Mahony, R.E. 1991. Consumers' perceptions of the quality of pork, with special reference to intramuscular fat in pork chops. M.Sc Thesis, University College, Cork, 1991.

Sahlin, K., Alverstrand, A., Brand, R. and Hultman, E. 1978. Intracellular pH and bicarbonate concentration in human muscle during recovery after exercise. J. Appl. Physiol. 45, 474-80.

Schaefer, A.L., Jones, S.D.M., Tong, A.K.W. and Young, B.A. 1990. Effects of transport and electrolyte supplementation on ion concentrations, carcass yield and quality in bulls. Can. J. Anim. Sci. 70, 107-19.

Smith, G.C. and Carpenter, Z.L. 1976. Eating quality of animal products and their fat content. Proceedings of the Symposium on Fat Content and Composition of Animal Products, National Academy of Sciences, Washington DC, pp. 147-82.

Staborsvik, E., Fretheim, K. and Froystein, T. 1984. Myosin denaturation in pale, soft and exudative (PSE) porcine muscle tissue as studied by differential scanning calorimetry. J. Sci. Food Agric. 35, 240-44.

Tarrant, P.V. 1989. The effects of handling, transport, slaughter and chilling on meat quality and yield in pigs - A review. Ir. J. Food Sci. Technol. 13, 79-107.

Taylor, A.A. and Tantikov, M.Z. 1989. Electrical stimulation and rapid chilling of pig carcasses. Proc. 35th International Congress of Meat Science and Technology, Copenhagen, pp. 1157-62.

te Nijenhuis, A.C.M., and Bon, M. 1988. Sensorisch onderzoek naar scharrelvarkensvlees en gervoon varkensvlees. Hogeschool Nijinegen, Voeding en Dietetiek, Nijinegen, The Netherlands, February, 1988.

Troy, D.J. and Tarrant, P.V. 1987. Changes in myofibrillar proteins from electrically stimulated beef. Biochem. Soc. Trans. 15, 297-8.

van der Wal, P.G. 1991. Free range pigs: carcass characteristics and meat quality. In: Proc. 37th International Congress of Meat Science and Technology, Kulmbach. 1, pp. 202-5.

von Mickwitz, G., 1982. In: Moss, R. (ed.) Transport of Animals Intended for Breeding, Production and Slaughter. Martinus Nijhoff, The Hague, pp. 45-56.

Warriss, P.D. 1982. Loss of carcass weight, liver weight and liver glycogen, and the effects on muscle glycogen and ultimate pH in pigs fasted pre-slaughter. J. Sci. Food Agric. 33, 840-6.

Warriss, P.D. 1991. A note on the influence of ambient temperature at slaughter on pig meat quality. Proc. 37th International Congress of Meat Science and Technology, Kulmbach, 2, 301-4.

Warriss, P.D. and Brown, S.N. 1983. The influence of pre-slaughter fasting on carcass and liver yield in pigs. Livest. Prod. Sci. 10, 273-82.

Warriss, P.D., Dudley, C.P. and Brown, S.N. 1983a. Reduction of carcass yield in transported pigs. J. Sci. Fd. Agric. 34, 351-6.

Warriss, P.D., Kestin, S.C. and Robinson, J.M. 1983b. A note on the influence of rearing environment on meat quality in pigs. Meat Sci. 9, 271-9.

Warriss, P.D. and Bevis, E.A. 1987. Liver glycogen in slaughtered pigs and estimated time of fasting before slaughter. Brit. Vet. J. 143, 354-60.

Williams, J.M. 1985. Causes and effects of porcine stress syndromes. The Veterinary Annual, 25th Issue, Bristol, Scientechnica, pp. 175-9.

Wood, J.D., Jones, R.C.D., Francombe, M.A. and Whelehan, O.P. 1986. The effects of fat thickness and sex on pigmeat quality with special reference to the problems associated with overleanness. 2: Laboratory and trained taste panel results. Anim. Prod. 43, 535-44.

Wood, J.D. and Warriss, P.D. 1991. The influence of the manipulation of carcass composition on meat quality. In: Proc. Nottingham Easter School, Butterworth-Heinneman (in press).

Wood, J.D., Warriss, P.D. and Enser, M.B. 1991. Effects of production factors on meat quality in pigs. Paper presented at a Symposium on The Chemistry of Muscle Based Foods, at The Queen's University, Belfast, on 9-11 September, 1991.

Discussion

Discussion after Dr. Tarrant's Presentation:

*Tornberg - I wonder about the fatty acid composition because we had that discussion in Sweden, about polyunsaturated fats. A study has shown that polyunsaturated fatty acids will cause oxidation in the body, and that can lead to atherosclerosis. So, the polyunsaturated fatty acids are not so favorable, and I think monounsaturated fatty acids are the most important, with regard to human health.

*Tarrant - You are absolutely right, and our dairy scientists have been feeding rapeseed to cattle to increase monounsaturated fatty acids in butter.

*Puolanne - One recommendation has been to have a 1:1:1 ratio of fatty acids, one third monounsaturated fatty acids, one third polyunsaturated, and one third saturated.

*Honikel - May I add another point to this question of polyunsaturated fatty acids? From the point of view of a meat scientist interested in meat quality for the purposes of processing, if there is too much unsaturated fatty acid, the fat gets soft. On the other hand, the fat oxidizes much easier. What we have done in the last ten or fifteen years, from a breeding point of view, and from the demands of the nutritionists, was not always good. I think we should change our point of view on the whole subject. Pork fat is about 50% unsaturated fatty acids, and most of it is monounsaturated fatty acids, especially oleic acid, and I think this is the best composition you can imagine. We should leave it as nature has intended.

*Puolanne - It is normally said that the post-mortem glycolysis is either fast or slow. But my idea, and it is shared by some others, is that the extent of pre-mortem glycolysis, is also important. At the time of slaughter, the pH may be 0.5 units below the normal level. Do you think that this is important, especially when the temperature of the body, and the denaturation are considered?

*Tarrant - I think it's very important. I was amazed when I read the results of Henckel (1992). They have found resting levels of pH, *in vivo*, down around 6.6. Previously, Sahlin and colleagues (1978), had shown that pH values of that magnitude were normally encountered in subjects after intense exercise, that is, that the pH *in vivo* would come down to about 6.5 as a result of exercise. But the Danish results now are suggesting that in both halothane positive pigs, and heterozygotes, the resting level of pH could be quite low. But I don't know why that was. I think that must be extremely important, because as soon as that muscle becomes anaerobic, and presumably when the pH gets down low, it's totally anaerobic, there will be muscle denaturation, whether the animal is alive or dead. The process has started. Once the circulation stops, the animal is already half way into rigor. So, the handling situation before slaughter is probably the critical point, and, slaughter intervenes to make the pH fall irreversible.

*Smulders - I'm a little bit worried when you say that feed deprivation before slaughter reduced the risk of salmonellosis, or at least salmonella shedding. I would argue that if you deprive an animal of feed, it is an extra stress factor, as has been shown by the people at Bristol University. It will increase the gut flow. It may decrease the volume of feces shed, but it will definitely make the salmonella crawl out of their niches. So, I think it is a little bit tricky to mention reduced salmonellosis as an advantage in feed deprivation.

*Tarrant - I accept your point about fecal release of salmonella. Feed deprivation may very well be an additional stress that increases a problem rather than reduces it, but, the risk of gut spillage during evisceration is definitely reduced.

Genetics of Porcine Stress Syndrome and Association with Oedema Disease

P. Vögeli, R. Fries, R. Bolt, C. Gerwig, P. Affentranger, N. Künzi, H.U. Bertschinger* and G. Stranzinger

Institute of Animal Sciences, Federal Institute of Technology (ETH), CH-8092 Zürich, Switzerland
* Institute of Veterinary Bacteriology, University of Zürich, CH-8057 Zürich, Switzerland

Introduction

Sudden, stress-induced death, pale soft exudative meat (PSE) and sensitivity to halothane-induced malignant hyperthermia (MH) are manifestations of the porcine stress syndrome (PSS)(Mitchell and Heffron, 1982; Louis *et al.*, 1990; Archibald, 1991). Deaths arise from uncontrolled skeletal muscle contracture with attendant hypermetabolic and hyperthermic reactions. PSS is characterized by sudden death in pigs subjected to stressors. It occurs most frequently at the time of marketing (when pigs weigh 100kg), although it may occur at other times, especially after physical exertion. As well as increasing PSS losses, the PSS gene adversely affects litter productivity. However, it also appears to give an economically important improvement in carcass yield. The opportunity now exists to manipulate the frequency of the gene to improve economic efficiency.

An abnormality in the Ca^{2+}-release channel of skeletal muscle sarcoplasmic reticulum (the ryanodine receptor) may account for the MH because, in skeletal muscle, both contraction and metabolism are regulated by the concentration of intracellular Ca^{2+} (O'Brien, 1986). The skeletal muscle ryanodine (RYR1) gene and the porcine MH (HAL) locus have been localized to pig chromosome 6p11-q21 suggesting linkage between porcine MH and RYR1 (Davies *et al.*, 1988; Harbitz *et al.*, 1990). Moreover, a point mutation associated with malignant hyperthermia in several breeds of swine has been identified in a putative important regulatory region of RYR1 (Fujii *et al.*, 1991). However, it must be noted that in contrast to swine, MH muscles in man may contain two populations of sarcoplasmic reticulum Ca^{2+}-release channels and therefore MH in humans does not arise from a single defect (Fill *et al.*, 1991).

This paper reviews current knowledge of the genetics of PSS with particular reference to halothane sensitivity and to the RYR1 gene, and examines some of the implications for pig breeding programs and for future research. Linkage between the porcine genes associated with bacterial colonization (oedema disease) and the HAL linkage group is documented. Detailed aspects of meat quality and physiological and biochemical aspects of PSS have recently been reviewed elsewhere (Lister, 1979; Mitchell and Heffron, 1982; Webb *et al.*, 1982; O'Brien *et al.*, 1990).

Inheritance of halothane sensitivity

Of the three phenotypes, only halothane sensitivity can be measured in the living animal through short, controlled, exposure to the anesthetic gas halothane (Eikelenboom and Minkema, 1974). Susceptibility to halothane-induced MH has been shown to be controlled by a recessive gene at a single autosomal locus (HAL) with both alleles exhibiting incomplete penetrance (Smith and Bampton, 1977; Vögeli *et al.*, 1988). Because sensitivity to halothane-induced MH is a recessive trait in pigs, the standard halothane challenge test fails to distinguish between the homozygous resistant animals (N/N) and the heterozygous carriers (N/n). Penetrance estimates for the recessive homozygote (n/n) range from 50 to 100% over a variety of breeds, and average roughly 89% (Webb *et al.*, 1982). Where pigs have received two or more consecutive halothane tests, estimates of both repeatability of the test result and the probability of correctly identifying a halothane positive (HP, n/n) pig on the first test fall within the range 0.90 to 0.95. Strong evidence of monofactorial inheritance is also provided by blood typing studies, which show that HAL is situated within a known linkage group (e.g. Vögeli, 1989).

Incidence of halothane sensitivity

Frequencies of HP reaction in world breeds range from 0 to 88% (Webb *et al.*, 1982). Large White or Yorkshire types have low or zero frequencies. Most Landrace strains have low to intermediate frequencies (5-20%), but those with extreme ham development (Pietrain and Belgium Landrace pigs) show higher incidences.

Methods for predicting liability to PSS

An economic benefit from manipulating the frequency of the halothane gene will depend on the ease with which a change can be achieved. In practice, the halothane test is quick, cheap, requires little training, and gives an immediate result. It carries a small mortality risk for the pig and a slight possibility of a health risk for the operator. The main disadvantage is that the heterozygote is not detected. A number of methods for predicting halothane genotype and liability to PSS are currently under investigation and/or are used.

Linkage between halothane and blood type genes

The HAL locus is situated within a linkage group of five other loci (Fig. 1). In most populations, strong linkage disequilibria give rise to associations of PHI, H and A1BG with halothane sensitivity and meat quality.

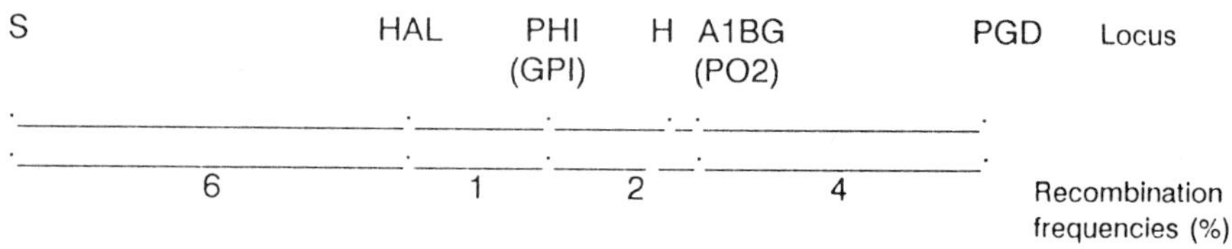

S: suppressor for A and O blood types (S and s)
HAL: halothane sensitivity (N and n)
PHI: phosphohexose isomerase (glucosephosphate isomerase)
H: H erythrocyte antigen locus
A1BG: α-1-B-glycoprotein
PGD: 6-phosphogluconate dehydrogenase

Figure 1. The halothane linkage group (adapted from Archibald and Imlah, 1985; Vögeli *et al.*, 1988; Vögeli, 1989).

Halothane and ryanodine receptor gene

The gene (RYR1) encoding the calcium-release channel of skeletal muscle sarcoplasmic reticulum (ryanodine receptor) has been proposed as a candidate for MH (MacLennan *et al.*, 1990). Fujii *et al.* (1991) found that the alteration of C1843 (C = cytosine) to T1843 (T = thymine) in porcine RYR1 leading to an alteration of arginine (Arg) 615 to cysteine (Cys) 615 in the ryanodine receptor (Fig. 2), was correlated with susceptibility to MH in five pig breeds.

In a recent study, we have amplified the 74-bp sequence between nucleotides 1811 and 1884 (Fig. 2) by the polymerase chain reaction (PCR). Because the mutation deletes a CfoI and creates an AspHI site (Fig. 2), we used these two enzymes for the detection of the C-T mutation in the PCR product (Fig. 3).

In our study, we have compared the accuracy of the DNA based test described above for the RYR1 C1843T mutation with the indirect halothane challenge test and S-PHI-H-A1BG-PGD haplotype analysis used to diagnose MH in informative families. Marker and HAL haplotypes were deduced by inspection of the phenotypes in informative litters and from knowledge of the parental and grand parental genotypes. A representative family is shown in Table 1.

Data came from 392 pedigreed pigs. The animals were either purebred or crossbred pigs with genetic contribution from Swiss Landrace, Piétrain, German Landrace, Norwegian Landrace, Danish Landrace and Belgian Landrace breeds.

The cosegregation of RYR1 genotypes, as determined by DNA-based analysis, and of phenotypes, as determined by halothane testing and examined by haplotype analysis, is summarized in Table 2.

Of the 68 animals reacting in the halothane test, 67 were confirmed as n/n by the DNA-based test. One reactor had the genotype N/n. Of the 324 nonreactors, 321 were confirmed as N/n or N/N, but three were shown to be homozygous (n/n) by the DNA-based test. Analysis of the marker haplotypes was consistent with the diagnosis of the three discordant animals as n/n nonreactors. The figure of 4.3% of the 70 n/n pigs that were nonreactors (false negatives) and the only animal of the 168 N/n pigs that was a reactor (false positive) lies within the range of incomplete penetrance recorded in other studies (Smith and Bampton, 1977; Gahne and Juneja, 1985; Vögeli *et al.*, 1988).

Table 1. Haplotype analysis of a representative family.
The loci in the haplotypes are written left to right S-RYR1-HAL-PHI-H-A1BG-PGD. C and T refers to nucleotide 1843 in ryanodine receptor gene. N and n alleles at the HAL locus are deduced from the results of the halothane test and examination of the marker loci. Pig O4 has received a recombinant haplotype s↑CNAcFB from the sow. S: sire; D: dam and O: offspring

Family	S	RYR1	Halothane reaction	PHI	H	A1BG	PGD	Haplotypes S-RYR1-HAL-PHI-H-A1BG-PGD
S42	ss	TT	+	AB	aa	SS	AA	sTnAaSA/sTnBaSA
D12	Ss	CT	-	AA	ac	FS	BB	sTnAaSB/SCNAcFB
								from sire/from dam
O1	Ss	CT	-	AB	ac	FS	AB	sTnBaSA/SCNAcFB
O2	ss	TT	+	AA	aa	SS	AB	sTnAaSA/sTnAaSB
O3	Ss	CT	-	AA	ac	FS	AB	sTnAaSA/SCNAcFB
O4	ss	CT	-	AA	ac	FS	AB	sTnAaSA/s↑CNAcFB
O5	Ss	CT	-	AB	ac	FS	AB	sTnBaSA/SCNAcFB
O6	ss	TT	+	AB	aa	SS	AB	sTnBaSA/sTnAaSB

Normal sequence

CRC1811

1811 5' GT TCC CTG TGT GTG TGC AAT GGT GTG GCC 1839
Val Ala

1843 CfoI CRC 1884

1840 GTG CG*C TCC AAC CAA GAT CTC ATT ACT GAG AAC 1872
Val Arg (615) Ser Asn Gln

1873 TTG CTG CCT GGC 3' 1884

Mutant sequence

CRC1811

1811 5' GT TCC CTG TGT GTG TGC AAT GGT GTG GCC 1839
Val Ala

1843 AspHI CRC 1884

1840 GTG TGC T*CC AAC CAA GAT CTC ATT ACT GAG AAC 1872
Val Cys (615) Ser Asn Gln

1873 TTG CTG CCT GGC 3' 1884

Figure 2. Nucleotide and deduced amino acid sequences of part of the exon in which the single Arg^{615} to Cys^{615} mutation is located (in the box). The forward primer used in PCR amplification corresponds to the sequence of nucleotides 1811 to 1834 (CRC 1811)(overlined) and the reverse primer is complementary to nucleotides 1861 to 1884 (CRC1884)(overlined). The alteration of the normal sequence G CGC TC starting at nucleotide 1842 to the mutant sequence G TGC TC leads to the deletion of a CfoI site and the creation of an AspHI site in the mutant sequence (adapted from Fujii *et al.*, 1991). * Indicates the cleavage site.

Table 2. Comparison of halothane challenge testing for phenotype and DNA-based analysis for genotype in the definition of the MH status in different purebred and crossbred pigs (number of animals in parenthesis)

A Halothane reaction	B Genotype deduced by marker and HAL haplotypes	C Genotype from DNA-based analysis	% Discordance	
			A-B	B-C
+ (67) - (3)	n/n (70)	n/n (70)	4.3	0.0
- (167) + (1)	N/n (168)	N/n (168)	0.6	0.0
- (154)	N/N (154)	N/N (154)	0.0	0.0

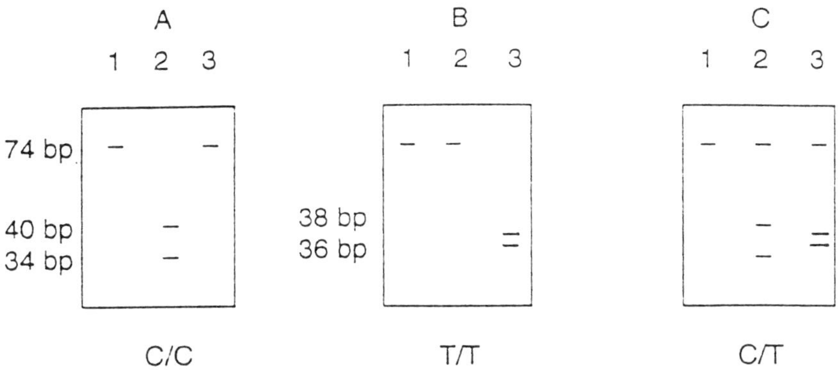

Figure. 3. Detection of the C to T mutation at position 1843 by CfoI (lane 2) and AspHI (lane 3) cleavage in the 74-bp PCR product from porcine genomic DNA. Lane 1 shows the PCR product undigested. CfoI digestion of the C-allele (HAL^N) generates 40- and 34-bp fragments (pigs A and C) and AspHI digestion of the T-allele (HAL^n) generates 38- and 36-bp fragments (pigs B and C). At the beginning the electrophoretic separation was conducted in a 6% NuSieve agarose gel stained with ethidium bromide. Because of the difficulties for visualizing the short fragments using this method we changed subsequently to the more sensitive radioactive PCR method. The length variation between fragments was then visualized on an acrylamide gel.

All of the four disagreements between the RYR1 genotype and the halothane test alone could be ascribed to incomplete penetrance of the HAL n allele. This conclusion is confirmed by examination of the marker haplotypes. Thus, there is complete correlation between MH status of 392 animals, as diagnosed through a combination of the halothane challenge test with S, PHI, H, A1BG, PGD haplotyping, and the DNA-based test. These data strongly support the identity of the C1843 to T mutation as a causal factor of porcine MH (Otsu *et al.*, 1991).

Associations between H blood-group system and the PHI (GPI) red-cell enzyme system and the locus specifying receptors for an *Escherichia coli* strain responsible for oedema disease

Oedema disease in the pig

Oedema disease of swine is a consistent problem in countries where modern animal husbandry is practiced. It occurs immediately after weaning and is caused in Switzerland mainly by *E. coli* strains of serotype

O139:K12(B):H1:F(107). In Switzerland, during the last 10 years, oedema disease was diagnosed in 13-21% of the dissected cadavers.

Oedema disease of pigs is an enterotoxaemia characterized by generalized vascular damage. The latter is caused by a toxin called Shiga-like toxin-II variant (SLT-IIv) produced by certain strains of *E. coli*. The bacteria have to be able to proliferate in the small intestine in order to produce an effective dose of toxin. Variable morbidity has been observed within, and between, individual litters in natural outbreaks, as well as under experimental conditions (Bertschinger *et al.*, 1986).

Description of the experiment

Susceptibility of weaned pigs to oedema disease was analysed in an experiment where pigs were orally inoculated with the *E. coli* strain 124/76 (serotype O139: K12(B): H1:F(107)) (Bertschinger *et al.*, submitted). Faecal shedding of the bacteria was monitored by daily rectal swabbing with cotton swabs. The numbers of characteristic hemolytic colonies were estimated and given a score from 0 to 8. A score of 2 corresponds to a count of 5.3 log CFU/g of wet feces, a score of 3 corresponds to 6.3 log CFU, and so forth. The extent of colonization was calculated as the mean of the two highest fecal scores. Pigs with a mean peak score of 3.5, corresponding to 6.7 log CFU/g or more, were considered susceptible to colonization. This limit was based on lack of mortality below this value, and on scores obtained from completely resistant litters. Abundantly colonized pigs were treated with an antimicrobial to save them for breeding and also for humane reasons.

Estimation of genotypes

Genotypes of pigs were estimated in order to test the hypothesis that susceptibility and resistance to colonization would be controlled by two alleles of the locus specifying receptors for the *E. coli* strain 124/76 (ECF107), and that susceptibility would be dominant over resistance. The genotypes of boars and sows were estimated based on their own colonization scores, and on results of progeny originating from at least one mating with a resistant partner. They were classified as resistant (ss), if there were no susceptible pigs in that litter, heterozygously susceptible (Ss), if approximately 50% were susceptible, and homozygously susceptible (SS), if approximately 100% of the progeny were susceptible.

Inheritance of susceptibility to bacterial colonization

Matings classified ss x ss produced 92 pigs, 89 of which were resistant to colonization (Table 3). One hundred and sixty-eight pigs descended from Ss x ss matings; 83 (49%) were found to be resistant. Of the 146 pigs originating from matings with at least one SS parent, 13 (9%) were classified resistant.

These results are considered compatible with a mode of inheritance with one genetical locus (ECF107), and with susceptibility being dominant over resistance.

Table 3. Susceptibility to colonization with E. coli strain 124/76 of 406 pigs originating from 50 matings (Bertschinger *et al.*, submitted). Colonization was assessed by monitoring fecal shedding of the bacterial strain after oral inoculation

Genotypes of parents[1] (ECF107 locus)	Numbers tested		Resistant pigs[2]	
	Litters	Pigs	Observed	Expected
ss x ss	12	92	89 (97%)	92 (100%)
Ss x ss	0	168	83 (49%)	84 (50%)
SS x ..	18	146	13 (9%)	0 (0%)

[1]Genotypic classification based on own colonization test results and on results from progeny.
[2]Pigs with fecal shedding score of 3.4 or less were considered resistant.

Estimates of recombination distances between the locus ECF107 and the loci H and PHI

Additional tests for blood factors of the H blood-group system as well as for the red-cell enzyme system PHI (GPI) were used to detect red cell antigens or electrophoretic variants respectively.

The logarithm of the odds (lod) score test is now widely used as a statistical test to determine whether the hypothesis of linkage can be accepted or must be rejected (Morton, 1955). The summary of lod scores is given in Table 4. The maximum likelihood estimates of the overall recombination frequencies ($\hat{o}$) for ECF107-H and ECF107-PHI linkage groups are 0.06 and 0.10 respectively. The lod scores are highly positive and are above the significance limit of 3. Therefore, the hypothesis of linkage between the ECF107 locus, and the PHI locus and the ECF107 locus and the H locus can be accepted. Since genes for H and PHI types are linked to those for HAL types, the ECF107 locus controlling susceptibility to colonization is, therefore, also assigned to this short chromosomal region on pig chromosome 6.

Linkage disequilibrium responsible for association between susceptibility to bacterial colonization and H and PHI types

Linkage disequilibrium designates a non-random association of alleles belonging to two linked loci. The coefficient of linkage disequilibrium ECF107-PHI (0.1479) is highly significant ($P<0.01$) and is nearly the maximum value (0.86) (Table 5). The linkage disequilibrium H-ECF107

(Table 5) is due to an excess of the haplotypes ($H^{a \text{ or } c}$, $ECF107^{s}$) and (H^{-}, $ECF107^{S}$) and a corresponding deficit of the haplotypes (H^{-}, $ECF107^{s}$) and ($H^{a \text{ or } c}$, $ECF107^{S}$). Pigs with the haplotypes (PHI^{B}, $ECF107^{s}$) and (PHI^{A}, $ECF107^{S}$) occurred more frequently than expected, whereas pigs with the remaining haplotypes occurred less frequently. The H^{a} and PHI^{B} alleles that are associated with resistance to colonization occur, however, with a higher frequency among halothane reactors (genotype HAL^{n} HAL^{n}) and among heterozygous nonreactors (genotype HAL^{N} HAL^{n}).

Table 4. Lod scores ($\hat{z}$) for linkage between the locus ECF107 and the loci H and PHI. $\hat{\theta}$ (recombination frequencies) and $\hat{z}$ are maximum likelihood estimates of $\hat{\theta}$ and $\hat{z}$

Loci	Number of informative parents	Number of informative offspring	$\hat{\theta}$	$\hat{z}$[1]
ECF107-H	3	48	0.06±0.03	10.84
ECF107-PHI	2	33	0.10±0.06	5.56

[1]Lod scores of + 3.0 or more are considered as sufficient evidence for linkage (Morton, 1955).

Table 5. Linkage disequilibrium (D), standardized linkage disequilibrium ($D_s = D/D_{max}$) and test statistics (k·w) between the linked loci ECF107-H and ECF107-PHI

Loci	D	k×w[1]	D_s
ECF107-H	0.0963	3.47	0.46
ECF107-PHI	0.1479**	9.64	0.86

**P<0.01

[1]The k-values were according to Hill (1974). This formula, however, is only valid for a random sample. A correction factor (w) must be used for family data (Vögeli and Schwörer, 1982). The k-values follow a X^2-distribution.

Conclusions

DNA-based technology has been used recently in studies of linkage with the MH (HAL) locus of both flanking and candidate genes. DNA clones for the linked markers, PHI and PGD, have been used to search for DNA polymorphisms linked to the HAL locus (Davies *et al.*, 1987, 1988; Brenig *et al.*, 1990). While these DNA markers do not give a direct test for the HAL alleles, they have allowed the HAL linkage group to be mapped on the long arm of porcine chromosome 6 by *in situ* hybridization (Davies *et al.*, 1988). In the present study, 392 informative meioses revealed cosegregation of the

C1843 to T mutation in the porcine RYR1 gene with the HAL n allele. This result provides evidence that this mutation is, indeed, the causal mutation for porcine MH.

The non-invasive DNA-based test for the C to T mutation at nucleotide 1843 can be used for diagnosis of the MH status of pigs. The test can distinguish three genotypes (N/N, N/n, n/n) eliminating the inaccuracy associated with current tests and resulting in an unambiguous definition of MH status of individual swine.

The frequency of halothane positive (HP) reaction varies from 0 to 88% in world breeds. In most breeds HP reaction is inherited as a recessive gene with adverse effects on PSS and reproduction, but beneficial effects on feed conversion index and lean content. Advantages in resistance to oedema disease also seem to be associated with the HAL^n allele. Hypersensitivity of the MH Ca^{2+}-release channel in muscle of MH susceptible pigs is a plausible explanation for the leanness. The mapping to the HAL linkage group of APOE (apolipoprotein E) or TGFB-1 (transforming growth factor B-1), with their known effects on myogenesis and adipogenesis (Massague, 1987), and by extrapolation from the human map of the LIPE (hormone-sensitive lipase) gene (Holm *et al.*, 1988), with potential effects on body fat content, supports an alternative explanation for some of the body composition phenotypes associated with the HAL locus. A challenge for the future will be to separate the beneficial effects of the halothane locus and loci in this region on carcass yield and resistance to oedema disease from the harmful effects on other traits in the population. Allele frequency can rapidly be changed by the DNA-based test. The cost-effectiveness of manipulating the frequency depends chiefly on economic balance between beneficial and harmful effects and the relative advantages of the heterozygote.

Summary

A single point mutation in the porcine gene for skeletal muscle ryanodine receptor (RYR1) is correlated with malignant hyperthermia (MH) in pigs. These results support the view that substitution of T for C at nucleotide 1843 is the causative mutation in porcine MH. The DNA-based test will provide the basis for manipulating the MH gene and its controlled inclusion in swine breeding programs. The case for manipulating the frequency of the MH gene depends on its economic advantages and disadvantages. Linkage is demonstrated between H and PHI (GPI), and a locus responsible for resistance and susceptibility to bacterial colonization (oedema disease), adding one locus to the previously identified linkage group S-HAL (RYR1)-PHI-H-A1BG-PGD.

Acknowledgements

These studies were supported in part by grants of the ETH-Zürich, the Commission for Support of Scientific Research, Berne, the Swiss Performance Testing Station Sempach (Drs. A. Rebsamen and D. Schwörer) and other public and private organizations in Switzerland. I would like to thank Mrs. U. Lenherr for typing the manuscript.

References

Archibald, A.L. 1991. Inherited halothane-induced malignant hyperthermia in pigs. In: Owen, J.B. and Axford, R.F.E. (eds), Breeding for Disease Resistance in Farm Animals, CAB International, Wallingford. UK. pp. 449-66.

Archibald, A.L. and Imlah, P. 1985. The halothane sensitivity locus and its linkage relationships. Anim. Blood Groups Biochem. Genet. 16, 253-63.

Bertschinger, H.U., Munz-Müller, M., Pfirter, H.P. and Schneider, A. 1986. Genetic resistance against *Escherichia coli* enterotoxaemia in the pig. J. Anim. Breeding. Genet. 103, 255-64.

Bertschinger, H.U., Stamm, M. and Vögeli, P. (submitted). Inheritance of resistance to oedema disease in the pig: Experiments with an *Escherichia coli* strain expressing fimbriae 107. Submitted to Vet. Microbiol.

Brenig, B., Jurs, J. and Brem, G. 1990. The porcine PHIcDNA linked to the halothane gene detects a HindIII and XbaI RFLP in normal and malignant hyperthermia susceptible pigs. Nuclear Acids Res. 18, 338.

Davies, W., Harbitz, I. and Hauge, J.G. 1987. A partial cDNA clone for porcine glucosephosphate isomerase: Isolation, characterization and use in detection of restriction fragment length polymorphisms. Anim. Genet. 18, 233-40.

Davies, W., Harbitz, I., Fries, R., Stranzinger, G. and Hauge, J.G. 1988. Porcine malignant hyperthermia carrier detection and chromosomal assignment using a linked probe. Anim. Genet. 19, 203-12.

Eikelenboom, G. and Minkema, D. 1974. Prediction of pale, soft,exudative muscle with a non-lethal test for halothane-induced porcine malignant hyperthermia syndrome. Netherlands J.Vet. Sci. 99, 421-6.

Fill, M., Stefani, E. and Nelson, T.E. 1991. Abnormal human sarcoplasmic reticulum Ca^{2+}-release channels in malignant hyperthermic skeletal muscle. Biophys. J. 59, 1085-90.

Fujii, J., Otsu, K., Zorzato, F., DeLeon, S., Khanna, V.K., Weiler, J., O'Brien, P.J. and MacLennan, D.H., 1991. Identification of a mutation in the porcine ryanodine receptor that is associated with malignant hyperthermia. Science 253, 448-51.

Gahne, B. and Juneja, R.K. 1985. Prediction of the halothane (Hal) genotypes of pigs by deducing Hal, Phi, Po2, Pgd haplotypes of parents and offspring: Results from a large-scale practice in Swedish breeds. Anim. Blood Groups Biochem. Genet. 16, 265-83.

Harbitz, I., Chowdhary, B., Thomsen, P., Davies, W., Kaufmann, U., Kran, S., Gustavsson, I., Christensen, K. and Hauge, J. 1990. Assignment of the porcine calcium-release channel gene, a candidate for the malignant hyperthermia locus, to the 6p11-q21 segment of chromosome 6. Genomics 9, 243-8.

Hill, W.G. 1974. Estimation of linkage disequilibrium in randomly mating populations. Heredity 33, 229-39.

Holm, C., Kirchgessner, T.G., Svenson, K.L., Frederikson, G., Nilsson, S.,Miller, C.G., Shively, J.E., Heinzmann, C., Sparkes, R.S., Mohandas, T., Lusis, A.J., Belfrage, P. and Schotz, M.C. 1988. Hormone-sensitive lipase: Sequence, expression, and chromosomal localization to 19cent-g13.3. Science 241, 1503-6.

Lister, D. 1979. Some physiological aspects of the stress cycle in relation to muscle function and meat quality. Acta Agric. Scand. Suppl. 21, 281-8.

Louis, C.F., Gallant, E.M., Remple, W.E. and Mickelson, J.R. 1990.Malignant hyperthermia and porcine stress syndrome: A tale of two species. Pig News Inform. 11, 341-4.

MacLennan, D.H., Duff, C., Zorzato, F., Fujii, J., Phillips, M., Korneluk, R.G., Frodis, W., Britt, B.A. and Worton, R.G. 1990. Ryanodine receptor gene is a candidate for predisposition to malignant hyperthermia. Nature 343, 559-61.

Massague, J. 1987. The TGF-B family of growth and differentiation factors. Cell 49, 437-8.

Mitchell, G. and Heffron, J.J.A. 1982. Porcine stress syndromes. Adv. Food Res. 28, 167-230.

Morton, N.E. 1955. Sequential tests for the detection of linkage. Am. J. Hum. Genet. 7, 277-318.

O'Brien, P.J. 1986. Porcine malignant hyperthermia susceptibility: Increased calcium-sequestering activity of skeletal muscle sarcoplasmic reticulum. Can. J. Vet. Res. 50, 329-37.

O'Brien, P.J., Klip, A., Britt, B.A. and Kalow, B.I. 1990. Malignant hyperthermia susceptibility: Biochemical basis for pathogenesis and diagnosis. Can. J. Vet. Res. 54, 83-92.

Otsu, K., Khanna, V.K., Archibald, A.L. and MacLennan, D.H. 1991. Cosegregation of porcine malignant hyperthermia and a probable causal mutation in the skeletal muscle ryanodine receptor gene in backcross families. Genomics 11, 744-50.

Smith, C. and Bampton, P.R. 1977. Inheritance of reaction to halothane anaesthesia in pigs. Genet. Res. 29, 287-92.

Vögeli, P. 1989. Position of the Phi and Po2 loci in the Hal linkage groupin pigs. Genet. Sel. Evol. 21, 119-25.

Vögeli, P., Kühne, R., Gerwig, C., Kaufmann, A., Wysshaar, M. and Stranzinger, G. 1988. Prediction of the halothane genotypes with the aid of the S, PHI, HAL,H, PO2, PGD haplotypes of parents and offspring in Swiss Landrace pigs. Züchtungskunde 60, 24-37.
Vögeli, P. and Schwörer, D. 1982. Linkage disequilibrium between susceptibility to the malignant hyperthermia syndrome (MHS, halothane sensitivity) and the phenotypes of the H blood-group system and the PHI enzyme in Swiss Landrace pigs. Züchtungskunde 54, 124-30.
Webb, A.J., Carden, A.E., Smith, C. and Imlah, P. 1982. Porcine stress in pig breeding. In: Proceedings of the 2nd World Congress on Genetics Applied to Livestock Production, Vol. VI, pp. 588-608.

Discussion

*Haley - I have a couple of questions on the oedema disease. What breed of pigs did you use?

*Vögeli - They were landrace breeds. We have selected them and first determined the fecal shedding. Then we found, with time, which of the animals were positive carriers, especially the parents. Then we began to select heterozygous animals, and homozygous recessives. Then we began to mate them. Then, we used the offspring for the fecal shedding test. We have also followed the different markers. We have now developed another test. We can now directly measure whether receptors are in the intestine or not. And I have seen that if you also use the other approach, you get the same results. And it seems that it is really very closely linked to the 'S' blood group locus. It's not closely linked to the halothane gene. It's linked much more to the 'S' blood-group locus, but it must be there somewhere. I don't think we really have the locus, directly, but it must be very close.

*Haley - It is quite surprising to see strong linkage disequilibrium over such a large distance in an outbreeding population. One possibility, from the data you present, is that there is obviously a small amount of misclassification on the basis of susceptibility. If penetrance is incomplete, that would tend to make the distances look larger than they actually are.

*Vögeli - Yes, it can also result from the test. This could be because it is a complicated test. The environment is not always so equal that you can analyse it. But, I have also used this analysis on pig-performance testing stations using randomly selected animals. And I have seen that only about 10 percent of our animals, these were Landrace, and Yorkshire animals, are resistant. Ninety percent are susceptible.

*Haley - So, you are seeing the same linkage disequilibrium in Large White as you do in Landrace?

*Vögeli - Yes.

*Russo - In our laboratory, we utilize PCR for the identification of halothane genotype using the mutation that you described. Now, I read the news in a pig magazine, that an international breeding company has patented this method of genotype determination. Do you know what exactly is patented. Is it the mutation, or the procedure? And, in general, is it possible to patent the results of discovery?

*Vögeli - I'm not a lawyer. We use the method now without paying any patent fees to anyone. But, we are doing research. When you use this method commercially, you must pay fees to Hoffman-LaRoche because they have the patented the method. The fees are in the range of about 10 percent of the price charged for analysing such a sample. And, the other question is patenting the probes. We don't use any probes for detecting the mutation, we only use PCR. You can use PCR directly to analyse this mutation. At the beginning, we used an approach without radioactive labelling, but then we have changed to use α-^{32}P phosphorus dATP, because the results were better. The bands were more sensitive. Especially the shorter bands.

*Lundström - I would like to ask if you have done any investigations on the halothane gene on wild boars, or other primitive domestic pigs? In the pig gene map project in Sweden, we have found the halothane gene in one of the wild boars, and it would be very interesting to see if it comes from there, or if it comes from the domestic pigs.

*Vögeli - That would be very interesting, but I don't know.

*Lingaas - I think I missed a little point about your linkage between the oedema disease gene, and the PHI group. Why didn't you do a direct comparison between the oedema disease-resistance gene, and the halothane gene?

*Vögeli - I didn't use the same animals for the DNA-based test that were used for the oedema test. It was not the same project. But, I have concluded, that because H and PHI are linked to the halothane gene, we can also say that Oedema disease gene, because it is also linked to H and PHI, is linked to HAL. This is an indirect conclusion, but, as far as locating the actual gene locus, I cannot really say where, precisely, it is, because the data is small. There must be families and recombinations, so that it can be located. There must be recombinations to determine how the gene is situated, and its position. You must use the same families, because different ones are not as informative. There may be recombination hotspots in some direction, then it is difficult to draw conclusions.

*Mäki-Tanila - It depend on the variation at these loci, that there is no linkage disequilibrium between H, and *E. coli* locus, whereas there is recombination between the PHI locus and the *E. coli* locus? Some things to explain!

Pig Gene Identification - Review

C.S. Haley and A.L. Archibald

AFRC Institute of Animal Physiology and Genetics Research, Edinburgh Research Station, Roslin, Midlothian, EH25 9PS, UK

Abstract

The identification of genes controlling variation in traits in the pig provides the opportunity to understand their function and to utilize them to improve performance and welfare. The genes of interest to the animal scientist are those where beneficial variants are already available or where there is the prospect of producing beneficial variants using molecular methods. Genes will continue to be isolated via the route of physiological study, supplemented by molecular biological techniques. In the future, further genes will be identified by the random cloning and sequencing of cDNAs followed by elucidation of gene function using such techniques as targeted mutagenesis in experimental species. Genetic mapping will allow the identification of some of the genes that contribute to genetic variation for traits of interest, but the subsequent cloning of these mapped genes is currently problematical. The mapping of genes of value also allows their manipulation via marker-assisted selection. The cloning of genes may allow their use in transgenic programs. For both transgenesis and marker-assisted selection, however, much further research is required before effective exploitation can be expected.

Introduction

Over recent decades, European breeders have been remarkably effective at enhancing the efficiency of pig production and the quality of pig carcasses. The conversion of feed into lean meat has been greatly improved and the levels of fat in the carcass have been drastically reduced, with the rate of genetic improvement in traits such as carcass lean percentage being more than 1% per annum (Mitchell *et al.*, 1982). Such changes have made pig meat both more acceptable and relatively less expensive to the consumer. These changes have been wrought by focusing selection efforts mainly on traits such as growth rate and ultrasonically measured fat depths which can be recorded relatively simply on the live animal.

Improvements in carcass lean content may have been accompanied by a slight decline in meat quality, as measured by objective indices (e.g. Christensen *et al.*, 1986). As a result, there is an increasing interest in meat quality, which is now included amongst the objectives of several breeding programs (Lundström *et al.*, 1989). It may be desirable to further increase the relative weight placed on meat quality as carcass lean content approaches an optimum level. Measuring the quality of meat is difficult and expensive, however, making large-scale recording impractical. Also, as such measures can often only be made on the slaughtered animal, selection may perforce be via an index of information from relatives.

Traditional methods of genetic improvement have relied on selection based upon measurements of performance at the whole animal level. These methods have been supplemented more recently by measurements of physiological variation that are correlated with the traits that we wish to improve and that bring us closer to the level of the genetic variation. It is now possible to contemplate the identification of individual genes that contribute to variation in traits of interest, and the direct manipulation of these genes for the improvement of animal performance, quality, or welfare. This approach may alleviate some of the problems associated with the improvement of traits that are difficult or expensive to measure, such as those associated with reproductive performance or meat quality. The focus of this paper is on methods for the identification of genes of value and on routes to their exploitation in animal improvement.

Which genes?

The genome of a mammal such as the pig may contain some 50000 to 100000 genes. In the context of animal improvement our primary interest is in a subset of these genes. The first category of genes of interest are those at which genetic variants (alleles) exist that influence animal performance. Such genetic variation may either be within a population (and thus contribute to the genetic variance of traits) or between populations (if populations are homozygous for different alleles), or both within and between populations (if alleles are at different frequencies in different populations). The second category of genes are those at which genetic variation may not currently exist (possibly because selection has already fixed the best allele), but at which directed changes could potentially produce new and better variants. Of course it may be possible to produce better variants for the genes in the first category also. Indeed, the presence of existing variation may be good evidence that such genes are amenable to further improvement.

Routes to gene identification - physiological genetics

The first route to the identification of genes of interest is essentially the tried and tested one of physiological study and experimentation, now supplemented by the techniques of molecular biology. Thus the study of a physiological/biochemical pathway leads to the identification of enzymes and peptide hormones in that pathway. This information can be used to isolate the DNA sequence responsible from a library of sequences, for example by use of a synthetic oligonucleotide probe predicted from a partial peptide sequence.

Gene identification by this route does not in itself show whether there is genetic variation already present for the gene or whether manipulation of the gene can be potentially beneficial. The latter part of this question can be studied by such methods as exogenous administration of the gene product, administration of agonists or antagonists to the gene product, or immunization against the gene product. The development of methods for gene transfer or manipulation in the live animal (broadly called "transgenesis", the techniques of which are summarized below) will provide even more powerful tools, for they potentially allow the effect of subtle changes to the amount, timing and site of gene expression to be studied in experimental animals.

Some indication of the potential role of identified genes in genetic variation within or between populations can be obtained by combining the methods of quantitative genetics with those of animal physiology. By comparison of lines differing for traits of interest, the genetic contribution of physiological variation to performance traits can be estimated. For example, selection for reduced activity of the NADPH-generating enzymes (the sum of G6PDH, 6PGDH, NADP-MDH and NADP-ICDH) has been shown to reduce backfat levels in the pig (Müller, 1986). This type of study identifies the genes encoding these enzymes as worthy of further study, but does not show that genetic variation at these loci themselves contributes to genetic variation in backfat levels. This is because variation in the activity of these enzymes may be due to variation at other loci, for example loci that control the level of expression of the enzymes, which may be located on other chromosomes.

Routes to gene identification - characterization of anonymous cDNAs

A cDNA library contains sequences that are complementary to messenger RNA and thus it contains essentially only expressed gene products. The complete or partial sequencing of randomly selected clones from a cDNA library is a very effective way of rapidly identifying genes (Adams *et al.*, 1991). Determining the role of the identified genes is, however, a major endeavor. Some of the genes may have a function suggested by their homology to previously identified genes, but for most, further study will be required before a function can be assigned. Again a powerful tool is provided by the ability to target mutations to individual gene loci in experimental species, for example allowing the effect of modifying or removing expression from particular genes. Studying the role of each individual gene in this way will be a major task, however. The characterization of anonymous clones from cDNA libraries does not seem practicable at present as a means of identifying the small proportion of genes that may influence any particular trait of interest in the pig. Nonetheless, genes characterized in this way and mapped in the human or murine genomes will, in the future, provide a supply of potential candidate genes for functional effects mapped in the pig.

Routes to gene identification - gene mapping

A third route to the identification of genes of interest is now being provided by the development of gene maps. In essence, a gene map shows the linear organization of genes on chromosomes. The distance between two gene loci that are linked together on the same chromosome is estimated (in centiMorgans - cM) from the frequency with which they recombine at meiosis. The aim of gene mapping is to produce a map with "marker" genes evenly spaced over every chromosome. In this way the whole genome is tagged and any new gene that is mapped will be found to be linked to one or more of the markers already on the map.

Marker loci

In order to estimate linkage distances between two gene loci it is necessary to record the genotypes of progeny from parents that are heterozygous for both genes. Only then is it possible to see the consequence of linkage - that a particular allele of one gene tends to be inherited with a particular allele of the second. Thus, in building gene maps, marker loci are selected which are both codominant (all genotypes at the locus can be distinguished from one another) and highly polymorphic, so that the probability of an individual being heterozygous for the marker gene is high. Often, however, marker loci have no function in themselves and the polymorphism they represent is in DNA which does not encode a functional product (such DNA represents a large fraction of the genome of mammals such as the pig). DNA technology allows the identification of large numbers of marker loci which are ideally suited for use in gene mapping. These loci fall into three basic types:

Restriction fragment length polymorphisms (RFLPs) These are often the result of a polymorphism in the DNA sequence (or site) recognized by a specific restriction endonuclease. The result is a polymorphism in the length of fragments produced when DNA is cleaved by the restriction enzyme and the fragments from a specific locus are visualized by Southern blotting. RFLPs are widespread and codominant and have proved very useful for mapping purposes, their level of polymorphism is usually low within a population because they generally only have two alleles, corresponding to the presence or absence of a particular restriction site.

Variable number of tandem repeats (VNTR) loci Some RFLPs are highly polymorphic, with many alleles, because between a pair of restriction sites there is a section of DNA that is tandemly repeated a number of times. The variability arises because the number of repeats, and thus the length of the restriction fragment, varies between alleles. One type of VNTR locus has core repeat units of ten or more base pairs and are known as minisatellite loci (Jeffreys *et al.*, 1985). Although minisatellite loci may be highly polymorphic and are widespread in the genome, their distribution may be clustered in certain regions (Royle *et al.*, 1988). A more recently discovered type of VNTR loci has a simple sequence repeat unit, commonly a dinucleotide CpA and these have been termed microsatellite loci (Litt and Luty, 1989). This type of locus is also codominant and highly polymorphic and appears to be very common and randomly spread in most genomes. It has the additional advantage that it is amenable to genotyping using the polymerase chain reaction (PCR). PCR is used to amplify the specific region containing the microsatellite repeats and then the length of the repeat length can be measured by running the PCR products on a polyacrylamide gel. The attractions of using microsatellites are the large number of loci in the genome and their high levels of polymorphism and the use of PCR, which only requires a small sample of DNA and which lends itself to future developments in automation.

Random amplified polymorphic DNAs (RAPDs) To isolate such markers, PCR primer pairs are selected essentially at random. If a pair of primers hybridizes to sequences in a DNA sample that are sufficiently close together and in the correct orientation, the PCR reaction will amplify the intervening sequence which can then be visualized on a gel. Polymorphism at the DNA level, however, means that there is often polymorphism in the PCR reaction - from some alleles a product is amplified whereas from others it is not. This type of marker has been used to rapidly produce maps in some plant species and has recently been applied to crosses between mouse lines (e.g. Woodward *et al.*, 1992). They are not ideal for general mapping purposes because there are often only two "alleles" (amplification or no amplification) and thus heterozygosity will not be high within a population. Furthermore, they are not codominant (it is not easy to distinguish an individual heterozygous for two alleles, one of which can be amplified and the other of which cannot, from an individual with two alleles which can both be amplified). Such markers may nonetheless be of use in line crosses, particularly to provide a skeleton map on which to hang more informative markers.

Mapping trait genes

Once a genetic map with markers spaced throughout the genome has been produced it can be used as a tool to map and identify trait genes. It must be noted that a trait gene must be polymorphic, and thus contribute to variation in the trait in the population under study, in order to be mapped. Many of the genes of interest will be contributing to variation in traits such as growth rate or leanness which vary quantitatively, not qualitatively, between individuals. Such genes have been called quantitative trait loci (QTL) and several or many QTL are likely to contribute to variation in each quantitative trait (although QTL may contribute to variation in several traits).

The principle of mapping a trait gene by linkage to a marker is relatively simple. In the progeny of an animal which is heterozygous for both a marker and a trait gene, one trait allele will tend to be inherited with one marker allele only if the trait gene and the marker are genetically linked. In this case there will be a phenotypic mean difference between the two groups of progeny inheriting the alternative marker alleles and such mean differences are diagnostic of linkage between a marker gene and a trait gene.

Types of population for mapping trait genes

It is possible to consider mapping trait genes, such as QTL, either within a breed or population, or in a cross between two breeds. In the former case evidence for genetic variation in traits of interest within the population is provided by non-zero estimates of the trait heritability. In the latter case a breed difference indicates that there would be genetic variation in the F_2 or backcross generations from the cross. In most cases a breed cross provides the best means of mapping trait genes. Firstly, it is more likely that QTL of large effect (i.e. those that are the easiest to map) will be segregating in a breed cross than within a breed. (Selection for a trait will tend to fix alleles of large effect first, removing the variation they cause from a population. In contrast, two breeds that have been selected in different ways may be fixed for different alleles, which will thus segregate in the breed cross.) Secondly, after a long separation, two breeds may have very different allele frequencies for marker loci due to the combined effects of genetic drift and selection. Within a breed, the maximum frequency of heterozygotes at a locus with two alleles is normally 0.5, whereas in an F_1 cross between two lines fixed for different alleles it will be 1.0. Therefore, for a marker and a linked QTL of this type, the frequency of the doubly heterozygous parents required to detect linkage will be $0.5^2 = 0.25$, whereas in the F_1 population it will be 1.0. Thus, a higher proportion of the families are expected to be informative for the detection of linkage in a breed cross than within a breed, making the breed cross a more powerful design.

The optimum strategy for the detection of QTL in the pig is thus to cross two breeds that differ for the trait or traits of interest. The larger the difference between the breeds the more likely it is that genes of large effect will be detected in their cross. The more genetically divergent the breeds are, the more informative the markers are likely to be in the cross. The F_1 individuals provide the heterozygous parents and genotyping for the markers and measurement of the traits in their offspring allows the trait genes to be mapped. The F_1 parents can either be crossed *inter se* to produce F_2 progeny or back-crossed to one or other of the parental breeds. Often the use of F_2 progeny will be optimum, because the F_2 is more powerful than either back-cross for detecting genes of additive effect and is the only one of the three crosses that can be used to detect genes that are dominant towards either parental breed.

The use of a breed cross facilitates the use of "interval mapping" (Lander and Botstein, 1989). In this procedure, instead of looking for a QTL linked to a single marker, it is looked for in the interval between two linked markers. Although the method can be applied to data collected within a breed (Knott and Haley, 1992a) its use in a breed cross is much simpler and more effective. In a breed cross, the method of interval mapping is only slightly more powerful for the detection of QTL than using a single marker, but it can provide much more accurate estimates of the effect and position of the QTL (Haley and Knott, 1992). Interval mapping also allows the evidence for QTL along individual chromosomes to be displayed graphically in the form of a likelihood surface on which the highest point indicates the most likely position of a QTL.

The disadvantage of interval mapping is that the statistical method (maximum likelihood) is relatively complex and thus it is difficult to extend the method. This is a problem because if two QTL are linked on a chromosome, the normal method, which effectively assumes that only a single QTL is present, can give biased estimates of their size and effects (Knott and Haley, 1992b). We have recently shown that a method of interval mapping using multiple regression can be used which produces very similar results to the method based on maximum likelihood (Haley and Knott, 1992). The regression method is easier to apply than maximum likelihood and can be easily extended to map two linked QTL.

Size of QTL detectable

Variation in any quantitative trait is likely to be controlled by a number of QTL. Evidence from plants and experimental species suggests that there will be a few QTL of moderate effect and many of small effect (Shrimpton and Robertson, 1988). Only the QTL of moderate to large effect can potentially be mapped in studies of reasonable size. As already noted, a cross between two breeds fixed for different alleles at both markers and QTL is the most powerful simple design for mapping and provides a conservative estimate of the number of animals that will need to be measured to detect QTL of a given size. For any trait, the absolute size of effect of QTL that can be detected will be related to the phenotypic variation in the trait and Table 1 gives examples for a number of traits.

Table 1. Size of genes detectable by mapping. The table shows the difference between the two homozygotes required for a gene of additive effect to be detected with 90% power at the significance level shown. Detection is by interval mapping with a map of 20cM spaced markers in an F_2 cross between two breeds fixed for different alleles

		Number of F_2 animals measured			
		100		1000	
		Significance level		Significance level	
Trait	Trait s.d.	5.0%	0.1%	5.0%	0.1%
Growth rate (g/day)	90.0	97.0	127.0	32.0	43.0
Backfat depth (mm)	2.5	2.7	3.5	0.9	1.2
Litter size (piglets)	2.5	2.7	3.5	0.9	1.2
Intramuscular fat(%)	0.70	0.76	0.99	0.25	0.33

Table 1 shows the detection of genes at two levels of significance. The normal 5% level is only appropriate when a single test is being performed. In practice, with many markers, many tests would be performed and the individual tests of significance will be required to be more stringent in order to achieve an overall level of false positive tests of 5%. Smaller genes are detectable as the sample size increases, with the size of gene detectable being approximately inversely proportional to the sample size. Thus as the sample size increases ten-fold, the size of gene detectable decreases by $\sqrt{10}$. The amount of variance in the F_2 population that these genes would explain varies from around 25% for a gene detected at the 0.1% level of significance with 100 individuals to 1.6% for a gene detected at the 5% level of significance with 1000 individuals. Thus gene mapping is not an easy option for gene identification, genes either have to have relatively large effects or large sample sizes are required. To detect the smaller sized genes shown in Table 1 requires 1000 animals to be measured for the trait and also genotyped for sufficient markers to cover the genome (*circa* 150 markers to cover the pig genome at 20cM resolution). This size of project can not be readily undertaken with today's technology, although more efficient and even automated genotyping methods may make it simpler in the future.

The Pig Gene Mapping Project (PiGMaP)

The recognition of the advantages of mapping the pig genome has resulted in it being the focus of the first multi-national co-ordinated program for genome analysis in a mammalian farm species. The Pig Gene Mapping Project (or PiGMaP) collaboration is centered in Europe and provides a focus for the research program and a framework for the collaboration (Haley *et al.,* 1990a; Haley and Archibald, 1991; Archibald *et al.,* 1991). PiGMaP is funded both from the EC BRIDGE program and from national sources. A wider grouping is provided by a "European Laboratory Without Walls", which currently includes 17 laboratories in Belgium, Denmark, France, Germany, Italy, the Netherlands, Norway, Sweden and the United Kingdom and may be further expanded by the inclusion of research groups outside Europe. PiGMaP was initiated in February 1991 and its major aims are: 1) to produce a moderate resolution linkage map with genetic markers spaced at approximately 20cM intervals covering at least 90% of the genome; 2) to identify the position and orientation of the linkage groups on the porcine chromosomes; 3) to align the porcine map with those of other mammals; and 4) to initiate studies to map QTL.

PiGMaP mapping strategy

Initially, map development and the study of QTL in the pig will rest upon crosses between divergent breeds. Most pig breeding, however, uses selection within breeds to improve performance, thus, ultimately a gene map will be required based on markers that are informative within breeds. The strategy adopted for genetic mapping is to build a map based upon crosses between divergent breeds and to incorporate markers into this which are also informative within breeds. The breeds of pig used for the reference crosses in PiGMaP are of three types; European commercial breeds, Chinese Meishan and European Wild Boar. In our own laboratory, Meishan and Large White pigs were crossed reciprocally to produce the F_1 generation, these animals were intercrossed to yield F_2 individuals for the linkage mapping.

Five laboratories are producing breed crosses, three of the laboratories are producing Meishan x Large White crosses and the other two laboratories are producing European Wild Boar x European commercial pig (either Large White or Pietrain) crosses. The linkage map is being developed by a group of laboratories within the PiGMaP collaboration, who all genotype the shared pool of animals. Stored DNA samples from reference family animals will allow new markers to be placed on the map as they are isolated. Markers selected from this map will provide the reference markers for further mapping studies.

Three of the types of molecular genetic markers described above (RFLPs and minisatellite- and microsatellite-based VNTRs) are being employed in the genetic mapping studies of the pig. Expressed sequences are being used in Southern blot analyses to detect diallelic RFLP loci (e.g. Couperwhite *et al.*, 1992; Mariani *et al.*, 1991). These expressed sequences provide the means to integrate the genetic and physical maps as well as to explore comparative aspects of gene mapping. In Edinburgh, DNA prepared from the Meishan and Large White animals has been digested with a panel of 12 restriction endonucleases and is being screened with a range of probes. The genetic divergence between the breeds means that many loci screened in this manner do indeed prove to be polymorphic. Several combinations of enzymes and probes have revealed optimum RFLP marker loci (i.e. ones for which the founder breeds are fixed for different alleles).

The RFLP loci that are optimum in breed crosses are not useful within breeds because they are fixed for one or other allele, i.e. they are not polymorphic within a breed. VNTR loci are polymorphic, and thus informative, both in a breed cross and within breeds. The presence of minisatellite loci in the pig have been demonstrated by the use of three different probes to reveal DNA fingerprint patterns (Georges *et al.,* 1988), but most of the effort on VNTR marker development in PiGMaP is focused on the microsatellite loci. Wintero *et al.* (1992) have isolated and sequenced over 100 loci containing simple tandem repeat (dCdA)n motifs. They have estimated that there are approximately 65000-100000 such loci in the porcine genome and they have also demonstrated that these loci are evenly distributed throughout the genome, with only the regions around the telomeres, centromeres and the nucleolar organizers showing any deficiency in such repeats. The porcine microsatellites, therefore, appear to have the same desirable characteristics as their human and murine equivalents. Over 200 porcine microsatellite loci have now been sequenced and several have been shown to be highly polymorphic, both between, and within breeds, in several of the PiGMaP laboratories, including our own (Johansson *et al.,* 1991; Archibald *et al.,* 1991; J. Brown and A.L. Archibald, unpublished results). In the future, multiplex PCR analysis of microsatellite loci and determination of the alleles with automated DNA sequencers will be essential for the execution of QTL mapping experiments involving hundreds or thousands of animals.

Genetic resources for QTL mapping

As argued above, crosses between breeds that are genetically distinct and phenotypically very different provide the most powerful resource for mapping QTL. The first results from the PiGMaP project are showing that a cross between the Meishan and Large White breeds is very informative as far as marker loci are concerned. However, as is shown in Table 2, the two breeds are also very different for a number of important traits including litter size, growth rate and fatness. For several of the traits the breed differences are several standard deviations in size, and thus there is the possibility that several genes of the magnitude shown in Table 1 could contribute to these breed differences and thus be detectable with a reasonably sized sample of F_2 animals. The table only shows a sample of the traits for which the breeds differ and other traits are being identified as the breeds are further characterized.

One reason for our focus on the Meishan x Large White cross in the first instance is the size of the breed differences for a number of traits. As the map is developed and the genotyping methods made more efficient it will be possible to consider mapping QTL which are responsible for smaller breed differences or in breeds differing for fewer traits. A number of breeds may provide useful resources for the mapping of QTL controlling particular traits, such as the Duroc or other breeds with a high intramuscular fat content and the Iberian pig or other breeds with specialized processing qualities. The qualities of the many different breeds available for study in China have hardly been assessed, and might include enhanced meat and eating quality, as well as particular disease resistance or hardiness characteristics.

There are also some opportunities for the mapping of genes within breeds. An obvious candidate as far as meat quality is concerned is the "rendement Napole" (RN) gene, which produces a difference between the two homozygotes at the locus of around two standard deviations in percentage yield after a standard pickling and cooking procedure (Le Roy *et al.*, 1990). A gene of such a size would be relatively easy to map. Further resources for mapping within breed are provided by divergent selection lines, such as those being selected for lean growth rate and efficiency and appetite at IAPGR. With sufficient divergence and a powerful enough study, it should be possible to consider the detection of some of the loci that mediate selection responses with a breed.

Table 2. European Large White and Chinese Meishan breed means and breed differences for commercially important traits. The breed difference is given in within-breed phenotypic standard deviations. The sex in which the measure was recorded is indicated. (Sources: Haley and Lee, 1990; Haley *et al., 1990;* Haley *et al.*, 1992; Serra *et al.*, 1992)

Trait	Meishan	Large White	Breed difference
Age at puberty (♀ days)	115	180	4.0
Number of teats (♀)	17	14	2.5
2nd parity litter size (♀, piglets)	15	10	1.5
2nd parity ovulation rate (♀ ova)	22	17	1.7
Prenatal survival (♀, %)	85	65	1.0
Piglet birth weight (♀ kg)	0.9	1.3	2.0
Growth rate (♂, g/day)	381	724	6.0
Subcutaneous fat depth (♀, mm)	26	9	4.5
Chop muscle area (♀, cm^2)	17	37	4.0
Age at 80kg (♂, days)	220	150	6.0

Cloning mapped genes

The use of gene mapping gives the possibility of identifying and locating some of the genes responsible for variation in a trait of interest. However, in order to study these genes in detail, it is necessary to identify and clone the DNA sequence that represents the gene. One approach, which has been used to clone disease loci such as the cystic fibrosis gene in man (Riordan *et al.*, 1990), is known as positional cloning. This approach requires the identification of DNA sequences representing genes that map at or near the known position of the trait gene. Each of these sequences must then be explored to see whether they contain a mutation that correlates with the trait variation. This approach requires a massive amount of work to identify and explore all of the potential trait genes and would not currently be possible for genes which are of agricultural interest only.

A second approach is to look at genes whose function suggests they may be associated with variation in the trait under study. It may be already known from previous studies that these "candidate" genes map in the same region as the trait gene, or if their position is not known, they can be mapped *de novo*. Study of candidate genes in the right map position may identify a variant that is likely to affect the structure or expression of the gene product and which correlates with the trait variation. This approach has been successful in a number of species and is essentially that which led to the identification of the Ryanodine receptor gene as a candidate for the halothane gene in pigs (Fujii *et al.*, 1991). The characterization and subsequent mapping of genes corresponding to "anonymous" cDNA clones is one way in which the number of potential candidate genes will be increased in future.

In either approach, the identification of a genetic variant that correlates with the trait variation does not prove conclusively that the trait gene has been identified. Genes that are very close together on the chromosome are often non-randomly associated (in linkage disequilibrium). Thus the correlation between the genetic variant and the trait may be fortuitous. One means of confirming the association as causal is to mimic the change in an experimental species such as the mouse (using targeted mutagenesis) and show that it produces the expected phenotypic effect.

The use of the candidate gene approach to identify genes contributing trait variation will become easier as more genes are cloned and characterized. Research on agricultural species such as the pig will benefit greatly from genome research in man and the mouse. This is because both DNA sequences of functional genes and linkage relationships (synteny) between genes show much conservation between different mammals. Once the gene map of the pig has been aligned with that of other mammals, the approximate position in other mammalian genomes of a gene mapped in the pig can be determined. A long-term aim of the human genome project is the identification, mapping and characterization of all functional DNA sequences in man. Thus it may, in the future, be possible to use the conservation of DNA sequence and synteny between species to identify potential candidate genes in man for trait genes that have been mapped in the pig.

A major drawback for any approach to the cloning of mapped genes is the accuracy with which they can be mapped. This is particularly applicable to genes controlling quantitative traits. The distribution of phenotypes associated with each of the genotypes has a large overlap and this means that they can only be mapped relatively imprecisely. For example, we have simulated an experiment using a map with markers spaced 10cM apart and 1000 F_2 animals from a cross between two breeds fixed for different alleles (Knott and Haley,

1992b). A gene which is responsible for one phenotypic standard deviation for the trait of interest between the breeds (and thus contributes 11% of the total variance for the trait in the F_2 population) can only be positioned to an accuracy of approximately ±4cM using interval mapping. The total map length of the pig is likely to be around 3000cM and, assuming it contains 100000 genes, the 8cM section to which the trait gene maps could possibly include 300 genes. Even if all of these can be identified and then unlikely candidates discarded, screening the remainder would be an extremely arduous task with current technology. Thus, routine application of the candidate gene approach to gene identification is likely to require both more accurate estimates of trait gene position and improved methods of screening candidate loci.

The value of gene identification

There are several arguments for the value of the identification of individual genes. In the first place, there is much to be learned about the control of pig performance. As already noted, past selection has largely been at the phenotypic level, with no knowledge of the action of the majority of individual genes contributing to this variation. The mapping of genes contributing to this variation will allow the study of the locations, effects and interactions of the more important loci. The mapping approach may also allow the integration of studies of phenotypic, physiological and genomic variation. With traditional approaches, if a breed difference in phenotype co-exists with one in a physiological variable it is difficult to decide whether the association is causal or casual, i.e. the physiological difference may be unrelated and irrelevant to the phenotypic difference. However, if variation for both traits proves to map in the same position this both provides some evidence of causality, significantly reducing the number of potential candidate loci that must be screened

The cloning of a gene of course permits a detailed study of its expression and action. The cloning of genes identified as QTL may be particularly valuable. By definition, at least two alleles with different phenotypic effect would already be available for study. The correlation of the difference at the DNA level between two alleles with their physiological and phenotypic effect will provide significant insight into the gene action. This in turn will aid the development of new variants with modified action for further study and perhaps exploitation via transgenesis.

Marker-assisted selection

The mapping of QTL provides the possibility of increasing the frequencies of a favored allele by selection of marker alleles known to be linked to the desired QTL allele. The use of markers to help the introgression of one, or a few, QTL of value from one line to another is probably where marker-assisted selection will see its greatest initial use in animal breeding. A QTL of value is mapped between two markers in an F_2 or back-cross population. One line (line A, for example) carries an allele of value at this QTL, but a second line (line B), is generally the better, so it is desired to introgress the QTL allele into this line. Animals from an F_2 or back-cross between the lines are selected which carry the alleles from line A for the markers flanking the desired QTL. From these, animals are selected with as many as possible of their remaining marker alleles from line B. These animals will thus carry the desired allele from line A, but have the majority of their genome from line B. The selected animals are back-crossed to line B and further rounds of marker selection and back-crossing performed until the desired proportion of genome derives from line B, then the selected animals are crossed with one another and animals homozygous for the desired region are identified.

In the traditional method of gene introgression, a cross between lines A and B followed by a similar program of back-crossing to line B with selection of individuals carrying the gene solely on the basis of their phenotype. There are several reasons why the use of markers can be advantageous. Firstly, it may be difficult to genotype animals for the gene of interest during the back-crossing phase. The gene to be introgressed may be recessive, i.e. it may not be possible to genotype animals at all in the back-cross, and in this case introgression using traditional techniques is difficult to contemplate. The gene of interest could be quantitative in effect, and thus difficult to genotype unequivocally. For example, a dominant allele that increased the litter size in pigs by two would be of major economic importance and yet this effect represents only about 0.8 phenotypic standard deviations. In a traditional backcrossing program it would be difficult to distinguish animals carrying the allele from those that did not because the distribution of their performance would substantially overlap. The litter size example demonstrates a further problem with the traditional approach, that is that many traits of interest are only expressed in one sex and that even obtaining phenotypic records on this sex may be expensive or time-consuming. Other traits that fall into these latter categories would include disease-resistance genes, which can only be measured by disease challenge, or meat-quality or carcass traits that can only be recorded post-slaughter.

Once a trait gene of interest has been accurately mapped, selection on markers during the back-crossing phase can help overcome the problems outlined above. Selection on flanking markers should be effective in retaining the gene even in the absence of good phenotypic information. A gene mid-way between two markers with each of which it has a recombination fraction of 0.1 will only be lost during introgression if a double recombinational event occurs in one cross-over on either side of the gene. In the absence of interference this event will only occur with a frequency of 0.01 and it will be less frequent if the gene is not mid-way between the markers. The marker genotype of individuals can also be measured in both sexes when they are born, removing the requirement for difficult or costly phenotypic measurements.

Selection on markers other than those flanking the introgressed gene can be used to rapidly increase the proportion of the genome derived from the recurrent parent (line B in our example) (Tanksley and Rick, 1980; Tanksley *et al.*, 1981; Hillel *et al.*, 1990). This is of value even when the gene being introgressed can be unequivocally genotyped without the aid of markers (an example here might be the introgression of a transgene into a line). Hillel *et al.* (1990) have shown that selection on markers under favorable circumstances can result in almost 100% of the genome being from the desired line by only the second back-cross.

Thus the use of marker-assisted introgression holds promise for the use of genes from genetically less-advanced breeds. For example, the mapping and transfer of one or more genes from the Meishan which control its prolificacy would have an appreciable economic impact. Further research is, however, required in order to determine the practicality of such programs and to explore their economic value relative to more traditional schemes.

Transgenesis

With the cloning of genes that may affect traits of interest, it becomes possible to contemplate the direct modification of the genome of an animal to enhance efficiency of production, quality of product or animal welfare. At present, the possibilities in large mammals are limited both by the available technology, and by the paucity of cloned trait genes. The only proven method of gene transfer in large mammals is the direct injection of DNA into the pronuclei of fertilized eggs, which are then transferred into recipient females. Often less than 1% of injected embryos produce transgenic offspring, however, and in those, the site of integration of the gene is random, leading to variable expression of the transgene and occasional disruption of endogenous genes. The method is also limited to the addition of genes, rather than the modification of genes already present. These limitations make direct injection a very unsatisfactory route for the production of transgenic farm animals.

In the future, a much improved route for the production of transgenic animals may be provided by the isolation of embryonic stem (ES) cells from farm species. These are cells that are isolated from the blastocyst inner-cell mass and grown in the laboratory, but retain the ability to differentiate into any cell type when incorporated into a recipient embryo (e.g. Bradley *et al.*, 1992). The value of such ES cells is that genetic manipulation can be carried out on large numbers of cells cultured in the laboratory. This means that subtle changes directed at individual genes (site-directed mutagenesis) can be performed, so endogenous genes can be altered as well as new genes added. Once the desired change has been made, the modified ES cells can be incorporated into recipient embryos. In some of the resulting chimeras, the modified cells may colonize the germ line and thus these animals may produce non-chimeric transgenic offspring.

Although the isolation of ES cells and their modification *in vitro* has been achieved in mice (reviewed by Bradley *et al.,* 1992), as yet there have been no reports of success in farm animals. When this technology is available, however, we may have the ability to make subtle changes, but in many cases our knowledge about the genes and the control of their expression will still be insufficient to identify potential applications. Often, successful exploitation of a gene may rely upon it being expressed in the right tissue, at the correct time, and in appropriate quantities. Thus, in addition to the identification of genes and the development of methods of gene transfer, further understanding of the control of gene expression is required for the full exploitation of these technologies.

An illustration of the value of the control of gene expression is provided by the published results of the production of transgenic pigs expressing additional copies of the growth hormone gene (e.g. Pursel *et al.,* 1989). Although the resultant animals do not show a dramatic increase in growth rate, they have reduced levels of subcutaneous backfat (7.5mm *versus* 21mm) and increased growth efficiency (by 16-18%) compared with non-transgenic litter-mates. The transgenic animals, however, have markedly reduced general fitness, with, for example, joint abnormalities and low reproductive performance. These abnormalities are associated with the over-expression of the growth hormone transgene as a result of its removal from the normal feedback control mechanisms.

Incorporation of genes into breeding programs

Although the production of individual transgenic animals is a major task, it is only part of the process for their use in breeding programs. The value of animals homozygous and hemizygous for any transgene must be assessed and they must be screened for undesirable side-effects (Smith *et al.,* 1987). This will require several generations to produce the numbers of transgenic animals needed for reasonably accurate assessment. During the period of assessment the normal breeding population will have been improved by the application of the traditional methods of selection. Thus, however genetically advanced the transgenic animals were at the time of their production, they will lag behind the normal population.

The incorporation of transgenes into advanced nucleus populations is probably best achieved by the back-crossing of transgenic females to nucleus males (Haley, 1991; Gama *et al.,* 1992). This reduces the genetic lag between the nucleus population and the transgenic population but cannot eliminate it altogether, it is likely to be equivalent at least to between one and two generations of selection (Haley, 1991; Gama *et al.,* 1992). Thus, for the introgression of a transgene into a nucleus population to be worthwhile, it must bestow sufficient benefit to overcome this genetic lag. As the economic merit of pig populations is improving at rates approaching 2% per annum (Mitchell et *al.,* 1982), it is likely that a transgene will have to improve net economic merit by at least 5% to be worth using.

These calculations have been performed with the introgression of transgenes into nucleus populations in mind. Very similar considerations, however, apply to the introgression of QTL from one breed to another. Genetic lag between the breed from which the QTL comes and the nucleus population will require to be minimized by the use of back-crossing. As noted earlier, the use of markers may speed this process (and also be of value in the introgression of transgenes). Nonetheless, it is likely that a QTL will have to have an appreciable benefit to make its introgression worthwhile.

Conclusions

The new technologies of molecular biology are facilitating the identification of genes of value for pig improvement. Gene-mapping technology has provided a new tool for the identification of genes that control trait variation. A complete gene map of the pig is currently being developed in PiGMaP. Nonetheless, improved efficiency of genotyping animals for genetic markers will be required before the methods can be applied on a sufficient scale to detect genes of all but the largest effect. The cloning of genes known only by their map position is currently a major hurdle. Genes identified through mapping may be utilized in breeding programs by marker-assisted selection, but further studies are required to show when this is worthwhile. Where genes can be cloned, they may be used in transgenic programs, but much greater knowledge of which genes to manipulate, better methods for the production of transgenic animals, and control of gene expression are required. In order to identify and assess the effects of genes the skills of animal physiology, and quantitative genetics are required. The new technologies complement other scientific skills, they do not replace them.

Acknowledgements

The authors are grateful for support for their work from the Ministry of Agriculture, Fisheries and Food, the Agricultural and Food Research Council and the European Commission BRIDGE program.

References

Adams, M.D., Kelley, J.M., Gocayne, J.D., Dubnick, M., Polymeropoulos, M.H., Xiao, H., Merril, C., Wu, A., Olde, B., Moreno, R.F., Kerlavage, A.R., McCombie, W.R. and Ventner, J.C. 1991. Complementary DNA sequencing: Expressed sequence tags and Human Genome Project. Science 252, 1651-6.

Archibald, A.L., Haley, C.S., Andersson, L., Bosma, A.A., Davies, W., Fredholm, M., Geldermann, H., Gellin, J., Groenen, M., Gustavsson, I., Ollivier, L., Tucker, E.M. and van de Weghe, A. 1991. PiGMaP: A European initiative to map the porcine genome. Anim. Genet. 22, Suppl. 1, 82.

Bradley, A., Hasty, P., Davis, A. and Ramirez-Solis, R. 1992. Modifying the mouse: Design and desire. Biotechnology, 10, 534.

Christensen, A., Sorenson, D.A., Vestergaard, T. and van Kemenade, P. 1986. The Danish pig-breeding program: Current system and future developments. Proc. 3rd World Congr. Genet. Appl. Livest. Prod. X, 143.

Couperwhite, S., Kato, Y. and Archibald, A.L. 1992. A TaqI RFLP at the porcine thyroid stimulating hormone, B-subunit locus (TSHB) Anim. Genet. (in press.)

Fujii, J., Otsu, K., Zorzato, F., de Leon, S., Khanna, V.K., Weiler, J., O'Brien, P.J. and MacLennan, D.H. 1991. Identification of a mutation in the porcine ryanodine receptor that is associated with malignant hyperthermia. Science 253, 448.

Gama, L.T., Smith, C. and Gibson, J.P. 1992. Transgenic effects, introgression strategies and testing schemes in swine. Anim. Prod. 54, 427-40.

Georges, M., Lequarre, A.S., Castelli, M., Hanset, R. and Vassart, G. 1988. DNA fingerprinting in domestic animals using four different minisatellite probes. Cytogenet. Cell. Genet. 47, 127.

Haley, C.S. 1991. Considerations in the development of pig-breeding programmes. Asia. J. Anim. Sci. 4, 305.

Haley, C.S., and Archibald, A.L. 1991. The Pig Gene Mapping Project. Commission of the European Communities DGXII, Brussels.

Haley, C.S. and Knott, S.A. 1992. A simple regression model for interval mapping in line crosses. Heredity (in press.)

Haley, C.S. and Lee, G.J. 1990. Genetic components of litter size in Meishan and Large White pigs and their crosses. Proc. 4th World Congr. Genet. Appl. Livest. Prod. XV, 458.

Haley, C.S., Archibald, A.L., Andersson, L., Bosma, A.A., Davies, W., Fredholm, M., Geldermann, H., Groenen, M., Gustavsson, I., Ollivier, L., Tucker, E.M. and Van de Weghe, A. 1990a. The pig gene-mapping project: PiGMaP. Proc. 4th World Congr. Genet. Appl. Livest. Prod. X111, 67.

Haley, C.S., Lee, G.J., Wilmut, I., Ashworth, C.J., Aitken, R.P. and Ritchie, W. 1990b. British studies of the genetics of prolificacy in the Meishan pig. In: Molenat, M. and Legault, C. (eds), Proceedings of the Chinese Pig Symposium, Toulouse, France INRA, Jouy-en-Josas, France.

Haley, C.S., D'Agaro, E. and Ellis, M. 1992. Genetic components of growth and ultrasonic fat depth traits in Meishan and Large white pigs and their reciprocal crosses. Anim. Prod. 54, 105.

Hillel, J., Schaap, T., Haberfeld, A., Jeffreys, A.J., Plotzky, Y., Cahaner, A. and Lavi, U. 1990. DNA fingerprints applied to gene introgression in breeding programs. Genetics 124, 783.

Jeffreys, A.J., Wilson, V. and Thein S.L. 1985. Hypervariable "minisatellite" regions in human DNA. Nature 314, 67.

Johansson, M., Ellegren, H., Edfors-Lilja, I. and Andersson, L. 1991. Cloning and characterization of highly polymorphic porcine microsatellites. Anim. Genet. 22: Suppl. 1, 64.

Knott, S.A., and Haley, C.S. 1992a. Aspects of maximum likelihood interval mapping in an F2 population Genet. Res. (submitted).

Knott, S.A. and Haley, C.S. 1992b. Maximum likelihood mapping of quantitative trait loci using full-sib families. Genetics (submitted).

Lander, E. S., and Botstein, D. 1989. Mapping Mendelian factors underlying quantitative traits using RFLP linkage maps. Genetics 121, 185.

Le Roy, P., Naveau, J., Elsen, J.M. and Sellier, P. 1990. Evidence for a new major gene influencing meat quality in pigs. Genet. Res. 55, 33.

Litt, M. and Luty, J.A. 1989. A hypervariable microsatellite revealed by *in vitro* amplification of a dinucleotide repeat within the cardiac muscle actin gene. Am. J. Hum. Gen. 44, 397.

Lundström, K., Karlsson, A., Lundeheim, L., Rydhmer, L. and Vestergaard, T. 1989. Current methods and perspectives in breeding for meat quality in pigs. 40th Annual Meeting of the EAAP, Dublin.

Mariani, P., Johansson, M., Ellegren, H. and Andersson, L. 1991. A TaqI restriction fragment length polymorphism at the porcine platelet-derived growth factor, B-receptor locus (PDGFRB). Anim. Genet. 22, 446.

Mitchell, G., Smith, C., Makower, M. and Bird, P.J.W.N. 1982. An economic appraisal of pig improvement in Great Britain. 1. Genetic and production aspects. Anim. Prod. 35, 215.

Müller, E. 1986. Physiological and biochemical indicators of growth and composition. In: Smith, C., King, J.W.B. and McKay, J.C. (eds), Exploiting New Technologies in Animal Breeding, Oxford University Press, UK, pp. 132-139.

Pursel, V.G., Pinkert, C.A., Miller, K.F., Bolt, D.J., Campbell, R.G., Palmiter, R.D., Brinster, R.L. and Hammer, R.E. 1989. Genetic engineering of livestock. Science 244, 1281.

Riordan, J.R., Rommens, J.M., Kerem, B., Alon, N., Rozmahel, R. *et al.* 1990. Identification of the Cystic Fibrosis gene: Cloning and characterization of complementary DNA. Science, 245, 1066.

Royle, N.J., Clarkson, R., Wong, Z. and Jeffreys, A.J. 1988. Clustering of hypervariable minisatellites in pro-terminal regions of human autosomes. Genomics 3, 352-60.

Serra, J.J., Ellis, M. and Haley, C.S. 1992. Genetic components of carcass and meat quality traits in Meishan and Large White pigs and their reciprocal crosses. Anim. Prod. 54, 117.

Shrimpton, A.E. and Robertson, A. 1988. The isolation of polygenic factors controlling bristle score in *Drosophila melanogaster.* 11 Distribution of third chromosome bristle effects within chromosome sections. Genetics 118, 445-59.

Smith, C., Meuwissen, T.H.E. and Gibson, J.P. 1987. On the use of transgenes in livestock improvement. Anim. Breed. Abstr. 55, 1.

Tanksley, S.D. and Rick, C.M. 1980. Isozyme gene linkage map of the tomato: applications in genetics and breeding. Theor. Appl. Genet. 57, 161.

Tanksley, S.D., Medina-Filho, H. and Rick, C.M. 1981. The effect of isozyme selection on metric characters in an interspecific backcross of tomato - basis of an early screening procedure. Theor. Appl. Genet. 60, 291.

Wintero, A. K., Fredholm, M. and Thomsen, P.D. 1992. Variable (dG-dT)n. (dCdA)n sequences in the porcine genome. Genomics 12, 281.

Woodward, S.R., Sudweeks, J. and Teuscher, C. 1992. Random sequence oligonucleotide primers detect polymorphic DNA products which segregate in inbred lines of mice. Mammalian Genome 3, 73.

Discussion

*Vögeli - How far has the QTL research progressed, and how will the QTLs be used in the future?

*Haley - Well, the first step is to produce the gene map, that is perhaps 2 years' work. The map then has to be applied to populations that are sufficiently large to find genes of interest. Our first interest will be to cross the Meishan and the Large White, because the differences are so large. Using this cross, genes can perhaps be found with only 500 or so individuals. So, we can contemplate finding genes on a time scale of about 4 years. Once we've found a few QTL, I would anticipate that the genotyping technology will have moved on sufficiently for us to contemplate using them we can begin integrating those genes into populations. So, I would anticipate that perhaps on a 5 year time scale we might have found a few genes of large effect which we could use by introgression in to our ordinary nucleus populations. That would be the first application, perhaps 5 years down the road.

*Lingaas - You have told us about these methods for QTL detection. They probably have the most potential for finding genes of great effect but for 30 years we have had intensive selection for production traits, or for economically important genes. If there were any genes with a really large effect, wouldn't you expect those genes to be fixed now?

*Haley - Yes, that is a very good point. I have focused here on using crosses between breeds. We can identify breeds that might have had different selection histories in the past, such as the Meishan and the Large White, which may be fixed for different alleles. If we go to a Large White population, for example, and look at that population which has, as you said, been selected for 20 generations for growth rate and leanness, it is unlikely that there will be any variance for genes with a large effect on growth rate and leanness in that population. The very genes easiest to map will be selected to fixation, and so, in those populations, we're not going to find genes of interest for the traits which have been under selection. The only ones we will find are ones that have arisen by recent mutation, or perhaps by recent introgression from another breed. But, there might be genes for traits which haven't been under selection particularly intensely. So that's why the preliminary focus is on breed crosses. It is much more difficult to find genes within breeds. You need much larger sample sizes to have any hope of doing it. So we can only contemplate doing it at a time when we have a machine that lets us take a blood sample from a pig, drop it in the machine, and it gives us a genotype for 150 loci on that animal, and it costs about 1 dollar. We can only contemplate looking within breeds when we have that sort of technology, and I don't know when we will have that sort of technology.

*Mäki-Tanila - Is there any difference, in principle, between a gene locus with a large effect on a quantitative trait, and a transgene which also has a large effect?

*Haley - I would say that the difference is, if you like, that we know that the gene of large effect works in the physiological context from which it came.

*Mäki-Tanila - Like the halothane gene.

*Haley - Yes, we know something of what it does. But, if you produce a transgene, it could produce myriad effects, which might actually be deleterious, as we have seen. So, there are differences between transgenes and QTL. But, in actually using those genes, if they did have the same effect, there would be no practical difference. Other than perhaps ethical concerns, that one is natural variation, and one is "unnatural" variation.

Breeding for Pork Quality

A.G. de Vries and P.G. van der Wal

DLO-Research Institute for Animal Production "Schoonoord"
P.O. Box 501, 3700 AM Zeist, The Netherlands

Abstract

Genetic parameters for pig meat quality traits in halothane-free populations were calculated and the prospects and strategies for genetic improvement of pork quality are discussed.

The required parameters were quantified in data from seven Yorkshire lines. Various traits were measured in the slaughterhouse and at the laboratory (among others (a.o.) pH, drip, cooking loss, color, tenderness and intramuscular fat). The genetic variance of water binding and color traits appeared to be equal to the variance due to day of slaughter (about 20%). Relationships between laboratory and slaughterhouse traits demonstrated opportunities for genetic improvement using selection based on inexpensive measurements like pH and reflectance. No unfavorable genetic correlations between carcass lean percentage and water binding or color traits could be found.

High heritabilities combined with a negative genetic relationship were found between carcass lean percentage and intramuscular fat percentage. Each further 1% increase of lean percentage will be accompanied by a reduction of 0.07% in intramuscular fat percentage.

Several strategies are discussed to breed for quality traits like color and water binding. In the short term, selection should aim for a decreased sensitivity to the PSE aberration. This can be achieved by selecting for low reflectance value and high pH values at 45 min post-mortem. For a more accurate evaluation, ultimate pH deviations below 5.6 and positive color scores deviations (PSE like) should be included. With this approach, no increased DFD incidence is expected.

Efficiency of selection for meat quality is probably rather low, when based only on phenotypic data from sibs. Future breeding programs may exploit DNA-markers as additional tools.

Introduction

In pig breeding, attention has primarily been directed to carcass quality (carcass lean percentage) and efficiency of production (Ollivier *et al.*, 1990). Improvements have been made not only by selecting for a better genotype, but also by providing better environmental conditions and feeding practices.

Until now, selection programs for pigs were mainly based on phenotypic performance data of individuals. This method of selection was successful for growth rate and lean percentage (Smith, 1984). For other traits, genetic improvement is more difficult to achieve, because of lower heritability, or difficulties with recording data.

A few developments which may challenge pig breeding organizations to shift selection pressure from growth characteristics to meat quality are: 1) increased attention of consumers to product quality; 2) growing interest in integrated production systems; and 3) increased possibilities to select for meat quality.

Apart from the effects of the halothane gene, knowledge about genetic effects on pork quality is limited. The same holds true for environmental effects. Quantification of genetic parameters is needed to evaluate the role of breeding organizations, and to find optimal strategies for integrating selection for meat quality into present genetic improvement programs. Since it can be assumed for today's breeding populations that the halothane gene is under control, either by gas testing (Eikelenboom and Minkema, 1974), or by DNA-probing (Fujii *et al.*, 1991), research should focus on the effects of other genes.

The aim of this paper is to quantify parameters of pig meat quality traits in halothane-free populations, and to discuss the prospects and strategies for genetic improvement of pork quality. Data from an experiment with seven Yorkshire lines was used for quantifying the required parameters. No attention is given to fat quality, although this may also be a trait of concern (Wood *et al.*, 1986; Cameron, 1990).

Materials and methods

Experimental data

The experiment was set up in co-operation with seven Dutch breeding organizations. From each organization, a random sample of one of their sire lines was evaluated. All animals were of Dutch Yorkshire origin. The performance test was done within each breeding organization's own test facilities, generally under *ad lib* feeding conditions. Each breeding organization also used their own slaughter facilities. Pigs were slaughtered in weekly batches. Comprehensive meat-quality evaluations were done, on average, on seven pigs per week. Some traits were also measured on about ten additional animals. Transport and handling was standardized within breeding organizations.

In the slaughter-line, leanness and different meat-quality traits were measured at 45min post-mortem, using the *M. longissimus dorsi* either between the 3rd and 4th lumbar vertebrae, or Hennessy Grading Probe (HGP) between the 3rd and 4th from last ribs, as:

- **Lean %**: carcass lean percentage estimated with the HGP (Walstra, 1988)
- **HGP-PSE**: PSE reflectance measurement by the HGP
- **pH1**: pH value.

The day after slaughter (about 20h post-mortem), the following measurements were taken:

-**FOP**: reflectance measurement by the Fibre Optic Probe (between 3rd and 4th lumbar vertebrae)

- **pH2**: pH value (same position)

-**Minolta L***: L* value (paleness) according to Cielab system with a tristimulus color analyser (Minolta CR 110) at the loin surface between 2nd and 3rd lumbar vertebrae after blooming for 30min.

For a more comprehensive evaluation of part of the carcasses, samples from the loin between the last and the 2nd lumbar vertebrae were transported to our institute. In the laboratory, the following measurements were performed after 24h post-mortem:

- **pH3**: pH value (between 3rd and 4th lumbar vertebrae)

-**Filter paper**: measurement of water-holding capacity with the filter paper test (Kauffman *et al.*, 1986)

-**Hunter L***: L* value (paleness) according to Hunter Labscan spectro-colorimeter LS-5000 after blooming for 30min

-**Swelling %**: measurement of water uptake during low speed centrifugation (900 x g) (Wierbicki *et al.*, 1962)

-**Drip loss**: drip loss, derived from the method of Honikel (1987); samples stored during 48h at 4°C on an absorption pad and covered with a packaging film

-**Cooking loss**: cooking loss, determined after drip loss, by packing the samples in vacuumized polyethylene bags and heated for 60min in a 75°C waterbath (Boccard *et al.*, 1981)

-**Shearforce**: tenderness measurement, using ten cores (diameter 1.26cm) from the heated samples (Boccard *et al.*, 1981); average of maximum forces required to shear the cores with a Warner-Bratzler blade

-**Intramuscular fat %**: intramuscular fat determined in a loin sample by the Fosslet extraction method (Olsman, 1978).

Statistical analysis

The following mixed model was used for analysing the data:

$$y = Xb + Ls + Zu + e$$

where y: vector with observations
b: vector with fixed effects
s: vector with random effects of day of slaughter
u: vector with random additive genetic effects
X: incidence matrix for fixed effects
L: incidence matrix for day of slaughter
Z: incidence matrix for additive genetic effects.

Fixed effects in the model were breeding organization (seven levels) and sex (three levels). Random slaughter day effects (257 levels) were nested within breeding organization. For the genetic effects, up to three generation relationships were used for fitting an animal model. Univariate REML estimates of the parameters were derived using a derivative-free algorithm (Meyer, 1991). A variance of the log likelihood values of 0.001% was taken as convergence criterion. Approximate standard errors were calculated for the variance components (Meyer, 1991).

Correlations were estimated, following the approach of Thompson and Hill (1990), using a canonical transformation and calculating covariances as half the difference between the variance of the sum of two traits and the summed variances of the individual traits (after scaling to equal phenotypic variances). In the first round, the original traits were analysed. In the following rounds, the traits were transformed to canonical variates using estimated variance/covariance components derived from the previous round. The procedure was stopped when the maximum change in correlation between successive rounds was less than 0.05. Standard errors on the correlations could not be derived with the procedure, but they might be approximated using the formula given by Falconer (1989).

Results

Levels and variance

Table 1 shows overall means and standard deviations. For comparison, the table also lists levels of traits for DFD, normal and PSE meat (based on visual scores) as found in previous studies at IVO-DLO (among others (a.o.) van der Wal *et al.*, 1988). The levels of most traits are within the range of normal meat. However, the standard deviations reveal considerable variation in water binding and color. As a result, considerable fractions of the experimental population tend toward DFD or PSE meat.

Table 1. Overall means, standard deviations (n = 1969), and levels of traits categorized as PSE, normal or DFD meat (based on previous studies at IVO-DLO)

Trait	Mean	S.D.	Category levels		
45 min pm			DFD	normal	PSE
HGP-PSE	47.2	7.3	<20	35-45	>60
pH1	6.36	0.33	6.5	6.3	5.8
20h pm					
FOP-2	32.7	13.3	<10	10-25	>35
pH2	5.71	0.25	6.6	5.6	5.4
Minolta L*	52.2	3.8	45	50-55	60
> 24h pm					
pH3	5.61	0.23	6.6	5.6	5.4
Filter paper (mg)	45.3	23.4	25	60	130
Hunter L*	55.6	3.4	45	50-55	60
Swelling (%)	40.7	28.4	130	40	20
Drip loss (%)	4.06	2.04	1	3.5	8
Cooking loss (%)	31.4	2.9	15	29	32
Shearforce (daN)	3.33	0.60	2.3	3.5	3.5
Intramusc. fat (%)	1.32	0.51		----- n.r.[1)]----	

[1)] not relevant

A low level of intramuscular fat content was found (1.32%), together with a rather high standard deviation (0.51%). This means that a high fraction of meat samples had an extremely low proportion of fat (<1%). Such meat is expected to be less tasty and not so tender (Bejerholm and Barton-Grade, 1986) as meat with more intramuscular fat.

Sources of variance

The variance components are given in Table 2. In addition to the meat quality traits, lean percentage is tabulated. The percentage of total variance for fixed effects (R^2) and accounted for by day of slaughter (s^2) were quite low for lean percentage and intramuscular fat. This indicates small differences between the breeding populations and between batches. The heritability of intramuscular fat percentage was quite high (0.41), which is in agreement with the literature (Schwörer *et al.*, 1990).

R^2 values for pH, color and water binding traits were around 0.20. These values can be explained by the differences in pre-slaughter management, e.g. one of the breeding organizations kept the pigs overnight at the slaughterhouse. Components of variance due to genotype (h^2) and day of slaughter (s^2) appeared to be of the same magnitude, around 0.20. Estimates of s^2 were higher than expected. These values were expected to be lower because there was not much variation in the (within breeding organization) transport or (pre-)slaughter conditions between batches.

Table 2. Percentage of total variance explained (R^2) by fixed effects (breeding organization and sex), phenotypic standard deviation (σ_p), relative part of phenotypic variance due to genotype (h^2) and day of slaughter (s^2)

Trait	R^2	σ_p	h^2(s.e.)	s^2(s.e.)	n
45 min pm					
Lean %	0.11	2.35	0.52 (0.05)	0.03 (0.01)	4055
HGP-PSE	0.11	6.52	0.18 (0.03)	0.27 (0.02)	4055
pH1	0.07	0.319	0.15 (0.03)	0.17 (0.02)	4055
20h pm					
FOP	0.41	10.2	0.15 (0.03)	0.13 (0.02)	4055
pH2	0.29	0.217	0.21 (0.03)	0.21 (0.02)	4055
Minolta L*	0.15	3.51	0.21 (0.05)	0.17 (0.02)	1881
> 24h pm					
pH3	0.19	0.208	0.15 (0.05)	0.22 (0.03)	1881
Filt. paper (mg)	0.07	22.5	0.20 (0.05)	0.21 (0.02)	1881
Hunter L*	0.22	3.02	0.27 (0.06)	0.11 (0.02)	1837
Swelling %	0.25	24.8	0.20 (0.05)	0.14 (0.02)	1881
Drip loss (%)	0.20	1.84	0.20 (0.06)	0.14 (0.02	1837
Cooking loss (%)	0.13	2.75	0.11 (0.05)	0.21 (0.03)	1837
Shearforce (daN)	0.12	0.559	0.20 (0.05)	0.17 (0.02)	1881
Intramusc. fat (%)	0.05	0.505	0.41 (0.07)	0.08 (0.02)	1837

n = number of records.

Correlations

Tables 3 and 4 show correlations between laboratory and slaughterhouse traits. Genetic correlations (r_G) appear to be stronger than correlations at slaughter day (r_S) or error level (r_E). Lower r_Ps compared with r_Gs were also found in other studies (a.o. Cameron, 1990; Hovenier *et al.*, 1993). They might be explained by uncorrelated random measurement errors. The high r_G values for Hunter L* and water-binding traits with pH and FOP indicate opportunities for genetic selection for meat quality based on cheap, non-destructive measurements at the slaughterhouse.

The r_Gs of lean percentage with pH, water binding and color appear to be close to zero in the present study. However, previous studies reported unfavorable genetic relationships. These differences in results might be explained by halothane status differences between the studied populations (Sellier, 1988).

The negative correlation (-0.37) between lean percentage and intramuscular fat percentage is in agreement with the literature (Schwörer *et al.*, 1990). Since intramuscular fat percentage is, at the moment, probably below its optimum, the negative correlation with lean percentage can be considered to be unfavorable.

Table 3. Correlations between slaughterhouse measurements at 45min and 20h pm due to genetic effects (r_G), day of slaughter (r_S), non-systematic environment (r_E), and the combination of these three factors (r_P) (n = 4055)

Trait		pH1 45 min	FOP 20h	pH2 20h	Carcass lean %
HGP-PSE	r_G	-0.43	0.40	0.01	-0.00
45 min pm	r_S	-0.07	0.01	0.08	-0.16
	r_E	-0.27	0.15	0.05	-0.08
	r_P	-0.25	0.16	0.05	-0.05
pH1	r_G		-0.55	0.10	0.10
45 min pm	r_S		-0.13	0.12	-0.10
	r_E		-0.30	0.25	-0.06
	r_P		-0.31	0.20	-0.01
FOP	r_G			-0.65	0.00
20h pm	r_S			-0.23	0.11
	r_E			-0.26	-0.03
	r_P			-0.32	-0.01
pH2	r_G				0.09
20h pm	r_S				-0.06
	r_E				-0.08
	r_P				-0.01

Principal components

Principal component analysis (PCA) was performed to examine underlying factors of variance. The PCAs were extracted from the raw correlation structure of a set of ten traits, covering color, water binding and pH development. The ten traits and their correlation with the PCAs are given in Table 5. The first PCA (PCA1) combines a high pH, dark color and a high water binding. The second PCA (PCA2) is associated with the decline of pH (+ pH1, - pH2, - pH3). with early *versus* late loss of water (- drip loss, + cooking loss). With these two PCAs, a large proportion of the variation (66%) in the 10 traits could be explained.

After rotation, based on the varimax criterion (SPSS, 1985), two interpretable components were found. The first (PCA1R) represented the characteristics of the DFD aberration (high final pH), whereas the second (PCA2R) represented the PSE aberration (low pH1). When factor analysis, based on Maximum Likelihood, was performed (instead of rotated PCAs), almost identical results were found.

Table 4. Correlations between laboratory and slaughterhouse measurements due to genetic effects (r_G), day of slaughter (r_S), non-systematic environment (r_E), and the combination of these three factors (r_P) (n = 1837)

Laboratory measurements		Slaughterhouse measurements HGP-PSE 45 min	pH1 45 min	FOP 20h	pH2 20h	Carcass lean %
Drip loss	r_G	0.14	-0.56	0.55	-0.69	-0.10
	r_S	0.03	-0.22	0.47	-0.48	-0.22
	r_E	0.17	-0.46	0.44	-0.45	0.17
	r_P	0.14	-0.44	0.46	-0.49	0.05
Cooking loss	r_G	0.26	-0.24	0.66	-0.82	-0.06
	r_S	-0.14	0.14	0.15	-0.39	-0.08
	r_E	-0.06	-0.12	0.13	-0.40	0.21
	r_P	-0.03	-0.09	0.19	-0.46	0.10
Hunter L*	r_G	0.14	-0.27	0.62	-0.69	-0.09
	r_S	0.02	-0.04	0.23	-0.56	-0.23
	r_E	0.19	-0.21	0.42	-0.48	0.04
	r_P	0.15	-0.20	0.43	-0.53	-0.03
Intramusc. fat	r_G	0.32	-0.12	0.32	-0.23	-0.37
	r_S	0.03	-0.22	0.03	0.10	-0.08
	r_E	0.11	0.05	0.07	0.16	-0.12
	r_P	0.15	-0.04	0.12	0.04	-0.23

Table 5. Raw correlations of original traits with unrotated principal components (PCA1, PCA2; R^2 respect. 0.507 en 0.156) and rotated principal components (PCA1R, PCA2R) (n = 1911)

Trait	PCA1	PCA2	PCA1R	PCA2R
45 min pm				
HGP-PSE	-0.32	-0.55	-0.01	0.63
pH1	0.36	0.70	-0.03	-0.78
20h pm				
pH2	0.80	-0.24	0.82	-0.18
Minolta L*	-0.71	-0.18	-0.54	0.50
> 24h pm				
pH3	0.80	-0.40	0.89	-0.04
Filter paper	-0.71	-0.46	-0.40	0.75
Hunter L*	-0.83	0.00	-0.73	0.40
Swelling %	0.82	-0.39	0.90	-0.06
Drip loss	-0.84	-0.22	-0.62	0.60
Cooking loss	-0.70	0.33	-0.77	0.05

The PCAs were submitted to the same (co)variance analyses as the original traits. The heritability estimates for the PCAs were close to 20% (Table 6). The variances due to slaughter day effects (s^2) were between 15 and 18%. The R^2 values reveal rather high contrasts between the breeding populations for PCA1 and PCA1R. This can be explained by differences in pre-slaughter conditions which affected the risk of DFD aberrations. On the other hand, the PCA representing the PSE aberration (PCA2R) as well as PCA2 showed only small contrasts between the breeding populations (low R^2 values).

Table 6. Percentage of total variance explained (R^2) by fixed effects (breeding organization), phenotypic standard deviation (σ_p), relative part of phenotypic variance due to genotype (h^2) and day of slaughter (s^2) for unrotated principal components (PCA1, PCA2) and rotated principal components (PCA1R, PCA2R) (n = 1859)

Variate	R^2	h^2	(s.e.)	s^2	(s.e.)
PCA1	0.25	0.21	(0.05)	0.16	(0.02)
PCA2	0.11	0.20	(0.05)	0.17	(0.02)
PCA1R	0.24	0.19	(0.05)	0.18	(0.03)
PCA2R	0.13	0.21	(0.05)	0.15	(0.02)

Table 7 gives the correlations between the PCAs at different levels. By definition, the raw correlation (i.e. before adjusting for fixed effects) between the PCAs is zero. The same holds true for the correlation between PCA1R and PCA2R. The figures in Table 7 show that the genetic correlation between PCA1R and PCA2R is close to zero. This means that a higher incidence of PSE in certain families was not associated with a higher or lower incidence of DFD meat.

Discussion

Prospects

The high heritability and large phenotypic standard deviation of intramuscular fat percentage imply a large genetic variance. From the estimates in Table 2 (h^2 and σ_P), a genetic standard deviation of 0.3 can be derived. In the same way, carcass lean percentage can be shown to have a genetic standard deviation of 1.7. Table 4 showed a genetic correlation of -0.37 between the traits. Thus, when attention is focused primarily on lean percentage, each percentage increase would be accompanied with a reduction of 0.07% for intramuscular fat percentage (-0.37 x 0.3/1.7). Simultaneous selection for both traits would considerably reduce the improvement of lean percentage. However, as soon as certain commercial products reach an optimal level of lean percentage, it might become worthwhile to start selection for intramuscular fat. Perhaps some commercial populations have already obtained an optimal lean content (Lundström, 1990).

Table 7. Correlations between unrotated (PCA1, PCA2) and rotated principal components (PCA1R, PCA2R) due to genetic effects (r_G), day of slaughter (r_S), non-systematic environment (r_E), and the combination of these 3 factors (r_P) (n= 1859)

Variate		PCA2	PCA1R	PCA2R
PCA1	r_G	0.11	0.84	-0.53
	r_S	-0.10	0.86	-0.36
	r_E	0.03	0.85	-0.48
	r_P	0.03	0.85	-0.48
PCA2	r_G		-0.43	-0.90
	r_S		-0.59	-0.89
	r_E		-0.49	-0.89
	r_P		-0.50	-0.89
PCA1R	r_G			-0.01
	r_S			0.15
	r_E			0.05
	r_P			0.05

In the present study, color and water binding traits had no unfavorable correlations with lean percentage. Research is underway to estimate the correlations with growth rate and feed intake, and to check for non-linearity of the relationships. In the absence of unfavorable relationships, breeding organizations could continue their present selection schemes with emphasis on growth and carcass characteristics. However, many of them are probably tempted to play an active role in the improvement of meat quality. The heritability of 0.20 offers possibilities for genetic improvement. With proper statistical techniques and large numbers of animals per batch, it will be possible to correct for the relatively large effects of the (pre)slaughter environment, and to estimate breeding values for individual animals.

Including selection for meat quality in a breeding program will imply reduced selection pressure for other traits. Each breeding organization has to counterbalance the possible gains to be made with meat quality and the reductions in gains for other traits. A choice for including meat quality will only be considered when efficient breeding strategies are found. Another requirement will be that the breeding organizations are able to share in the financial returns from improved quality. This need can be satisfied by integrated production systems. Recently, many such production systems have been initiated (e.g. Cöp, 1989; van der Sluis, 1992).

Strategies

The first topic to be addressed is the most favorable direction of selection for meat-quality traits. The level of intramuscular fat should probably be raised. The optimal breeding objective for the other traits is debatable (Lundström *et al.*, 1989). To reduce the incidence of PSE meat, selection should be for more uniform, less pale (darker) meat with lower drip loss. However, at the same time, care is needed avoid raising the chance of DFD meat. If an increase of DFD frequency is inevitable, a good approach could be to define optimal ranges for the levels of each individual trait, and then to derive economic values of trait changes from their effects on the fractions of carcasses categorized as optimal (Hovenier, 1992). Important assumptions for such an approach are that the variances of traits are not affected by selection, and that the effects of individual traits on consumer's attitude are independent. When the latter assumption does not hold, the procedure might be combined with the approach of de Vries (1989), who defined a model demonstrating the feature that the weakness in one trait might not be fully compensated for by the strength in another trait.

Instead of changing the level, one could also try to reduce the variation in color and water binding. Breeding organizations could try to provide populations with reduced genetic variance. However, this will only reduce a small part of the variance, approximately 80% will remain, which implies a reduction of 11% in the standard deviation. A better alternative is to breed animals that are more resistant to pre-slaughter stress, thus having a smaller risk of aberrations.

As recently demonstrated by Karlsson (1992), principal component analysis appears to be a useful method for meat-quality studies. The genetic correlation between PCA1R and PCA2R being close to zero (Table 7) implies that selection against the incidence of PSE aberration would not necessarily be accompanied with a higher chance of DFD meat. However, in practice, not all ten traits in Table 5 can, or will, be measured.

A relatively inexpensive option is to measure only pH1, HGP-PSE (45 min post-mortem), pH2 and a color score (24h post-mortem). Selection should be then for low HGP-PSE and high pH1. Deviations of pH2 should only be included in the evaluation, when they are below 5.6. Similarly, only negative color score deviations (PSE-like) should be taken into account. In this way, DFD carcasses are not favored.

Like lean percentage, meat quality can not be evaluated directly on selection candidates. Muscle biopsies, as an indirect measurement on live animals, may have predictive value for meat quality. This technique can be a useful tool, especially for selection against aberrations due to altered biochemical composition of the muscles, e.g. the acid meat aberration found in Hampshire pigs (Monin and Sellier, 1985).

When meat quality has to be assessed on carcasses, performance data are not available and thus selection will rely only on family information. This implies selection between full-sib or half-sib families, since most pig-breeding organizations apply short-generation intervals. Artificial insemination boars used in nucleus populations might, at the same time, be used in subnucleus or commercial field populations. When simple quality measurements are done on large numbers of progeny from these boars, selection between half-sib families will be very accurate. However, effectiveness of selection also depends on intensity of selection. Because of inbreeding considerations, a fixed number of families should be maintained. Therefore, substantial genetic improvement can only be made in large nucleus populations with a large number of half-sib families available for selection. In general, nucleus populations are small due to the high costs per animal.

Apart from the halothane gene and the RN-gene, there might be more genes with large effects on pork quality. The unfavorable alleles of such "major genes" might directly cause an aberration or they might induce a higher risk for aberrations. Quantitative and/or molecular genetics should be able to detect such genes. DNA-probes should then be developed to eliminate unfavorable alleles. Instead of direct probes (like the probe for the halothane mutation), DNA-markers close to major genes can be utilized for selection, this technique is referred to as marker-assisted selection (MAS; Beckman and Soller, 1983). The European project for a pig gene map (PiGMaP; Haley *et al.*, 1990) can play an important role in providing such markers. Genotype scoring in MAS schemes is not restricted to sibs, but it can also be done on selection candidates. Therefore, marker scores combined with phenotypic carcass measurements (sibs) enable selection between, as well as within, families.

Conclusions

Water binding and color

1. The genetic variance of color and water binding traits is of the same magnitude as the variance between batches (about 20%).
2. Color and water binding have no unfavorable correlations with carcass lean percentage.
3. In the short term, selection in most breeding populations should aim for decreased sensitivity to the PSE aberration.
4. Selection for meat quality can be based on relatively inexpensive measurements taken in the slaughterhouse.
5. DNA-markers can become important additional tools.

Intramuscular fat

1. The heritability of intramuscular fat percentage is high (40%).
2. Intramuscular fat percentage is unfavorably related with carcass lean percentage.

References

Beckman, J.S., and Soller, M., 1983. Restriction fragment length polymorphisms in genetic improvement: methodologies, mapping, and costs. Theor. Appl. Genet. 67, 35-43

Bejerholm, C. and Barton-Grade, P.A. 1986. Effect of intramuscular fat level on eating quality of pig meat. In: Proc. of 32nd Eur. Meeting of Meat Research Workers., August 24-29, 1986, Ghent, Belgium, pp. 389-91.

Boccard, R., Buchter, L., Casteels, E., Cosentino, E., Dransfield, E., Hood, D.E., Joseph, R.L., MacDoughall, D.B., Rhodes, D.N., Schön, I., Tinbergen B.J. and Touraille, C. 1981. Procedures for measuring meat quality characteristics in beef production experiments. Report of a working committee in the commission of the European Communities (CEC) Beef Production Research Program. Livest. Prod. Sci. 8, 385-97.

Cameron, N.D. 1990. Genetic and phenotypic parameters for carcass traits and eating quality in pigs. Livest. Prod. Sci. 26, 119-35.

Cöp, W.A.G. 1989. Ensure quality by controlling the whole production process. Pigs 5 (6), 18-19.

de Vries, A.G. 1989. A method to incorporate competetive position in the breeding goal. Anim. Prod. 48, 221-7.

Eikelenboom, G. and Minkema, D. 1974. Prediction of pale, soft, exudative muscle with a non-lethal test for the halothane-induced porcine malignant hyperthermia syndrome. Neth. J. Vet. Sci. 99, 421-6.

Falconer, D.S. 1989. Introduction to quantitative genetics. Third edition. Longman, Essex, p. 317.

Fujii, J., Otsu, K., Zorzato, F., de Leon, S., Khanna, V.K., Weiler, J., O'Brien, P.J. and MacLennan, D.H. 1991. Identification of a mutation in the porcine ryanodine receptor that is associated with malignant hyperthermia. Science 253, 448-51.

Haley, C.S., Archibald, A., Andersson, L., Bosma, A.A., Davies, W., Fredholm, M., Geldermann, H., Groenen, M., Gustavsson, I., Ollivier, L., Tucker, E.M. and van de Weghe, A. 1990. The pig gene-mapping project: PIGMaP. Proc. 4th World Congr. Genet. Applied to Livest. Prod., Edinburgh, Vol. XIII, pp. 67-70.

Honikel, K.O. 1987. How to measure the water-holding capacity of meat? Recommendation of standardized methods. In: Tarrant, P.V., Eikelenboom, G. and Monin, G. (eds) "Evaluation and control of meat quality in pigs", a seminar in the CEC Agric. Research Progr., Dublin, Ireland, 1985. Martinus Nijhoff Publishers, Dordrecht, pp. 129-42.

Hovenier, R. 1992. Personal communication.

Hovenier, R., Kanis, E., van Asseldonk, Th. and Westerink, N.G. 1993. Genetic parameters of pig meat quality traits in a halothane negative population. Livest. Prod. Sci. (in press).

Karlsson, A. 1992. The use of principal component analysis (PCA) for evaluating results from pig meat quality measurements. Meat Sci. 31, 423-33.

Kauffman, R.G., Eikelenboom, G., van der Wal, P.G., Merkus G. and Zaar, M. 1986. The use of filter paper to estimate drip loss of porcine musculature. Meat Sci. 1, 191-200.

Lundström, K. 1990. Genetics of meat quality. Proc. of the 4rd World Congr. on Genetics applied to Livest. Prod., Edinburgh, July 23-27, 1990, Vol. XV, 507-10.

Lundström, K., Karlsson, A., Lundeheim, N., Rydhmer L. and Vestergaard, T. 1989. Current methods and perspectives in breeding for meat quality in pigs. In: 40th Annual Meeting of the EAAP, Dublin, Ireland, August 27-31, 1989.

Meyer, K. 1991. Estimating variances and covariances for multivariate animal models by restricted maximum likelihood. Genet. Sel. Evol. 23, 67-83.

Monin, G. and Sellier, P. 1985. Pork of low technological quality with a normal rate of muscle pH fall in the immediate post-mortem period: The case of the Hampshire breed. Meat Sci. 13, 49-63.

Ollivier, L., Gueblez, R., Webb, A.J. and van der Steen, H.A.M. 1990. Breeding goals for nationally and internationally operating pig-breeding organisations. Proc. of the 4rd World Congr. on Genetics applied to Livest. Prod., Edinburgh, July 23-27, 1990, Vol. XV, 383-94.

Olsman, W.J. 1978. De betrouwbaarheid van de bepaling van het vetgehalte in vlees met het Foss-Let apparaat. Ware-n-chemicus, 8, 109-14.

Schwörer, D., Morel, P., Prabucki, A. and Rebsamen, A. 1990. Genetic parameters of intramuscular fat content and fatty acids of pork fat. Proc. of the 4rd World Congr. on Genetics applied to Livest. Prod., Edinburgh, July 23-27, 1990, Vol. XV, pp. 545-8.

Sellier, P. 1988. Aspects génétiques des qualités technologiques et organoleptiques de la viande chez le porc. Journèes Rech. Porc. en France, 20, 227-42.

Smith, C. 1984. Rates of genetic change in farm livestock. Res. and Devel. in Agric. 1, 79-85.

SPSS 1985. Manual. SPSS Advanced Statistics Guide. McGraw-Hill, New York.

Thompson, R. and Hill, W.G. 1990. Univariate REML analyses for multivariate data with the animal model. Proc. of the 4rd World Congr. on Genetics applied to Livest. Prod., Edinburgh, July 23-27, 1990, Vol. XIII, 484-7.

van der Sluis, W. 1992. Thoughts. Pigs 8 (1), 4.

van der Wal, P.G., Bolink, A.H. and Merkus, G.S.M. 1988. Differences in quality characteristics of normal, PSE and DFD pork. Meat Sci. 24, 79-84.

Walstra, P. 1988. Automated grading probes for pigs currently in use in Europe, their accuracy, costs, and ease of use. In: O'Grady, J.F. (comp.) New techniques in pig carcass evaluation. Pudoc, Wageningen, pp. 16-27.

Wierbicki, E., Tiede M.G. and Burrell, R.C. 1962. Determination of meat swelling as a method for investigating the water-binding capacity of muscle proteins with low water holding-forces. Fleischwirtschaft, 42, 948-51.

Wood, J.D., Jones, R.C.D., Fracombe, M.A. and Whelehan, O.P. 1986. The effects of fat thickness and sex on pig meat quality, with special reference to the problems associated with overleanness. 2: Laboratory and trained taste-panel results. Anim. Prod. 43, 535-44.

Discussion

*Kallweit - In your closing remarks, you mentioned that we should increase the intramuscular fat to an optimum. What would you suggest the optimum is?

*van der Wal - As far as I know, consumers can see the amount of intramuscular fat when it rises above about 2%. So, when it comes above 2% IMF, chemically determined, there will be problems with consumer acceptance. We have to raise the amount from about 1.3-1.5%, to about 2%, but not over it. But, for taste, and perhaps tenderness, there are indications that it has to be a little bit higher but then we would have to counter the complaints of the meat consumers.

*Lundström - Do you know if you have the halothane gene in the animals in your experimental material?

*van der Wal - These were all halothane-free animals, but I can't give a guarantee that there were no heterozygotes.

*Lundström - The heritability for the Hunter-L value was a bit higher than the other reflectance values. Could that be due to an influence of the pigment content ?

*van der Wal - Perhaps, it is possible. But, normally, in pigs, there is not a big difference in the amount of hemoglobin and myoglobin present.

*Demeyer - You stated in your conclusions that slaughtering conditions were very important but in your text, I'm reading that each breeding organization had its own slaughterhouse. Can you dissociate its effect then?

*van der Wal - It is very difficult, and we were not so happy that every breeding organization had its own slaughter house. There was also a complication in one of the breeding organizations, who kept the animals overnight in the slaughterhouse. In the pigs coming from that company, we found indications of DFD. It was most likely not caused by the slaughter procedure, but rather by the long period the animals remained in the stables, because we didn't find it in the other animals.

*Demeyer - I noticed, and maybe it's a bit related to the former question, that the Minolta value was a little different from the Hunter-L value. Can you comment on this?

*van der Wal - Yes, there was a little difference, but the correlation between the two is rather high.

*Demeyer - But it's measured at the same position?

*van der Wal - Not exactly at the same position. The Minolta lightness value was measured in the slaughterhouse, on the lumbar part of the loin. After that, a few hours later, and after making a new cut, the Hunter-L value was determined in the laboratory after a bloom.

*Demeyer - So, it is not the apparatus, it is a real difference.

*van der Wal - Between the apparatus there is a little difference, but they show a very high correlation. In this case, there are also differences in the position in the loin.

*Kallweit - At what time post-mortem did you measure color brightness with the Minolta?

*van der Wal - About 20h post-mortem, in the slaughter house.

*Kallweit - Not directly after slaughter, within 40 min or so?

*van der Wal - No, it was impossible. The Hennessey Grading Probe PSE scattering values were made 45 min post-mortem, together with the Fiber

Optic Probe measurements in the slaughter house. In the lab, the Hunter-L values were measured at approximately 24h post-mortem

*Lister - I'm a little bit concerned about your recommendation to increase the IMF content to 2%. I think there are very good reasons to avoid promoting high fat meat products. I'm also thinking about Vivion Tarrant's comment this morning. In the study he reported, people couldn't determine the difference between 0.8 and 2% anyway. So, I can see the logic, to get nearer to the optimum, but I think, in principle, it's wrong to try to attain higher IMF content, and I don't think it's going to do any good anyway.

*van der Wal - We are starting, in the next study, with a taste panel on the amounts of IMF, and I hope that it can give much more information about the Dutch situation but there are complaints from consumers that the meat, the pork, is too dry, and less tender. We also found in the literature, the Swiss literature, that a higher level of IMF could improve both tenderness and taste. But, we are not sure. The opposite details are also available, as Dr. Tarrant showed us this morning.

*Tornberg - Tomorrow, I will talk about the investigation where we looked at IMF content from 0.5 to 3%, and normalized all other factors to be standardized and to vary very little. We have looked at the tenderness, and juiciness, and flavor of meat and found no influence at all. So, I don't think the IMF content is that important.

*Essén-Gustavsson - I am also going to talk about IMF tomorrow, but I think that what we have to really realize is that IMF is all kinds of fat. It's between the fibers, it's within the fibers, and it's between the fascicles. I think, that maybe the fat within the fiber is important. For sensory properties yes, but the fact is, that it is important because that muscle fiber can grow, or work in a better, more efficient way, without giving rise to low pH values due to a rapid glycogenolysis.

*Demeyer - We made cooked ham made from halothane-positive and -negative hams, and measured all kinds of quality criteria. The consumers, or taste panels, definitely preferred the halothane-negative cooked ham over the halothane-positive cooked ham. The only measurement of the fresh meat criteria that was related to that was the fat content.

*Tornberg - Yes, but my measurements were on loin. That could be a difference.

*Demeyer -That's cooked meat, of course, but the fat content is definitely related to the quality of the hams.

*Mäki-Tanila - What are your doing about IMF now in the Netherlands? Are you trying to improve it? I mean, what is "improvement" in the Netherlands?

*van der Wal - At first, we want to see what information we can get from the taste-panel analysis, that is the first point of research. We will start that project soon.

Pig Behavior

M. Harri

Department of Applied Zoology, University of Kuopio, PO Box 1627, SF-70211, Kuopio, Finland

General biology

The order *Artiodactyla* includes several domesticated livestock species, including cattle, sheep, goat, camels, and deer. The suborder *Suina* includes pigs, peccaries and hippos. The *Suidae* (pigs) is an Old World family, although swine have become widespread in the New World, since domestic, feral and wild stocks have been imported into the Americas, New Zealand, Australia, Indonesia, Hawaii, and practically all over the world (Epstein and Bichard, 1984; Graves, 1984). The ancestors of the domesticated pigs are to be found among the wild pigs of the species *Sus scrofa* which ranges from the western coast of Eurasia to the Pacific Ocean and from southern Sweden to Atlas countries and Sudan, in North Africa. *Sus scrofa* is classified into a large number of subspecies. *S.s. viffatus* originates from Malaysia and Indonesia, while *S.s. scrofa* has a European origin (Epstein and Bichard, 1984). The former subspecies is smaller than the northern races, its height at shoulder being in the range 70-90cm and the weight 90-135kg. The shoulder heights of the European boar often exceed 100cm and weights of adult animals between 24 and 350kg have been recorded. The skull of *S.s. vittatus* is short and high, whereas that of *S.s. scrofa* is elongated. The other species of *Sus* may, and often do, cross with *S. scrofa*; hybrids are fertile in both sexes (Corbet and Mill, 1980).

Pigs are characterized by rooting behaviour, and the associated anatomical and physiological features. Pigs utilize their muzzle, which is flattened into a tough, rounded disk, in an unending search for food on, and under, the substrate. Wild pigs have a complete dentition of 44 teeth, while in some domestic breeds the number of teeth is reduced (Epstein and Bichard, 1984). The stomach is two-chambered, simple, and non-ruminating. Pigs are omnivorous. In the wild state they feed on roots, plant stems, fungi, acorns, chestnuts, beechmast, grain, insects, larvae, worms, eggs, frogs, reptiles, and carrion. Although vegetative material composed a major part of their overall diet, during a drought, when pigs had nothing but grass to eat, many died of starvation (Graves, 1984).

Pigs wallow in mud to plaster their skin with mud as a protection against flies and mosquitoes, to rid the hog of ectoparasites, or for thermoregulation during the hot season (Conley *et al.*, 1972; Eisenberg and Lockhart, 1972). However, there are reports of instances of hogs breaking through ice to wallow, or wallowing while being chased by hunters (Stegeman, 1938). Rubbing may be performed to remove ectoparasites (Conley *et al.*, 1972) but also as a comfort behavior. Honing of tusks on trees may be performed by boars in preparation for a battle with other males, or as a territorial behavior. Wild hogs spend most of their inactive hours in day beds (Conley *et al.*, 1972). Dens are used only for farrowing, but in cold weather more elaborate beds are made (Stegeman, 1938).

Pigs are social animals. Piglets stand within minutes of birth and begin to form dominance relationships with litter-mates within hours. Several females may combine litters, so that interactions between individuals within, and between, litters start early in life and persist into adulthood. In the same way, early associations with humans persist, making establishments of wild stocks from pen-reared hogs difficult (Lewis, 1966). The central social unit in wild swine is based around one to several females and their offspring. Adult wild boars tend to remain solitary. They associate with the female(s) whenever the female(s) exhibits sexual receptivity (Epstein and Bichard, 1984; Graves, 1984). However, the number of individuals per group varies seasonally and geographically (Graves, 1984).

Sows come into heat every 21 days, and the duration of estrus is 2-3 days. The litter of wild sows, farrowed 112-120 days after mating, numbers 4-6 (1-9). Wild boars reach sexual maturity at approximately 18 months. In warm regions, farrowing may continue throughout the year. However, reproduction is strongly associated with the availability of food, rainfall and other climatic conditions. European wild sows were in an anestrus condition during years of mast failure (Matschke, 1964).

About 1-3 days pre-partum, the pregnant sow separates from the herd and searches for a quiet place, in a thicket, under a bush, etc. (Graves, 1984; Jensen, 1989). She constructs a concave farrowing nest by rooting a shallow depression in the soil or by rooting soil and vegetation into a pile (Graves, 1984). The size, location and construction of the nest depends on climatic conditions, being more protected and containing more material in winter (Dellmeier and Friend, 1986; Jensen, 1989). Farrowing typically lasts for a few hours. The sow may, or may not, eat blood, other fluids, and after births expelled (Signoret *et al.*, 1975), but does not assist piglets cleaning themselves. Lactating sows can be successfully rebred.

The precocial piglets are very active and stand up soon after birth. They find the teats by using auditory, olfactory, visual, and tactile stimuli immediately after birth (Rohde *et al.*, 1991). Aggressive interactions are rare during the initial teat-seeking phase. After finding a teat, they vigorously massage the udder and then move from teat to teat (Signoret *et al.*, 1975). During this phase, they display aggressive behavior by biting their littermates with needle teeth or by pushing with nose or shoulder. After this teat-sampling phase the piglets limit their activities to a specific area of the udder and develop preferences for suckling from a specific teat or pair of teats during the short milk letdown period, which occurs about once an hour (Fraser, 1978). Thereafter, the piglets settle into a teat maintenance phase, during which they have access to their specific teat(s), and alternate sleeping with suckling (Graves, 1984). The first few piglets born may have lower mortality rates than those piglets born later (Harstock and Graves, 1976). These piglets also tend to be heavier. Thus, they are able to win more fights for the most productive anterior teats, and gain more weight than other piglets. They are also less likely to be injured or crushed by the sow (Fraser, 1978). Dominance continues to play major role in settling disputes over access to food, mates, and other limiting resources, and allows greater freedom of movement and less restricted personal space (McCort and Graves, 1982).

The habitat, use of space, and daily activity, are strongly dependent on local environmental conditions. It is mainly the availability of food that determines the home range. During summer dryness, swine move from grassland to marsh area. Generally, most of the activity of feral or wild pigs is carried out at intervals throughout the night. However, depending on disturbances, weather conditions and even season, etc. the pigs may be active during daylight, or only during the darkest night (Hanson and Karlstad, 1959; Nichols, 1963; Barrett, 1978). Swine are not strictly territorial, but they do have distinct home ranges (Graves, 1984).

From wild to domesticated pigs

The earliest remains of domesticated pigs have been excavated at Çayönü, in southeast Anatolia, dated to *c.* 7000 BC (Reed, 1977). Later remains of pig bones have been found all over the world. Several breeds have been developed. The British Islands had breeds of their own, while in the European Continent the four basic breeds - Large White, Danish Landrace, German Yorkshire and German Landrace - had a major influence on further breed development. These breeds have since evolved into further subbreeds. Additional local breeds have added genetic variance of their own into all these basic breeds (Epstein and Bichard, 1984). All this means that there is no single domestic pig. In addition, the management methods up to the 1950s and 1960s were rather extensive, today's intensive housing systems having thus a rather short history, compared with the history of the pig as a domestic animal. As a result of this, the long domestication history has not provided pigs with enough time to adapt themselves to intensive husbandry systems.

In terms of productivity, changes from the practically wild state in the farmyard to modern intensive husbandry system have been enormous. While the litter of wild sows numbers 4-6, that of the domestic pig is 8-12, or even more. Many intensive sow herds are now producing over 22 weaned piglets per sow per year. Just to mention a few figures, the pigs are growing to 90kg in around 165 days, with a conversion rate of under 3kg concentrated feed to 1kg live weight gain (Epstein and Bichard, 1984).

Todays' paradox stems from the changed attitudes of the consumers towards husbandry systems. It is no longer important that the price of the pork is low. The consumers want organic meat from happy animals, returned from piggery back to nature. However, this idea suffers from serious misinterpretations of natural life, animal husbandry, and pure and applied ethology.

1. It should be borne in mind that none of the domestic pig breeds are now the same animal as the wild hog. In the process of domestication, the pig has changed in morphology and physiology perhaps more than any other animal, with the exception of the dog. The backfat thickness of wild hog is approximately three times that of a Yorkshire, and activities of adipose tissue lipogenesis enzymes and gluconeogenesis enzymes are several times higher (Wangsness *et al*., 1980). All these features help the wild hogs to store energy for the time of starvation that they, sooner or later, will encounter in the nature. The growth potential of domesticated pig, with its increased capacity to lay down lean tissue instead of fat reserves, would not be beneficial in the wild, when availability of food is limited and attainable only after hard, long-term work (Graves, 1984). With its round body form, and hairless skin, the domestic hog would succumb in the first thicket, after an intensive attack of mosquitoes, burning sunshine, or autumn cold. With the altered cranial structure, the domestic pig is capable of moving straw from one place to another. However, if it had to root worms or roots from sun-dried ground, its chances of survival would be minimal.

2. Although it is true that during the course of evolution wild animals have been adapted to their particular environment, adaptation does not equal happiness (Dawkins, 1980). If a sow constructs a farrowing nest, she does it not in order to be happy, but rather in order to guarantee that at least some of its piglets will survive to reproduce and to transfer her genes to the next generation. While observing the play of wild piglets during the summer sunshine, we easily forget their litter-mates that are starving to death because of the poorer teats they have received from their high-ranking litter-mates. Marian Dawkins (1980) warns that we should not fall into the trap of thinking that a natural life is better simply because it appears to be more romantic to us as outsiders. The "goals" of nature are different from those of man. In terms of evolution, it would be satisfactory if only one single piglet out of all the litters one sow produces, during her whole life span of about 20 years, would survive to reproduce. This guarantees that the heritage of that particular sow will remain in the population. On the other hand, man's goal is to grow all the piglets a sow produces in a year, often more than 20, to mature size, and not only the litters of one year, but also those of several consecutive years.

3. Despite major changes in morphology and physiology during domestication, the behaviour of pigs has changed very little. The degeneracy hypothesis claims that domestic animals have "lost" many of the biologically adaptive traits of their wild ancestors (Kavanau, 1964). However, despite many behavioural differences between wild and domestic stocks of many species, there is no evidence that domestication has caused behaviours to be eliminated from the species repertoire, or that even the basic structure of the motor patterns for such behaviours has been changed (Hale, 1969; Miller, 1977). In nearly all cases, behavioural differences between wild and domestic stocks are quantitative in character and are best explained by shifting stimulus thresholds of the behavioural patterns (Price, 1984). Atrophy, or the reduction of disappearance of any innate activity, is found in some characters. The constant attention and vigilance of the wild pig is said to be largely missing in the domestic pig (Hemmer, 1990). The social structures of the wild pig is relaxed in the domestic pig and its spontaneous motility is lower, as are its nest-building activities, and its defense of young. On the other hand, the sexual behaviour is hypertrophied to be more intense, and the number of young is higher. Aggression is generally associated with sexual behaviour, especially in males. Since fear of humans has greatly diminished, man will often become the target of sexually active animals (Hemmer, 1990). Wild pigs synchronize their behaviour within the group more than domestic ones, and make social contacts more often. They are also more aggressive.

Very often, comparisons between wild and domestic pigs have been carried out by transporting stall-housed pigs to outdoor pens (Jensen, 1986, 1991; Wood-Gush and Stolba, 1982; Petersen *et al.*, 1990). However, Robert *et al.* (1987) transferred wild pigs from natural free-range enclosures into intensive housing conditions. Under these circumstances, differences between wild and domestic pigs persisted. The distribution of rest and activity patterns during day-time was different between wild and domestic pigs, although bimodal in the two groups. Wild pigs exhibited diurnal activity. They were resting less often than domestic pigs during day-time, this lower percentage being reflected in a higher locomotor activity. This comparison confirms the hypothesis that impoverished environment is not the only factor influencing the behavior of animals. Domestic pigs have been selected for their ability to maximize their productivity. Long durations of rest, associated with low levels of motor activity, favors the conservation of energy that can then be directed to production. These differences between wild and domestic breeds, in spite of their quantitative features, confirm the well-known fact that quantitative traits can also be genetic, and thus subject to selection. It is interesting to note that the wild pigs kept in intensive housing spent much more time resting than wild pigs living in natural free-range environments (Robert *et al.*, 1987), an indication of the dependence of behaviors on environmental factors as well.

Pure versus applied ethology

Historically, the scientific comparison of wild and domestic situations has been retarded by the separation between the animal scientists and veterinarians, who study improving domesticated animal production and health, and the ethologists who tend to study the behavior of undomesticated species (Kilgour, 1976). The latter group is interested in cost-benefit comparisons of different behavioural strategies in terms of survival and ecological fitness. As an example, let us formulate a hypothesis that the more the wild hog roots the better are its chances of survival, and lifetime reproduction. To test this hypothesis we observe amount of rooting of different individuals and compare the results with lifetime reproductive data of the same animals. If it appears that individuals who root more than lazier ones can also find more food in times of food shortage, and, accordingly, on average leave more offspring that survive to reproduce, we can accept that rooting behavior has a fitness-value to an animal. In evolutionary time, then, the individuals with a higher rooting activity will replace ones with a weaker activity, and leave their genes, including this behavioral pattern to the population. These results, however, do not give an explanation of whether or not sows in one particular intensive husbandry system produce more meat of better quality, are healthier, or suffer less from stress and behavioral abnormalities than ones in another intensive system.

On the other hand, applied ethologists usually do not pay any attention to the two fundamental questions of pure ethology. These questions are "how?" and "why?". How questions require answers about how the proximate mechanisms within an individual - its genetic development and physiological-psychological systems - enable the individual to behave. Why questions require answers about why these systems have evolved, and why the behavioral patterns of an individual help it overcome the obstacles to reproduction in its environment (Alcock, 1989). The applied ethologists compare behaviors of animals in different husbandry system in an attempt to use this information to decide which one of the systems is better for productivity, health and welfare of animals. In addition, the methods used by pure and applied ethologists are as different as are their ways of formulating hypothesis. They do not speak the same language. Despite the apparently same field of study, the pure and applied ethologists have very little in common.

Ethology and animal welfare

Since domestic animals in captivity have much less freedom and experience less variation in their environment than those in the wild, are they suffering, since they are prevented from displaying and experiencing the entire range of their possible behavioral repertoire? Furthermore, when returned back to nature, the sow builds up a nest as skillfully as her wild counterparts despite having no early experience of nest building. On the other hand, in the stall, the same sow may give birth to 12 piglets, all of which will survive with the help of a heat lamp provided by man, while in the nature the same sow, with its low stores of fat energy and hairless skin, would not survive the winter, and even less give birth to and nurse piglets, however skillfully she is able to construct the nest. We can also ask, why should she build, or even have an ability to build the nest inside a heated house with an extra heat lamp whenever needed? Does she suffer from the lack of opportunity to engage in a behavior that is not necessary for survival in that particular environment?

In nature, animal needs are, to a great extent, based on survival and reproduction. Behaviors that are important for the animal's survival have come to have a physiological structure that causes the animal to seek out the goals associated with these behaviors.

Since all behavioral patterns have evolved through natural selection, and have physiological consequences, it is logical to think that behavior is needed by the animal. The psychohydraulic model by Konrad Lorenz (1981) assumes that the non-performance of a behavior brings about a damming up of some energy or drive, which finally, if that particular behavior is not released by a specific stimuli, leads to the occurrence of that behavior as a vacuum activity. The Lorenzian model fosters the idea that the inability to perform a particular behavior implies disturbed welfare. In another model, behavior is conceived of as a tool by which animals regulate their internal or external state. If the goal, e.g. warmth, or a nest, is achieved by other methods, then the behavior is no longer needed. Homeostasis and regulation are key concepts (Wiepkema, 1982). Attempts to fit all behavioral patterns into either of these opposite hypotheses are unsuccessful. Behavioral patterns can be divided into three main categories (Hughes, 1973):

1. Behaviors that require a certain releasing stimulus to be elicited. If the stimuli are not present the behavior is not released and no frustration is evident (e.g. anti-predator behavior).
2. Behaviors that are dependent on internal stimuli. If the stimuli are not present welfare is impaired (e.g. hunger).
3. Behaviors that are more frequent in the presence of external stimuli but which are elicited even if these stimuli are not present. Absence of the stimuli may lead to welfare problems, or may not, because they are released in the absence of the appropriate stimuli (e.g. dust-bathing in caged hens). Behaviors exhibited in the absence of appropriate stimuli are called vacuum activities. These behaviors are considered to be abnormal, but may fulfil the animals need to exhibit these behaviors.

Observation of animal behavior in nature gives us an idea about the behavioral repertoire the animal has. However, the behavior of a domesticated animal when returned to nature is only raw material for the generation of hypotheses or questions (Jensen, 1991). If the animal in confined conditions does not perform the same behavioural patterns as it did in nature, two interpretations are possible:

1. The animals does not have any need to perform the behavior (e.g. nest building), because man has already made it, or
2. The impoverished environment does not allow the animals to perform the behavior. It is apparent that the welfare problems are associated with the latter interpretation only.

On the other hand, if the animal performs a behavior, e.g. nest-building, despite the presence of man-made nest and warm environment, again, two opposite interpretations can be drawn:

1. The nest building is pre-programmed into the pig's brain. During the course of evolution, the only those sows to produce offspring were those which have constructed the protective nest. The sow performs the appetitive phase of nest-building because it its programmed to do so, but without consciously paying any attention to the meaning of that particular behavior, or,
2. The animal is consciously motivated to build a nest when its physiology (hormonal state) demands the performance of that behavior. Welfare problems are associated with the latter interpretation only. Ironically then, both the performance or non-performance of a particular behavioral pattern can be interpreted as a sign of poor welfare.

Behavioral methods of assessing welfare

It is generally agreed that in any assessment of welfare, all the available evidence should be taken into account, including that of the animals' health, production, physiology and biochemistry as well as behavior (Dawkins, 1980; Duncan, 1987; Smidt, 1983). There are problems associated with all these classes of evidence. There have been several different approaches that have used behavior to assess welfare, and some have been more successful than others.

The call of the wild

One method has been to compare the behavior of an animal in an environment that is assumed to be ideal, with that of an animal in a confined housing system. However, domesticated species have changed in many respects in the course of domestication. It is difficult to decide what actually constitutes an "ideal environment". As stated above, it is almost certainly not the natural environment in which their wild progenitors evolved. With regard to farm species, perhaps "traditional" husbandry systems would be nearly ideal since, until about 40 years ago, they had changed only slowly, which meant that the species involved could also evolve under an artificial selection pressure to suit the system. However, traditional systems have obvious failings. In fact, the main reasons why the traditional systems have been replaced by more intensive ones is the low productivity of pigs in the traditional system, due to diseases, fights between individuals, accidents, high-maintenance energy costs resulting from cold, exercise, etc.

Another possibility is that feral populations of farm animals that are living and breeding in the wild may serve as a reference for behavior in an ideal environment. However, the same criticism can be raised as with animals under traditional systems. The animals are often severely stressed and malnourished, which undoubtedly reduces their welfare. Selection of the natural environment as a reference for an ideal environment would mean a ban of the pig industry in many northern countries e.g. in Finland where the wild boar cannot survive to such an extent that reproducing stocks could exist.

The third suggestion has been that the "ideal environment" is one that elicits a rich repertoire of behavior (Hughes, 1980). This can rapidly develop into a circular arguments. Thus, if the "ideal environment" has been chosen because it evokes a wide variety of behavior, then it should hardly be surprising that when compared with this, all other environments appear to be "poorer", in that they stimulate a more meager range of behavior.

Apart from the problem of deciding what to use as the reference environment, the comparative method provides results that are difficult to interpret. It should be no surprise that animals behave differently in different environments. In fact, why should animals display "natural behavior" in an unnatural environment? Differences in behavior may simply demonstrate how adaptable animals are. As discussed above, the interpretation of results from this method is particularly problematic when a behavioral pattern is missing in the test environment.

While taking all the weak points of the naturalistic method into account, it seems somehow surprising how much pig scientists have been attracted to this class of method. Attention has been focused on behavioral patterns, and the weak points of the method have been recorded, but not thoroughly discussed. For example, the piglet mortality rate, recorded by Jensen (1989) in a free-range natural housing for domestic pigs would hardly be acceptable in an intensive pig husbandry. In addition, environmental conditions, especially in the north, add additional problems. One of these is sunshine, which, even in northern latitudes, easily burns the unpigmented and hairless skin of the domesticated pig. Another problem is small in size but severe in magnitude. It is the mosquito.

Behavior in seminatural environment

All kinds of housing systems include some confinement, restriction of total freedom of movement. There have been several attempts to construct systems that contain benefits of the natural free-range system, and that of the very intensive one, without their disadvantages. In these systems, partially because of environmental demand, the behavior of the animals is something between that in nature and that in the confined system. Stable social groups, communal nests, certain defecation places, and plenty of rooting activity belong to the daily routine of the pigs (Stolba and Wood-Gush, 1984; Newberry and Wood-Gush, 1986, 1988; Stolba, 1981; Kerr *et al.*, 1988). The family pen system by Stolba and Wood-Gush was developed by identifying some key features of free-range pigs and by adapting these to the intensive unit (Stolba, 1981; Stolba and Wood-Gush, 1981). The basic unit of the system was a series of interconnecting pens to accomodate four sows and their followers. The pens were partially roofed and partially open, to simulate a forest border where many activities took place in the large enclosure. Each sow was provided with a nest site, 2-3m in diameter. Straw for building the nest was provided from a rack. A corridor was provided for defecating 4.5-11m away from the nest. Activity areas containing straw, peat and bark for rooting and wallowing, horizontal bars that the pigs could lever with their snouts, sufficient space for all pigs to feed simultaneously, and refuges for any pigs coming under social pressure were also provided. Although no absolute measure is available, all the symptoms that were suggestive of reduced welfare were absent. All types of reduced welfare stereotypes were absent, tail-biting and cannibalism were absent, aggression was at a low level, more time was used in manipulating and exploring objects, and fear responses were not exaggerated. The family pen system replaces dry sow stalls, farrowing crates and the modern fattening house, the three areas of pig production that have attracted most criticism. The reproductive performance of the sows has been very good, and the growth rate and feed conversion rate of the piglets satisfactory. The drawbacks, from a commercial point of view, have been high mortality of the newborn piglets, above average labor costs, and, for the large-scale producer, an increased use of space (Kerr *et al.*, 1988). The latter feature becomes even more important in the North, where heating cost represents a major part of the total costs of any piggery.

Behavior in different situations

One method has been to study how animals behave in various states of suffering such as frustration, fear and pain. This method of stressing animals experimentally, recording their responses and comparing these responses with what occurs under commercial practice, has been used with variable results in assessing welfare of domestic animals. With regard to animal suffering, these results are relevant only when coupled with the results of physiological and biochemical measurements. These measurements describe how animals behave in different situations; however, they do not directly give information about how the animals feel about the situation.

Instrumental methods for assessing animal welfare

Operant conditioning has been widely and successfully used to assess animals' opinions about their physical environment. Pigs placed in a cold chamber are able to press for bursts of infrared heat or switch off a fan to keep the effective chamber temperature at about 25°C (Baldwin and Ingram, 1967). This is close to the thermoneutral zone for the pig. However, the response was dependent on nutritional state, presence of neighbors, time of day, type of floor, type of heat reward, amount of bedding etc. (reviewed by Kilgour *et al*., 1991). Pressing for heat has been put to practical use in the USA (Morris and Curtis, 1982). Pigs able to break a light beam to switch lights on and off preferred light for 72% of each 24h. If they were required to work for lights they did this for only 0.5% of the 24h (Baldwin and Meese, 1977). Vibrations of a machine that mimicked a road transporter were more aversive to pigs than was noise (Bailey *et al*., 1983; Stephens *et al*., 1985). Matthews and Ladewig (1985) showed that it was possible to quantify the relative value of different environmental features for pigs. Social contact was more valued than a simple stimulus change (door-opening), and food had the highest value of all.

In addition to operant conditioning, preference tests have been used to assess pigs' opinions about their environment. For example, van Rooijen (1983) showed that pigs would make a large number of responses in order to get from a concrete floor to an earthen floor. However, preference tests have been widely criticized (Duncan, 1978; Dawkins, 1983). They give only relative values, minority choices are difficult to interpret, and an animal's short-term preference may not be in the best interests of long-term welfare. In addition, the history of an animal, and the other simultaneously influencing factors, affect the result. For example, the pig's preference for straw was markedly dependent on ambient temperature (Fraser, 1985).

Behavior in intensive housing

It should be borne in mind that the commercial intensive housing systems have been developed for a reason. Attempts have been made to solve any problem encountered by the producer. After having solved each individual problem, however, only an impoverished environment was left. Many management problems with domestic pigs stem from our inattention to providing occupation for the pig's muzzle, which, over an evolutionary time scale, has developed for continuous employment (Harstock, 1982). Addition of occupational material such as straw, chains, rubber rings, wood pieces, toys etc. in addition to noise and music have been shown to reduce abnormal behavior, aggression, stress, and fear of humans in pigs (van Putten, 1980; Warnier and Zayan, 1985; Grandin *et al*., 1983, 1987; Grandin, 1989a; Pearce *et al*., 1989).

Domestic pigs kept in groups soon form hierarchies, and the common procedure of feeding them from open troughs presents the dominant pigs with a defendable food supply which they are able to monopolize. In more natural conditions, pigs are usually feeding on dispersed food. Stolba (1985), based on careful behavioral observations, first increased the length of the troughs, and then tried various ways of screening off the feeding pigs from each other. The most cost-effective method was to divide the trough into individual sections using small screens placed along its length (bin stalls). The design of stall division affected both the spatial use of the stall, aggressive behavior, and cortisol concentration. The behavioral data suggest that pigs in the stall constructed of horizontal bars were more actively avoiding their neighbors than the animals housed in vertical stalls (Barnett *et al*., 1991). The addition of mesh to the sides of the tether stall virtually eliminated aggressive interactions (Barnett *et al*., 1987). In these experiments, however, hormonal data was not correlated with measures of aggressive behavior, an indication of how difficult it is to measure animal welfare. Straw on the floor, and hessian covers over the standard farrowing crate increased nesting behavior and piglet-directed behavior in sows. The incidence of piglet mortality was lower in the modified crate than in the standard crate. Thus a small modification of the farrowing environment can affect maternal behavior, resulting in improved piglet survival (Cronin and van Amerongen, 1991). However, the preference for a certain environment can also change. Thus less-experienced sows preferred an enclosed farrowing crate, but this preference largely disappeared with repeated experience of open crates (Phillips *et al*., 1991).

Handling the animals

Modern production systems require pigs to be handled during preventive medicine programs, mixed socially into new groups for ease of management, and moved into new accommodation several times during their growth to slaughter weight, and finally transported to the slaughter plant. Three behavioral problems are associated with these procedures:

Mixing of unfamiliar animals:

The mixing of unfamiliar animals will result in a range of aggressive and other behavioral patterns until a new dominance hierarchy has been established (Rasmussen *et al*. 1962; Meese and Ewbank, 1973). The reactions are variable and individual, but usually the aggression peaks within the first 24h and stable dominance hierarchies are established by 48h (Meese and Ewbank, 1973). Aggression generally increases with increasing group size and decreasing space per pig (Ewbank and Bryant, 1972; Meese and Ewbank, 1973). When unfamiliar sows are brought together, the enclosure in which they are initially mixed should enable individuals to distance themselves at least four to five body lengths from one another (Baxter, 1985). Fencelike contact prior to mixing will markedly help the establishment of the dominance hierarchies. Extreme crowding often occurs in a gathering pen or during transport. Also, the introduction of one strange sow into a established group can result in serious active fights (Fraser, 1980). Hanging clothing strips in the pen (Grandin, 1989c), straw (Kelly *et al.*, 1980), sedative drugs (Symoens and van der Brande, 1969), and aerosolized androsterone (McClone and Morrow, 1988) have been used to reduce aggression. Recently Grandin and Bruning (1992) modified the last method by adding sexually mature boars into a group of slaughter-weight pigs. This procedure effectively hampered fighting.

Sudden change in environment:
Man controls the environment and social life of the pigs. In evolutionary time, however, pigs are adapted to survive in an environment that is far from perfect. They are adapted to pleasant and unpleasant changes in their environment. Ironically then, their production environment is too good. In addition, the artificial selection pressure for certain production characteristics has produced animals that are more adapted to a barred environment than are their wild counterparts. One aim of the multi-activity pens is to accustom pigs to the variation of life. It has been even suggested that the pigs should be subjected to stress, including noise and vibration during their growth phase so that they could then withstand the stress of transport and slaughter when the moment unequivocally comes (Grandin, 1989b).

Socializing sows to people:
Pigs are a contact species (Hediger, 1968), and they frequently touch each other in greeting. They will extend this gregariousness to people to whom they are positively socialized, which greatly facilitates their handling. Scratching by people, usually during the feeding time, usually accelerates positive socialization to humans. It is often better to allow the pig to take initiative in approaching, rather than actively approaching the animal (Hemsworth *et al.*, 1986a). Unpleasant handling only increases the stress and fear of humans (Hemsworth *et al.*, 1986b; Pearce *et al.*, 1989). However, the effects of handling, either pleasant or unpleasant, on performance characteristics are either minimal, absent, or controversial (Hemsworth and Barnett, 1991; Hemsworth *et al.*, 1986a, 1986b, 1986c; Gonyou *et al.*, 1986). Furthermore, the unpleasantly handled pigs did not show evidence of chronic stress (Hemsworth and Barnett, 1991). This result, and the finding that group housing in many respects induces similar unfavorable changes in production performance support the concept that the effects of socialization are not so straightforward as anticipated (Hemswoth and Barnett, 1991). In terms of practical pig husbandry, it is more important to avoid unpleasant treatment of the animals rather than to routinely try to pet each individual animal of the herd. In addition, the trait "fear of humans" has a high heritability of $h^2 = 0.38$ (Hemsworth *et al.*, 1990). Thus gilts could be bred to be less fearful of humans.

Pig behavior and meat quality

It is expected that behavioral problems are reflected in meat quality. However, although pigs in several management systems can suffer from stress and behavioral problems, the relationship between behavior and meat quality is established in a few well-known situations only. Most unfavorable changes in meat quality do not stem from the growth period but rather during the transport of pigs from piggery to the slaughter plant and the conditions preceding the stunning. Management conditions such as the size of crates or pens, type of floor, physical environment, feeding system, and handling have an effect on the welfare of pigs, and for this reason their development should be encouraged. Of course, poor management systems can affect meat quality indirectly: poor ventilation can increase diseases, and poor flooring can cause traumas. Natural feed sources from a natural enclosure can affect meat taste and flavor, and lack of exercise can affect the consistency of muscles. But it is difficult, or even impossible to show that the meat quality is dependent on the size of crate, group size of fattening pigs, early handling by humans, or corticosteroid levels in plasma if the animals are otherwise healthy (Immonen, 1984). It is difficult to conclude that selection for pigs that adapt to the intensive management systems can have a greater effect on meat quality than modification of the systems according to the requirements of the pigs.

Some problems associated with meat quality may be based on behavioral factors. A large number of slaughter pigs suffer from leg weakness, most probably because of lack of locomotor activities (van Putten, 1989). Leg weakness is not only a result of poor management systems, but also due to the fact that the fastest growing pigs in pens are also the ones with largest locomotor problems.

Porcine stress syndrome (PSS) develops in pigs subjected to stress associated with unfavorable environmental conditions, or social stress. Death of the animals may result, and, of course, meat from a dead animal is worthless. The piglets with the PSS trait, which is inherited as an autosomal recessive gene, develop muscular rigidity when exposed to halothane anesthesia. The PSS is thus a trait that has clear effect on meat quality, induces behavioral changes, is hereditary, and can be detected by using a rather simple pharmacological test. It is natural that a question of the use of behavioral tests in order to detect behavioral disorders has attracted scientists. Indeed, there are some behavioral differences between stress susceptible and normal piglets, but these differences are small and observed for a certain time of day only (Robert and Dallaire, 1986; Schaefer *et al.*, 1989). Unexpectedly, stress susceptible sows did not show more stereotypes, a behavior generally considered as abnormal, than did normal sows; only the type of stereotypes showed some differences (Ödberg *et al.*, 1991). As with the leg weakness, PSS is also associated with fast growth and a lean, muscular body composition. Thus, despite some behavioral differences between stress-susceptible and normal pigs, the PSS problem, which manifests itself by behavioral features, can still be best detected by other than behavioral tests. Because of its hereditary nature, the trait can be more easily culled out off the population (that is exactly what has been done to a large extent), rather than treated by using behavioral methods.

Acute stress situations result in the release of adrenaline from the adrenal medulla. If the pig does not have time to adapt to the situation, anaerobic glycolysis leads to utilization of glycogen stores, resulting in excessive lactic acid production before the slaughter. This process leads to the formation of pale, soft, exudative PSE-pork. If the stress state is prolonged, as a result of long transport and the pre-slaughter process, the glycogen stores are exhausted long before the stunning, and no lactic acid is formed. The pH of the meat remains high, leading to dark, firm, dry DFD-pork. Under these circumstances, good pig handling improves pork quality. The underlying behavioral principles responsible for improved pork quality are simple: i) gentle handling; ii) a rest period of at least 2h prior to stunning; iii) showering pigs during hot weather; and iv) shorter stunning to sticking intervals (Grandin, 1986).

Transport and holding

Behavioral principles have a place in animal handling. Physiological drives are forces that generally activate animal behavior toward appropriate goals, with some variation among individuals and large temporal variation (Fraser and Herchen, 1979). Often, however, a group behavior exhibits a collective drive of higher motivational force that is normally observed in individual animal behavior; such behavior can be beneficial (e.g. following a leader), or detrimental (e.g. a stampede). Behavior becomes a balance of the interacting driving forces: for newly mixed animals, aggression is dominant, but it soon diminishes as the social order becomes established and the feeding drive becomes dominant. Relationships between animal handling and behavior provide bases for improved handling techniques and facilities (Grandin, 1980, 1987a, 1989a, 1990; Ewbank, 1968) (Table 1).

The tendencies of animals to "follow the leader", and to circle around and maintain visual contact with handlers are particularly important. Shadows that fall across an alley can cause animals to balk. On the other hand, animal movement is facilitated by eliminating harsh contrasts of light and dark and sudden changes in floor level or texture. All livestock have a tendency to move toward a more brightly illuminated area (Grandin, 1987a).

The flight zone of an animal is that space surrounding the animal that will elicit avoidance or escape when encroached upon. When a person enters an animal's flight zone, it will move away. Flight distance can be influenced by previous handling experience and by gentle handling. Handlers can increase their control over animals by exerting dominance, that is, acting like a dominant conspecific. If possible, the dominant animal should be handled first (Grandin *et al.*, 1983).

Table 1. Behavioral principles that may facilitate or make difficulties in transport and slaughter

1.	Unestablished dominance hierarchies
2.	Follow the leader
3.	Contact with neighbors
4.	Solid walls around lanes and holding facilities
5.	Individual fright zone
6.	Frightened pigs tend to move forward, or into the center of the herd
7.	Pigs move from dark to light
8.	No sudden turns in lanes
9.	Gentle handling
10.	Pigs find vibration and noise aversive
11.	Aversive floor texture or angle

Pigs move more readily in single-lane races (Grandin, 1982). However, in abattoirs slaughtering more than 500 or more pigs per hour, jamming at the single-lane race can be reduced by installing two lanes side by side, with a wedge-shaped partition centered between the lanes. The lanes should have solid sides but the fence separating the two single lanes should be constructed so that the pigs can see each other. This encourages the following behaviour.

When frightened, pigs tend to move forward. This can lead to jamming and bruises. Jamming can be partially avoided by rolls on the walls, and by constructing a stepwise narrowing of the lane (Hoenderken, 1976). When two lanes join, a funnel angle of 30° is the most effective (van der Wal and Eikelboom, 1981). Curved lanes and rounded holding pens with rounded corners are recommended (Grandin, 1982; van der Wal and Eikelboom, 1981).

With the abovementioned general principles in mind, the loading and transport stress can be minimized, although never totally eliminated. The pigs should not be fed prior to loading, since feeding increases heat production, which, when combined with the transport stress, can lead to exhaustion. Ideally, a holding or piggery facility should be built at truck deck level to eliminate ramps, or the pigs can be lifted by using a hydraulic lift. If a ramp cannot be avoided, its slope should ideally be 15°, and not steeper than 20-24° (Fraser *et al.*, 1986; van Putten, 1981). Holding and loading facilities must have non-slip flooring. Flooring that moves causes swine to balk (Kilgour, 1988). Optimum spacing of the cleats is 5-10cm depending on the size of the pig (Phillips *et al.*, 1988). The last part of the corridor leading to the truck should be straight with no obstacles in the path of the pigs. At any one time, only a small herd of 5-10 pigs should be loaded. Preferably, pigs from different piggeries are separated from each other. Since they are standing for the whole time the truck is moving, densely, as well as inadequately crowded loading, as well as sudden accelerations and decelerations, add the stress of the transport. Optimal conditions for temperature and relative humidity are 16-20°C and 50-60%, respectively (Mothes, 1981; Lambooy, 1988). During a hot day, the detrimental effects of high temperature on meat quality can be partially decreased by artificial ventilation (Lambooy, 1988). Pigs find vibration in itself aversive, and do not easily habituate to it. The aversiveness of vibration is increased by feeding a large meal before the transport. On the other hand, excessive hunger also adds to the aversive nature of vibration (Stephens and Perry, 1990).

In some modern abattoirs, electrically driven belt conveyors transport the pigs to stunning (Grandin, 1982). They work well, and reduce the stress of the pigs. Improved V-shaped conveyors have been installed in some abattoirs with variable success (Grandin, 1987b; Lambooy, 1986). The principle has been developed further by Grandin (1987b).

Not all stresses may be eliminated by optimal treatment, but it is possible at least to reduce the stress level to a minimum. Harsh handling, and electric prods are unnecessary stressors to the pig (Grandin, 1986). However, the personnel in the slaughter plants and pig-transport services should first receive more information about pig behavior. This information is the best way to motivate them for good animal handling. The design of the equipment is important, but the person handling the pigs is always the most important (Braathen, 1981). As to the meat quality, difficult-to-handle pigs produced darker meat with a higher pH (Weeding *et al.*, 1991). To which extent the result is due to genetic differences of the pigs, or to their handling history is not known.

Stunning

All stunning methods used in modern slaughter plants cause unconsciousness in a shorter time than any way of death encountered by the animals in the nature. Even carbon dioxide stunning, after an initial phase of excitement, causes unconsciousness within about 10s. High-voltage electric shock leads to unconsciousness within one to two seconds (van der Wal and Eikelenboom, 1981). Therefore, acceptance or non-acceptance of a stunning method by the society is more an ethical phenomenon, rather than related to physiology or behavior of the animal.

References

Alcock, J. 1989. Animal Behaviour. An Evolutionary Approach. 4th edn. Sinauer Inc. Sunderland, MA.

Bailey, K.J., Stephens, D.B., Ingram, D.L. and Sharman, D.F. 1983. The use of a preference test in studies of behavioural responses of pigs to vibration and noise. Appl. Anim. Ethol. 11, 197.

Baldwin, B.A. and Ingram, D.L. 1967. Behavioural thermoregulation in pigs. Physiol. Behav. 2, 15.

Baldwin, B.A. and Meese, G.B. 1977. Sensory reinforcement and illumination preference in the domestic pig. Anim. Behav. 25, 497.

Barnett, J.L., Hemsworth, P.H. and Winfield, C.G. 1987. The effects of design of individual stalls on the social behaviour and physiological responses related to the welfare of pregnant pigs. Appl. Anim. Behav. Sci. 18, 133.

Barnett, J.L., Hemsworth, P.H., Cronin, G.M., Newman, E.A. and McCallum, T.H. 1991. Effects of design of individual cage-stalls on the behavioural and physiological responses related to the welfare of pregnant pigs. Appl. Anim. Behav. Sci. 32, 23.

Barrett, R. 1978. The feral hog on the Dye Creek Ranch, California, Hilgardia. Agr. Sci. Publ. Univ. California. 46, 283.

Baxter, M. 1985. Social space requirement of pigs. In: Zayan, R. (ed.) Social Space for Domestic Animals. Martinus Nijhoff, Boston, MA. pp. 118-19.

Braathen, O.S. 1981. Ideas about methods and equipment for transport and lairage, stunning and debleeding in order to reduce the incidence of PSS meat. Norwegian Meat Res. Lab. Rep., Oslo, Norway. pp. 306-27.

Conley, R.H., Henry, V.G., and Matschke, G.H. 1972. European hog research project W-34. Tennessee Game and Fish Comm., Nashville. Final Report p. 259.

Corbet, G.B. and Mill, J.E. 1980. A World list of Mammalian Species. British Museum (Natural History), London.

Cronin, G.M. and van Amerongen, G. 1991. The effects of modifying the farrowing environment on sow behaviour and survival and growth of piglets. Appl. Anim. Behav. Sci. 30, 287.

Dawkins, M.S. 1980. Animal Suffering: the Science of Animal Welfare. Chapman and Hall, London.

Dawkins, M.S. 1983. The current status of preference tests in the assessment of animal welfare. In: Baxter, S.H., Baxter, M.R. and MacCormick, J.A.C. (eds) Farm Animal Housing and Wellfare. Martinus Nijhoff, The Hague. pp. 20-6.

Dellmeier, G.R. and Friend, T.H. 1986. Behaviour of sows farrowing on unmaintained pasture. J. Anim. Sci. 63 (Suppl. 1), 168.

Duncan, I.J.H. 1978. The interpretation of preference tests in animal behaviour. Appl. Anim. Ethol. 4, 197.

Duncan, I.J.H. 1987. The welfare of farm animals: an ethological approach. Sci. Prog., Oxf. 71, 305.

Eisenberg, J.F. and Lockhart, M. 1972. An ecological reconnaisance of Wilpattu National Park, Ceylon. Smithsonian Inst. Press, Washington DC.

Epstein, H. and Bichard, M. 1984. Pig. In: Mason, I.L. (ed.), Evolution of domesticated Animals. Longman, London, New York. pp. 145-162.

Ewbank, R. 1968. The behavior of animals in restraint. In: Fox, M.W. (ed.), Abnormal Behaviour in Animals. W.B. Saunders, Philadelphia. pp. 159-78.

Ewbank, R. and Bryant, M.J. 1972. Aggressive behaviour amongst groups of domesticated pigs kept in various stocking rates. Anim. Behav. 20, 21.

Fraser, D. 1978. Observation of the behavioural development of suckling and early weaned piglets during the 1st 6 weeks after birth. Anim. Behav. 26, 22.

Fraser, A.F. 1980. Farm Animal Behaviour. Balliere Tindall, London, pp. 179-87.

Fraser, D. 1985. Selection of bedded and unbedded areas by pigs in relation to environmental temperature and behaviour. Appl. Anim. Behav. Sci. 14, 117.

Fraser, A.F. and Herchen, H. 1979. The behaviour of liberated livestock. Appl. Anim. Ethol. 5, 95.

Fraser, D., Phillips, P.A., and Thompson, B.K. 1986. A test of a free-access two-level pen for fattening pigs. Anim. Prod. 42, 269.

Gonyou, H.W., Hemsworth, P.H. and Barnett, J.L. 1986. Effects of frequent interactions with humans on growing pigs. Appl. Anim. Behav. Sci. 16, 269.

Grandin, T. 1980. Designs and specifications for livestock handling equipment in slaughter plants. Int. J. Stud. Anim. Probl. 1, 178.

Grandin, T. 1982. Pig-behaviour studies applied to slaughter-plant design. Appl. Anim. Behav. Sci. 9, 141.

Grandin, T. 1986. Good pig handling improves pork quality. Proc. 32nd European Meeting of Meat Research Workers. Ghent, Belgium, 1, 105-109.

Grandin. T. 1987a. Animal handling. Vet Clin. N. Amer. Food Anim. Pract. 3, 323.

Grandin, T. 1987b. High-speed double-rail restrainer system for stunning and ritual slaughter. Proc. Int. Congr. Meat Sci. and Technol. Helsinki, Finland, 1, 101.

Grandin, T. 1989a. Pre-slaughter treatment and transportation research in the United States. Proc. Int. Congr. Meat. Sci. and Technol., Copenhagen, Denmark, 1, 133-9.

Grandin, T. 1989b. Environmental enrichment reduces excitability in confinement reared hogs. Proc. Int. Congr. Meat Sci. and Tehnol., Copenhagen, Denmark, 3, 971-4.

Grandin, T. 1989c. Effect of rearing environment and environmental enrichment on the behavior and neural development of young pigs. Dissertation, University of Illinois, Champaign/Urbana, IL, USA.

Grandin, T. 1990. Design of loading facilities and holding pens. Appl. Anim. Behav. Sci. 28, 187.

Grandin T., and Bruning, J. 1992. Boar presence reduces fighting in mixed slaughter-weight pigs. Appl. Anim. Behav. Sci. 33, 273.

Grandin, T., Curtis, S.E. and Greenough, W.T. 1983a. Effects of rearing environment on the behaviour of young pigs. J. Anim. Sci. 57 (suppl. 1), 137.

Grandin, T., Curtis, S.E. and Greenough, W.T. 1983b. Objective measurements of the effects of environmental complexity on young pigs. University of Illinois Swine Research Reports.

Grandin, T., Curtis, S.E. and Taylor, I.A. 1987. Toys, mingling and driving reduce excitablity in pigs. J. Anim. Sci. 65 (suppl.), 230.

Graves, H.B. 1984. Behaviour and ecology of wild and feral swine (Sus scrofa). J. Anim. Sci. 58, 482.

Hale, E.B. 1969. Domestication and the evolution of behaviour. In: Hafez, E.S.E. (ed.), The Behaviour of Domestic Animals, 2nd edn., Baillière Tindall and Cassell, London. pp. 22-42.

Hanson, R.P. and Karlstad, L. 1959. Feral swine in the southestern United States. J. Wildl. Manage. 23, 64.

Harstock, T.G. 1982. Behavioural responses to management systems - Swine. In: Woods, W. (ed.), Proc. of the Symp. on Management of Food Producing Animals. Purdue University, West Lafayette, IL, pp. 185-93.

Harstock, T.G. and Graves, H.B. 1976. Neonatal behaviour and nutritional-related mortality in domestic swine. J. Anim. Sci. 42, 235.

Hediger, H. 1968. The Psychology and Behaviour of Animals in Zoos and Circuses. Dover, New York. p. 66.

Hemmer, H. 1990. Domestication. The Decline of Environmental Appreciation. 2nd edn. Cambridge University Press, Cambridge.

Hemsworth, P.H. and Barnett, J.L. 1991. The effect of aversively handling pigs, either individually or in groups, on their behaviour, growth and corticosteroids. Appl. Anim. Behav. Sci. 30, 61.

Hemsworth, P.H., Barnett, J.L. and Hansen, C. 1986a. The influence of handling by humans on the behaviour, reproduction and corticosteroids of male and female pigs. Appl. Anim. Behav. Sci. 15, 303.

Hemsworth, P.H., Barnett, J.L., Hansen, C. and Gonyou, H.W. 1986b. The influence of early contact with humans on subsequent behavioural responses of pigs to humans. Appl. Anim. Behav. Sci. 15, 55.

Hemsworth, P.H., Gonyou, H.W. and Dziuk, P.J. 1986c. Human communication with pigs: The behavioural response of pigs to specific human signals. Appl. Anim. Behav. Sci. 15, 45.

Hemsworth, P.H., Barnett, J.L., Treacy, D., and Madgwick, P. 1990. The heritability of the trait fear of humans and the asscociation between this trait and subsequent reproductive performance of gilt. Appl. Anim. Behav. Sci. 25, 85.

Hoenderken, R. 1976. Ein verbessertes System für das Treiben von Schlachtsschweinen zum Restrainer. Fleischwirtschaft 56, 838.

Hughes, B.O. 1973. Animal welfare and the intensive housing of domestic fowls. Vet. Rec. 93, 658.

Hughes, B.O. 1980. Behaviour of the hen in different environments. Anim. Reg. Stud. 3, 65.

Immonen, M. 1984. Kasvatusolosuhteiden vaikutus sianlihan laatuun. Sika 3/1984, 10.

Jensen, P. 1986. Observations on the maternal behavior of free-ranging domestic pigs. Appl. Anim. Behav. Sci. 16, 131.

Jensen, P. 1989. Nest site choice and nest building of free-ranging domestic pigs due to farrow. Appl. Anim. Behav. Sci. 22, 13.

Jensen, P. 1991. Back to nature: the use of studying the ethology of free-ranging domestic animals. In: Appleby, M.C., Horrell, R.I., Petherick, J.C. and Rutter, S.M. (eds) Applied Animal Behaviour: Past, Present and Future. SVE, UFAW, Herts. pp. 62-4.

Kavanau, J.L. 1964. Behaviour: Confinement, adaptation and compulsory regimens in laboratory studies. Science 143, 490.

Kelly, K.W., McGlone, J.J. and Gaskin, C.T. 1980. Porcine aggression: measurement and effects of crowding and fasting., J. Anim. Sci. 50, 336.

Kerr, S.C.G., Wood-Gush, D.G.M., Moser, H. and Whittemore, C.T. 1988. Enrichment of the production environment and enhancement of welfare through the use of the Edinburgh family pen system of pig production. Res. Dev. Agr. 4, 119.

Kilgour, R. 1976. The contributions of psychology to a knowledge of farm animal behaviour. Appl. Anim. Ethol. 2,197.

Kilgour, R. 1988. Behaviour in the pre-slaughter and slaughter environments. Proc. Int. Congr. Meat Sci. and Technol. Part A. Berisbane. Qld 29, August -2 September, 1988. Livestock and Meat Authority of Queensland, Brisbane, Qld. pp. 130-8.

Kilgour, R., Foster, T.M., Temple, W., Matthews, L.R. and Bremner, K.J. 1991. Operant technology applied to solving farm animal problems. An assessment. Appl. Anim. Behav. Sci. 30, 141.

Lambooy, E. 1986. Automatic electrical stunning of veal calves in a V-type restrainer. Proc. Eur. Meeting of Meat Res. Workers., Ghent, Belgium.

Lambooy, E. 1988. Road transport of pigs over a long distance: some aspects of behaviour, temperature and humidity during transport and some effects of the last two factors. Anim. Prod. 46, 257.

Lewis, J.C. 1966. Observation of pen-reared European hogs released for stocking. J. Wildl. Manage. 30, 832.

Lorenz, K. Z. 1981. The Foundations of Ethology. Springer-Verlag, New York. pp. 176-88.

McClone, J.J. and Morrow, J.C. 1988. Reduction of pig antagonistic behaviour by androsterone. J. Anim. Sci. 66, 880.

McCort, W.D. and Graves, H.B. 1982. Social dominance relationship and spacing behaviour of swine. Behav. Processes 7, 169.

Matthews, L.R. and Ladewig, J. 1985. Operant conditioning: theory and practical application in farm animal welfare research. Aktuel. Arb. Artgemassen Tierhalt., KTBL, 311, 134.

Matschke, G.H. 1964. The influence of oak mast on European wild hog reproduction. Proc. of The 18th Ann. Conf. Southeastern Assoc. of Game and Fish Commissions, October 18-21. Clearwater, FL. pp. 35-39.

Meese, G.B. and Ewbank, R. 1973. The establishment and nature of the dominance hierachy in the domesticated pig. Anim. Behav. 21, 326.

Miller, D.B. 1977. Social displays of mallard ducks (*Anas platyrhynchos*): Effect of domestication. J. Comp. Physiol. Psychol. 91, 221.

Morris, G.L. and Curtis, S.E. 1982. Let the pig turn on the heat. Pig. Int., Jan. pp. 30-32.

Mothes, E. 1981. Technologie der Tierproduktion. Lehrbuch für Veterinärmediziner. VEB Gustav Fischer Verlag, Jena.

Newbery, R.C. and Wood-Gush, D.G.M. 1986. Social relationships of piglets in semi-natural environment. Anim. Behav. 34, 1311.

Newberry, R.C. and Wood-Gush, D.G.M. 1988. The development of certain behavioural patterns in piglets under semi-natural conditions. Anim. Prod. 46, 103.

Nichols, L.Jr. 1963. Job Completion Report. Research Project Segment. Hawaii, Ecology of the wild pig. July 1, 1962-June 30, 1963. Project No.: W-5-R-14. Job No. 46(14). p 11.

Ödberg, F.O., Gijselbrecht, L. and Bouquet, Y. 1991. Stereotypy frequency in halothane+ and halothane- Landrace sows. In: Appleby, M.C., Horrell, R.L., Petherick, J.C. and Rutter, S.M. (eds) Applied Animal Behaviour: Past, Present and Future. SVE, UFAW, Herts. p. 93.

Pearce, G.P., Paterson, A.M. and Pearce, A.N. 1989. The influence of pleasant and unpleasant handling and the provision of toys on the growth and behaviour of male pigs. Appl. Anim. Behav. Sci. 23, 27.

Petersen, V., Recen, B. and Vestergaard, K. 1990. Behaviour of sows and piglets during farrowing under free-range conditions. Appl. Anim. Behav. Sci. 26, 169.

Price, E.O. 1984. Behavioural aspects of animal domestication. Quart. Rev. Biol. 59, 1.

Phillips, P.A., Thompson, B.K. and Fraser, D. 1988. Preference tests of ramp designs for young pigs. Can. J. Anim. Sci. 68, 41.

Phillips, P.A., Fraser, D. and Thomson, B.K. 1991. Preference for a partially enclosed farrowing crate. Appl. Anim. Behav. Sci. 32, 35.

Rasmussen, O.G., Jenks, E.M., Berry, T.H. and Becker, D.E. 1962. Social dominance in gilts. J. Anim. Sci. 21, 520.

Reed, C.A. (ed.) 1977. Origins of Agriculture. Mouton, Paris.

Rooijen, J. van 1983. Operant preference tests with pigs. Appl. Anim. Ethol. 9, 87.

Robert, S. and Dallaire, A. 1986. An exploratory sudy of behavioural differences between young pigs susceptible and non-susceptible to stress syndrome. Appl. Anim. Behav. Sci. 16, 335.

Robert, S., Dancosse, J. and Dallaire, A. 1987. Some observations on the role of environment and genetics in behaviour of wild and domestic forms of Sus scrofa (European wild boars and domestic pigs). Appl. Anim. Behav. Sci. 17, 253.

Rohde Parfet, K.A. and Gonoyou, H.W. 1991. Attraction of newborn piglets to auditory, visual, olfactory and tactile stimuli. J. Anim. Sci. 69, 125.

Schaefer, A.L., Sather, A.P., Tong, A.K. and Lepage, P. 1989. Behaviour in pigs from three genotypes segregating at the halothane locus. Appl. Anim. Behav. Sci. 23, 15.

Signoret, J.P., Baldwin, B.A., Fraser, D., and Hafez, E.S.E. 1975. The behaviour of swine: In: E.S.E. Hafez (ed.), The Behaviour of Domestic Animals. Ballière Tindall, London. pp. 295-29.

Smidt, D. 1983. Advantages and problems of using integrated systems of indicators as compared to single traits. In: Smidt, D. (ed.) Indicators Relevant to Farm Animal Welfare. Martinus Nijhoff, Boston. pp. 201-07.

Stegeman, L.C. 1938. The European Wild boar in the Cherokee National Forest, Tennessee. J. Mammal 19, 279.

Stephens, D.B. and Perry, G.C. 1990. The effects of restrain, handling, simulated and real transport in the pig (with reference to man and other species). Appl. Anim. Behav. Sci. 28, 44.

Stephens, D.B., Bailey, K.J., Sharman, D.F., and Ingram, D.L. 1985. An analysis of some behavioural effects on the vibration and noice components of transport in pigs. Q. J. Exp. Physiol. Cogn. Med. Sci., 70, 211.

Stolba, A. 1981. A family system of pig housing. In: Symposium on Alternatives to Intensive Husbandry Systems. SVE, UFAW, Herts. pp. 52-67.

Stolba, A. 1985. Minimizing social interference during feeding in pig groups. In: Abstracts of the 19th International Ethological Conference, Toulouse, France, Vol. 2. p. 460.

Stolba, A. and Wood-Gush, D.G.M. 1981. The assessment of behavioural needs of pigs under free-range and confined conditions. Appl. Anim. Ethol. 7, 388.

Stolba, A. and Wood-Gush, D.G.M. 1984. The identification of behavioural key features and their incorporation into housing design for pigs. Ann. de Recherch. Vet. 15, 287.

Symoens, J. and van der Brande, M. 1969. Prevention and cure of aggressiveness in pigs using the sedative azaperone. Vet. Rec. 85, 64.

van Putten, G. 1980. Objective observations on the behaviour of fattening pigs. Anim. Reg. Stud. 3, 105.

van Putten, G. 1981. Handling slaughter pigs prior to loading and unloading on a lorry. Paper presented at the Seminar on Transport. C.E.C. Brussels, July 7 - 8 1981.

van Putten, G. 1989. The pig: A model for discussing animal behaviour and welfare. Appl. Anim. Behav. Sci. 22, 115.

van der Wal, P.G. and Eikelboom, G. 1981. Stunning of pigs for slaughter; recent development in the Netherlands. In: Froystein, T., Slinde, E. and Standal, N. (eds) Porcine Stress and Meat Quality- Causes and Possible Solutions to the Problem. Agricultural Food Research Society, Norway, pp. 298-305.

Wangsness, P.J., Gobble, J.L. and Sherritt, G.W. 1980. Feeding behaviour of lean and obese pigs. Physiol. Behav. 24, 407.

Warnier, A. and Zayan, R. 1985. Effects of confinement upon behavioural hormonal responses and production indices in fattening pigs. In: Zayan, R. (ed.), Social Space for Domestic Animals. Martinus Nijhoff, Dordrecht. pp. 128-50.

Weeding, C.M., Hunter, E.J., Gruise, H.J. and Penny, R.H.C. 1991. The effect of slaughter handling system and pig behaviour on welfare and meat quality. In: Appleby, M.C., Horrell, R.L., Petherick, J.C. and Rutter, S.M. (eds) Applied Animal Behaviour: Past, Present and Future.SVE, UFAW, Herts. p. 157.

Wiepkema, P.R. 1982. On the indentity and significance of disturbed behaviour in vertebrates. In: Bessei, W. (ed.) Disturbed Behaviour of Farm Animals. Hohenheimer Arbeiten. pp. 7-17.

Wood-Gush, D.G.M. and Stolba, A. 1982. Behaviour of pigs and the design of a new housing system. Appl. Anim. Ethol. 8, 583 (abstract).

Discussion

*Smulders - A phenomenon was presented to the meat science community a couple of years ago by Temple Grandin. She said that there might be a genetic component in the behavior of pigs. Then she illustrated that by saying that in some areas in Australia, for instance, there were whole populations of what she calls crazy pigs. And, she says, that in America it's the same situation. There are some areas, I think it was the midwest, where the populations were extremely wild, and stressful. But if I look at the slide you showed of Roberts and Dallaire's work, it can be a day effect, can't it? If you watch the animals in certain surroundings, and use the same surroundings on another day, the same population may give you an entirely different view of what the situation is. I'd like to have your comments on that. The second question has to do with the angle that you suggested for the funnel-shaped tunnel. I would say that, based on the experience of the people at the Institute of Animal Production, that funnel-shaped tunnels are a "no-no". There shouldn't be any funnels involved in lairage, there should be a more staircased-type of wall.

*Harri - May I answer the last question first, because it is easier. I actually agree that the stepwise narrowing of the lanes would be more beneficial. But if there must be some funnel, there must be some angle, and then it should be 30°. OK then, let's agree, no funnels. What about the crazy pigs? Yes, that is a problem. There is always a very strong genetic component in pig behavior, for example, crazy pigs. And there are also results in which those pigs that were difficult to handle produced meat of lower quality when they were brought to the slaughter plant. But, of course, under certain surroundings, crazy pigs, are more beneficial. That's actually the real problem, we can't solve all those problems. We can't have pigs that are alert, and aggressive, and can be active, and at the same time they are calm, and docile, and so on. And, I think that, maybe, this is the most difficult thing. If we solve one problem, we create a new problem, and if we solve the new problem, we create another problem.

*Smulders - But, Dr. Grandin suggests that, by selection, we have actually introduced aggressiveness into some lines of pigs. My question is: when looking at the slide you showed of the study by Robert and Dallaire, it may well be that under another circumstance, these populations are not crazy or aggressive at all. So, it is simply a matter of when exactly the populations should be observed. The reason why I stress this once again, is because it is considered a big problem, and Dr. Grandin's advice has been taken very seriously. So, I would like to have your opinion on whether is it really a genetic effect, or is it rather a matter of methodology, how to observe the population?

*Harri - That is very difficult for me to answer, because I have not personally studied that particular problem. I have been studying the selection of pigs in a temperature gradient for quite a long time. I don't know, but I think there are both components, environment seems likely, and genetics seems likely too. When we select animals, we are selecting for several traits at the same time.

*Tarrant - I would like to know, in relation to mixing different groups of pigs, this is always a problem, how important is the sense of smell in the aggression, and in the sexual interaction between groups of strangers? I ask this because it's something that would be open to the possibility of correction or camouflage.

*Harri - I think the smell would be very important. Just a couple of days ago I read a paper in applied animal behavioral science where a sexually active boar was put in with a group of fattening pigs. That settled down all aggressions. So, of course the smell is very important. Androsterone has been used, for example, and that has settled down some aggressions.

*Tarrant - So, in your opinion, it should be possible to treat groups of strangers to reduce the level of aggression, and sexual interaction.

*Harri - Yes, it is, but to what extent it is possible, that I don't know, but we can reduce it, at least to a certain amount.

Physiology and Pork Quality - Review

D. Lister

CAB International, Wallingford, Oxon., OX10 8DE England

The experimental investigation of the link between the general physiology of pigs and the quality of their carcasses and meat is plagued with many difficulties. There is a well-established consensus that today's pigs grow faster, leaner and use feed more efficiently than their forebears. Some would claim that this is mostly due to improved management, feeding and health of the average pig. Others would argue that genetic selection and breeding for performance has been equally, if not more, important, especially in recent years.

The down side of this progress has been the greater awareness and apparent incidence of problems with the quality of meat. Over the last 30 years retailers and consumers have become increasingly vocal about PSE and DFD meat, soft fat and eating quality. The Porcine Stress Syndrome gives rise to PSE and DFD, and in its more extreme form, the death of animals and consequent cost to the producer when animals are mixed, transported, or even mated.

There is no doubt that the highest incidence of extreme reactions is to be found in breeds of pig, such as Pietrain, Poland China, Belgian Landrace and certain other Landrace types, all being noted for their lean and muscular carcasses. The question that still baffles scientists is whether selecting and breeding pigs to become ectomorphic (i.e. muscular) automatically predisposes them to become stress-prone and to produce meat of inferior quality. Or, is it that ectomorphism and meat quality problems are controlled by specific genes which, though separate and distinct, are closely linked and tend, therefore, to be inherited as a group of characters? If that were the case only gene importation or mutation would introduce the "stress syndrome" into an improved line which previously had been lean and stress-resistant. Unfortunately practical pig breeding frequently chooses to improve leanness by using those sires that carry genes for both leanness and stress susceptibility. Indeed, many of today's meat-line sires have been identified for their role by tests that identify them as candidates for PSS e.g. the halothane test.

All this has caused researchers to look at the general objective to secure carcass and meat quality and its physiological implications from a number of viewpoints. The most important will now be described.

The physiology of muscle function and its relation to whole animal physiology

This approach starts from a knowledge of the physiological and biochemical basis of the control of meat quality and PSS and seeks to identify which mechanisms, if any, also exercise control over growth and development.

The underlying hypothesis is that the syndrome known as malignant hyperthermia (MH) provides a useful model of the porcine stress syndrome. MH can be induced in susceptible pigs, such as the Pietrain and Belgian Landrace, by the anaesthetic halothane and the muscle relaxant suxamethonium (see Lucke *et al.*, 1979).

It was recognized very early in these studies, though not proven, that some impairment of the mechanism regulating calcium fluxes in muscle and hence its contraction and relaxation could account for the muscle rigidity and heat production that were common consequences of the reaction. It was only later that convincing evidence for calcium's involvement began to appear as the successful control of MH reactions became possible with the use of dantrolene sodium, which is a specific sequestering agent for calcium (Hall *et al.*, 1980).

Calcium was further implicated by Cheah and Cheah (1981, 1984) who proposed that enhanced mitochondrial calcium efflux might give rise to a cascade phenomenon resulting in an enduring, calcium excited, muscle contraction and the fatal metabolic sequelae.

Postulating a role for calcium in the PSS, in meat quality control or in MH is one thing, discovering how its damaging effects are triggered and maintained is quite another matter.

Stress reactions have been characterized classically in terms of the stimulation and role of the autonomic nervous system, adrenal medullary and cortical hormones. It is not surprising, therefore, that this aspect has been under close investigation for some time, though it must be said that the clinical and biochemical techniques and assays for detailed and critical investigations in farm animals were not available to the early researchers in this area. It was, however, eventually shown that one of the most dramatic physiological events in developing MH was the rise in circulating catecholamines and associated mobilization of energy substrate to fuel the metabolic changes. Further investigations revealed more specific detail of these changes.

Blocking the α- but not the β-adrenergic pathway of the sympathetic nervous system (SNS) can prevent the development of MH in Pietrain pigs, and on occasion, α-adrenergic stimulation can induce it (Hall *et al.*, 1977).

The range of meat quality from PSE to DFD is usually attributed to the effects of acute and chronic stress. Lister and Spencer (1983), however, suggested that it may well be more completely described in terms of the balance of α- and β-adrenergic stimulation and the consequent availability of metabolic substrates (see also Lister, 1987).

Whatever the interpretation, there would seem to be clear involvement of the autonomic nervous system in the various metabolic emergencies to which animals like Pietrains are prone. Gregory and Lister (1981) have provided further supporting evidence. Using a valsalva-like manoeuvre, together with intravenous tyramine, noradrenaline and phenylephrine, they were able to examine various components of the sympathetic nervous responses of lean, stress-prone Pietrain and fat, stress resistant Gloucester Old Spot pigs. The SNS of Pietrains was found to be more responsive to stimulation because of greater preadrenoreceptor responsiveness. They suggested that this could account for the Pietrain's greater leanness, susceptibility to myocardial failure and propensity to stress-induced metabolic acidosis.

Another mechanism to which this characteristic can contribute is in fat mobilization. Wood *et al.* (1977) have shown that fat mobilization can be stimulated by noradrenaline more readily in Pietrains than in Large Whites. For this to be an effective route *in vivo,* Pietrain adipose tissue would need to be exposed to higher levels of noradrenalin, or to possess a greater concentration of receptors, or receptors of greater sensitivity. The lower threshold of sympathetic nervous sensitivity and greater responsiveness of Pietrains might provide a mechanism for at least one of these to occur.

The fat-mobilizing rather than fat-depositing potential of Pietrain pigs is further sustained by other hormonal pathways. Insulin secretion and sensitivity are lower in Pietrains than in Large White pigs (Wood *et al.*, 1977). A further link between the SNS, catecholamines and metabolism was established by Moss (1987), who showed that the PSE condition is associated with a faster rate of utilization of thyroxine than of secretion.

These physiological features of changed responsiveness to stressors amongst different breeds of pig not only have implications for body composition, but clearly for how animals respond to environmental conditions (e.g. heat/cold) and physical (e.g. handling/transport), interventions.

The bulk of this work ended in the early 1980s. Since then, a great variety of new approaches for inducing greater growth performance and efficiency have been established. Many of these have been made possible by the enormous advances that have been made recently in biotechnology. The precision with which such techniques can be used allows much of the previous concern about genetically linked or associated physiological characteristics to be by-passed. The concerns now are with characterizing the effects and precision of the new methodologies, how such effects are achieved, and what disadvantages exist.

Fundamental mechanisms controlling growth and development and their manipulation

Genetic selection and breeding have been used for generations with more or less scientific precision. In the last 10 years or so it has been possible to identify specific genetic control mechanisms and ways in which they, or the hormonal messenger system involved could be manipulated (see Table 1, for example).

Table 1. Control of growth and product quality via the somatotropin (ST) axis

Genome modification	(a) Selection for higher ST levels (b) Insertion of ST gene
Affecting messengers	(c) Potentiation of ST with monoclonal antibodies or vaccines (d) Administration of ST or growth hormone releasing factor

The somatotropin axis has perhaps received most attention because there is potential benefit from its actions to both the meat and milk sectors of animal agriculture. As Table 1 points out, an array of approaches is open for manipulating this axis. Selection for higher ST levels (a) is the process by which traditional breeding for increased rate of growth may have achieved its success, but there is no clear evidence to support this contention. The most recent and novel approach (b) is to transfer appropriate controlling genes into developing fertilized eggs. The genes of interest are those that code for the hypothalamic, pituitary and liver polypeptides stimulating protein anabolism. Transgenesis is the procedure that allows the chosen genes to be incorporated into a new genome without risk of transferring undesirable genes which is, of course, the major concern in conventional animal breeding. It also permits the transfer of novel genes, even from other species, and in so doing, gives access to greater genetic variation and scope for selection.

Gene transfer in farm animals (for a complete review of this, and associated biotechnology, see Lister *et al.*, 1991, and van der Wal *et al.*, 1991) has been achieved so far only through direct injection of DNA into one of the pronuclei of fertilized eggs. The injected ova can be returned to a recipient potential mother that has been maintained at an appropriate stage of the oestrous cycle. Only a small proportion (on average about 25%) of the ova survive to term and of these, only 1-2% will have incorporated the injected DNA into their genomes. In addition to these limitations, there is no control over the site of incorporation or the number of copies of the gene incorporated. There is, therefore, considerable variation in the expression of the transgene across a range of recipients. Whilst this may not be a problem in rodents, it is not tolerable in farm animals and the technique is unlikely to find a place in practical agriculture.

Alternative methods for gene transfer are still at an early stage of consideration for farm animals though they have been evaluated in mice. These methods include retroviral and embryo stem-cell-based gene transfer. Retroviruses are double-stranded RNA viruses which, after entering the host, are converted to DNA and incorporated into the host's genome. Foreign DNA has been inserted into hosts via this route though the success rate is approximately the same as that which is achievable by direct injection methods, and there are limits to how much extra DNA can be incorporated.

A greater problem is, perhaps, the risk of undesirable viral sequences being formed and infective agents being produced.

Embryo stem cells retain an ability to develop into a variety of tissue cells. During their culture *in vitro,* new genes can be inserted and the cell mass can be used to recolonize the inner cell masses of blastulae to become both somatic and germ cells. The transfer gene may be present in only a proportion of the body cells i.e. the transgenic animals are chimeras, and the germ cells may not be amongst them. To overcome this in farm animals attempts are being made to use nuclei from embryo stem cells as the vehicle for transfers for this would provide an ideal system for use in these species.

A further approach (c) for manipulating the ST axis relies on immunology. The rationale for immunological interference with an animal's hormonal system is based on the notion of modifying endogenous hormone production using antibodies raised in the animal itself (active immunization) or in a donor (passive immunization) in the same way that disease resistance may be conferred to man and animals. The hope is that a short period of treatment, perhaps even a single injection, may be responsible for a prolonged effect on a production characteristic.

One of the first attempts at using such a technique in farm animals was that of Spencer *et al.* (1983). In their work, Spencer *et al.* used antibodies against somatostatin, which regulates the release of ST from the pituitary. The work was carried out in sheep, but its initial success in increasing their rate of growth was not always confirmed by other researchers and there is no consensus on the value of this approach.

When a "foreign" substance is injected into an animal, the immune reaction raises a number of antibodies, each combining with specific sites on the antigenic substance. Such antibodies are referred to as polyclonal antibodies. Monoclonal antibodies specific to one site may also be cultured and cloned. Such antibodies have many uses, but in the present context they may be complexed with certain sequences of ST to enhance its biological activity. Pell *et al.* (1991), have reviewed this procedure and its extension to the use of polyclonal antibodies of restricted specificity, which again increase the activity of the hormones of the ST axis. This work has been done mainly with lambs, but the authors argue that the principle is applicable to other farm species.

The third immunological approach requires the use of anti-antibodies or anti-idiotypes. If, for example, bovine ST is injected into rabbits, the rabbits will respond by making antibodies to the bovine ST. Such antibodies can be separated from rabbit serum and returned to a ruminant that will make further antibodies to them. These anti-idiotypes are capable of binding to ST receptors and acting in effect as the original hormone. A potential advantage of such anti-idiotypes is that they may be prepared to be receptor specific, which could allow them to create specific effects e.g. raising milk production, growth or fat mobilization. The added advantage over ST administration is the need, in prospect, for only infrequent injections of the first antibody to produce a prolonged response.

All of these techniques for manipulating with the ST axis are still, for the most part, research procedures. This is not the case for the administration of porcine ST (pST) or its releasing factor (d). The topic has been extensively reviewed by a variety of authors including Bonneau (1991a, 1991b). For pigs in particular, there is debate only on the size of the positive response to these treatments. Moreover, there is now ample evidence for effects on growth and also on carcass composition and meat quality.

Dramatic progress became possible in research with pST when recombinant pST possessing consistent biological effects that mimic native pituitary pST became available. Many groups of workers throughout the world recognized that daily (usually) injections of pST could reduce the daily feed intake of pigs by 2-22%, fat deposition by 7-44%, increase growth up to 47% and improve feed efficiency by 3-38%. The size of some internal organs e.g. liver, kidney and heart, is increased and contributes to a lower dressing percentage in treated pigs. Growth hormone-releasing factor (GRF) has comparable effects to ST, but at the same dose level as ST, they are significantly smaller.

Once the beneficial effects of treatment with pST became apparent, researchers turned their attention to establishing what further influences there might be on responses. These included the form, dose and treatment schedule of pST, the role of diet and nutrition, stage of development and the genotype of treated stock.

The dose response to pST varies from one character of interest to another. The pST is injected in doses related to body weight (μg per kg) or as mg per day. Daily feed intake falls linearly with increasing dose but growth rate is maximum at doses between 3 and 6mg per day and feed efficiency is then not further improved beyond 6mg per day. Carcass fat declines in a linear fashion whereas lean increases, in some cases linearly to doses reaching 200μg/kg/day. It is most common to administer single boluses of pST.

It is generally considered that pST is best administered at the stage of development when fat deposition starts to accelerate, although benefits have been observed from treatment at both earlier and later stages and may continue to weights in excess of 100kg.

The improved performance is evident in both *ad lib,* and restricted fed pigs and there are signs that at modest levels of feed restriction i.e. less than 25%, treated pigs are less affected by feed restriction than are control animals. But, it is not known with certainty whether dietary protein and amino acid requirements are affected by pST. Certainly if protein is limiting in the diet the growth of treated animals is greatly reduced. Improvements in feed efficiency suggest that total energy requirements are lower in treated pigs.

It is generally found that extremes of type e.g. slow-growing fat Chinese breeds respond most markedly to treatment, whereas the lean Pietrain's response can be marginal. A wide range of commercial genotypes tend to respond to a similar degree. The same association between body type and response can be seen in comparisons of castrates, females and males, the males responding least.

There are now many reported studies on the effects of pST on various aspects of meat quality. The general conclusion is that the eating quality of meat is unimpaired except at high and unpractical dosage levels when significant toughening, reduced juiciness and flavor have been observed. There seem to be no deleterious effects on processed products such as hams and sausages.

These practical observations are supported by research at a more fundamental level. Thus, there are signs of an increase in the size of muscle fibers in redder muscles and an accompanying change of enzyme profile towards that which is found in more glycolytic fibers. There is, however, little indication that post-mortem biochemical changes are dramatically altered and such changes as there are, tend to be small and non-significant, all of which would confirm the findings on sensory qualities.

The conclusion to be drawn from all of these studies would confirm that manipulation of the appropriate and specific pathway, namely the ST axis, is likely to achieve the benefits that are sought by classical selection and breeding. Moreover the benefits are evident in one generation and are clear of the problems of the fresh and cooked meat that have been evident for some time in the better-performing animals. There are presently practical, though not sufficiently refined, approaches for providing ST exogenously and, as we have seen, in the process of development are other far-reaching techniques that could offer advantages over the former. The problems over these methodologies are, however, not to do with efficacy and benefit, but whether or not they will be practices that are acceptable to consumers.

The identification and use of physiological mechanisms to monitor or correct specific practical deficiencies

Animal scientists have for many years searched for a test that, when applied to living pigs, might provide an indication of their likely meat quality after slaughter. The most common of those considered have relied on blood parameters which can be measured in samples of blood taken from restrained pigs which may or may not have suffered a period of stress or exercise before the sample was taken. The biochemical profile might include enzymes such as creatine phosphokinase, lactate dehydrogenase and others directly involved in muscle metabolism and function. Other scientists looked for changes in the levels of circulating stress hormones and metabolites e.g. adrenalin, noradrenalin, cortisone, lactate, glucose and electrolytes. Yet others examined blood gases, PO_2, PCO_2 or base excess phenomena. It must be said, however, that these parameters define the "resting state" and the metabolic consequences of stimulation. They may even describe some physiological deficiencies e.g. "leaky" membranes, but rarely are they descriptive of causal mechanisms. This should present no surprise since even now there is no clearly identified cause of the stress syndrome.

The most widely used "test" for both stress susceptibility and potential leanness is, of course, the halothane test. Though this is regarded by the majority of researchers and practical pig breeders and producers as an appropriate model for the stress reaction, the view is not unanimous (Mormede and Dantzer, 1987). Few, however, would disagree that the most effective way of introducing stress proneness into a population or conversely removing it is to change the incidence of the halothane gene in it.

It is evident from what has gone before that specific mechanisms in the hypothalamic-pituitary axis can be manipulated to improve growth performance and carcass quality without significantly affecting animal behavior and the biochemical processes that, post-mortem, can control the fresh and eating quality of meat. In other words the two major concerns of the pig world, carcass and meat quality, are not always inextricably linked as they might appear to be from the known linkage of genes contained at the *Hal* locus which conveys halothane sensitivity to pigs. This linkage (Archibald and Imlah, 1985) is well recorded to be associated with increased lean content of carcasses but also with disadvantages in stress susceptibility, meat quality and litter productivity. The practical difficulty for geneticists and animal breeders has centered on the problem of extracting the deleterious components whilst retaining the desirable. Now, it seems, that a specific biochemical mechanism and its controlling gene have been discovered that may accomplish this.

As indicated at the start of this paper, the fundamental problem in MH has always been thought to rest within those mechanisms regulating calcium fluxes in muscle. If this were to prove to be the case and that its genetic control lay in a single locus, investigations of the etiology of the halothane reaction and perhaps the porcine stress syndrome would undoubtedly benefit from the use of techniques in molecular biology and, perhaps, genetic engineering. This is now a prospect as a result of recent discoveries concerning a calcium-channel protein of the sarcoplasmic reticulum in muscle, which could be the site of the primary defect in MH (Louis *et al.,* 1990). This story is developed at length elsewhere (see contributions by Mickelson and Vögeli) but it is perhaps worth a brief comment to close this paper.

The process of excitation-contraction coupling in muscle is ultimately triggered by the ability of the sarcoplasmic reticulum to release and take up Ca^{2+} ions at concentrations in muscle around 10^{-6}M and in so doing allow the contractile proteins of the muscle to contract or relax.

There are several sites in muscle that regulate the concentrations of cytoplasmic calcium in addition to the sarcoplasmic reticulum. These include mitochondria (see Cheah and Cheah, 1981, 1984) and surface membranes including the plasma membrane and transverse tubule membranes. It is now considered that the role of mitochondrial calcium efflux may be in promoting an MH reaction that has already been initiated and that previous suggestions of defects in the plasma membranes might have been the result of deficiencies in their preparation (see Louis et *al.,* 1990).

Ervasti *et al.* (1989), however, identified a defect in the dihydropyridine receptor of the transverse tubule which is coupled to the calcium-channel protein of the sarcoplasmic reticulum. Once the mechanism to release the calcium that initiates muscle contraction has been stimulated the defect will not allow this release to be curtailed, the muscle is maintained in the contracted state and a full-blown MH episode may follow (Fill et *al.,* 1990).

It is now thought that an altered region or even a single amino acid in the release channel may be the focus of the defect and the result of a mutation in a single protein.

The ryanodine receptor by which the calcium-release channel is characterized (see O'Brien, 1986) has now been associated, via a DNA-based test, with a particular sequence - C 1843 in the *Hal* gene, and offers the basis for a specific test for MH susceptibility (Fujii *et al.,* 1991).

We have now therefore come full circle in this paper. What started out as an exercise in whole animal physiology where a pattern of discrete physiological events is characterized and a unifying hypothesis proposed has turned into one where one particular aspect of the enquiry has become dominant to allow the whole process to be short-circuited and a practical result to be within reach. There would seem to be great merit in pursuing this along a variety of biotechnological channels, though the pursuit of the big biological questions of the relations between growth, development, metabolism and behavior may ultimately prove more rewarding both to classical physiologists and the newer breeds of molecular biologist who need to know which are the important mechanisms to look for before they can practice their art.

References

Archibald, A.L. and Imlah, P. 1985. The halothane sensitivity locus and its linkage relationships. Animal Blood Groups and Biochemistry Genetics 16, 253-63.

Bonneau, M. 1991a. Administration of elements of the somatotropin axis: effects in pigs and poultry. In: van der Wal, P., Weber, G.M. and van der Wilt, F.J. (eds), Biotechnology for Control of Growth and Product Quality in Meat Production: Implications and Acceptability. Pudoc, Wageningen, pp. 67-89.

Bonneau, M. 1991b. Regulation of pig growth by somatotropic hormones: II. The effect of exogenous GRF or pST administration on performance and met quality. Pig News and Information 12, 39-45.

Cheah, K.S. and Cheah, A.M. 1981. Skeletal muscle phospholipase A2 and the interaction of mitochondria and sarcoplasmic reticulum in porcine MH. Biochemica et Biophysica Acta 638, 40-49.

Cheah, K.S. and Cheah, A.M. 1984. Endogenous calmodulin Ca^{2+} and phospholipase A2 activity and their relationship to halothane sensitivity in young and adult pigs. In: Proceedings of 30th European Meeting of Meat Research Workers, Bristol. pp. 106-107.

Ervasti, J., Claessen, M.T., Mickelson, J.R., and Louis, C.F. 1989. Altered transverse tubule dihydropyridine receptor binding in malignant hyperthermia. J. Biol. Chem. 264, 2711-17.

Fill, M., Coronado, R., Mickelson, J.R., Vilveen, J., Ma, J., Jacobson, B.A., and Louis, C.F. 1990. Abnormal ryanodine receptor channels in malignant hyperthermia. Biophys. J. 50, 471-75.

Fujii, J., Otsu, K., Zorzato, F., Dehern, S., Khanna, V.K., Weller, J.E., O'Brien, P.J., and McLennan, D.H. 1991. Identification of a mutation in porcine ryanodine receptor associated with malignant hyperthermia. Science 253, 448-51.

Gregory, N.G. and Lister, D. 1981. Autonomic responsiveness in stress - resistant and stress-sensitive pigs. J. Vet. Pharmacol. and Ther. 4, 67-75.

Hall, G.M., Lucke, J.N. and Lister, D. 1977. Porcine malignant hyperthermia v. Fatal hyperthermia in the Pietrain pig associated with the infusion of alpha adrenergic agonists. Brit. J. Anaes. 49, 855-63.

Hall, G.M., Lucke, J.N. and Lister, D. 1980. Malignant Hyperthermia: Pearls out of swine. Brit. J. Anaes. 52, 165-71.

Lister, D. 1987. The physiology and biochemistry of the porcine stress syndrome. In: Tarrant, P.V., Eikelenboom, G. and Monin, G. (eds), Evaluation and Control of Meat Quality in Pigs. Martinus Nijhoff, Dordrecht, pp. 3-16.

Lister, D. and Spencer, G.S.G. 1983. Energy substrate provision *in vivo* and the changes in muscle post mortem. Meat Sci. 8, 41-51.

Lister, D., Archibald, A.L. and Flint, D.J. 1991. Biotechnology in meat production. In: Smulders, F.J.M. (ed.), The European Meat Industry in the 1990s, Audet Tijdschriften, Nijmegen, pp. 1-24.

Louis, C.F., Gallant, E.M., Remple, E. and Mickelson, J.R. 1990. Malignant hyperthermia and porcine stress syndrome: A tale of two species. Pig News and Information 11, 341-44.

Lucke, J.N., Hall, G.M. and Lister, D. 1979. Malignant hyperthermia in the pig and the role of stress. Ann. NY Acad. Sci. 317, 326-35.

Mormede, P. and Dantzer, R. 1987. Implications of the porcine stress syndrome for animal welfare. In: Tarrant, P.V., Eikelenboom, G. and Monin, G. (eds), Evaluation and Control of Meat Quality in Pigs, Martinus Nijhoff, Dordrecht, pp. 89-98.

Moss, B.W. 1987. Thyroid function in the aetiology of the PSE condition. In: Tarrant, P.V., Eikelenboom, G. and Monin, G. (eds), Evaluation and Control of Meat Quality in Pigs, Martinus Nijhoff, Dordrecht, pp. 51-68.

O'Brien, P.J. 1986. Porcine malignant hyperthermia susceptibility: hypersensitive calcium-release mechanism of skeletal muscle sarcoplasmic reticulum. Can. J. Vet. Res. 50, 318-28.

Pell, J.M., Aston, R., Flint, D. J. and James, S. 1991. Immunomodulation of hormones of the somatotropin axis. In: van der Wal, P., Weber, G.M., and van der Wilt, F.J. (eds), Biotechnology for Control of Growth and Product Quality in Meat Production: Implications and Acceptability, Pudoc, Wageningen, pp. 51-66.

Spencer, G.S.G., Garssen, G.J., and Hart, I.C. 1983. A novel approach to growth promotion using auto-immunization against somatostatin. I. Effects on growth and hormone levels in lambs. Livestock Prod.Sci.10, 25-37.

van der Wal, P., Weber, G.M. and van der Wilt, F.J. (eds) 1991. Biotechnology for Control of Growth and Product Quality in Meat Production: Implications and Acceptability. Pudoc, Wageningen.

Wood, J.D., Gregory, N.G., Hall, G.M. and Lister, D. 1977. Fat mobilization in Pietrain and Large and White pigs. Brit. J. Nutr. 37, 167-86.

Discussion

*Haley - Going back to that serendipity question, the problem with your story is that the ryanodine receptor was not isolated by anyone that was working on the halothane problem, it was isolated by a group working on a completely different problem. The tie up, is the fact that it had been found, and people knew something about what he was doing, to calcium-release channels, in fact all this work had been done on calcium release allowed the two to be tied together. But the actual isolation of the gene itself was the serendipitous event that provided people "in the know," the ability to tie the two things together.

*Lister - I would agree totally. That's usually the way these things happen isn't it? Someone in an entirely different field discovers something that is of vital importance to what is going on elsewhere. The secret is to know what is happening elsewhere, and to be able to combine it with our other current thoughts. And that is exactly what was done.

*Haley - Yes, I quite agree with that. I suppose that the point I was trying to make in my talk was that the very targeted way of going towards the gene was not easy to achieve, and probably not the way that most of them would be found, and tied up to traits.

*Smulders - I am a little bit confused about your statement on the effects of porcine somatotropin on meat quality. If I am informed correctly, I think it is quite dose-dependent, what you can actually expect. If Beerman is right in his old review, I think the book is called "Designing Foods", he states something, I'm not sure about the exact levels, but at low doses, there is no reason to fear any meat-quality defects. But, at high doses, yes, there is an inflection point where it starts to hurt meat quality. So, I don't think it's a black-and-white situation.

*Lister - Well, others have also discussed this, and, the general consensus is that at practical levels, there is no risk to meat quality. I agree, that if you overdose them, then you get a bad effect.

*Smulders - Well, it's interesting, because in ruminants it's a completely different situation.

*Lister - People are apt to think that pST and **β**-agonists have similar effects. They are, of course, entirely different drugs and the mechanisms they employ are quite different. There is no question about the damaging effects of **β**-agonists on meat quality. pST, on the other hand, when used at the recommended dose levels appears to have no such effect.

*Demeyer - I would like to come back to what Dr. Smulders said, because, 2 years ago, we did some meat quality measurement on stress-susceptible Belgian pigs, treated with pST. There was a significant rise in carcass temperature between 1 and 6mg per day (we had three doses) and there was clearly an indication, that with these pigs, pH_1 was still lowered. Also, there was a very clear indication that there was some damaging effect on meat quality. There was no effect on texture however. Whereas the **β**-agonists affect texture, pST, did not.

*Lister - Yes, I accept that some problems may occur. But it is interesting that you were dealing with Belgian Landrace pigs. It could be that the effects on pH, were secondary to those that affected carcase temperatures rather than there being direct effects on muscle as would be the case for **β**-agonists.

*Møller - I also have a comment to this pST discussion. In a recent report, it was stated that the fibers get larger, and there was a higher frequency of giant fibers with pST treatment. The question about whether the meat quality being changed or reduced to the PSE direction were not measured. But, this change in fibers could be an indicator that we have to be a little bit careful. And, although, as I remember, some of the very first records of pST showed very small changes in meat quality, in some of the more recent records, some toughness problems have been observed.

*Tornberg - We are coming to this toughness, because when you are talking about meat quality, you mean PSE. But, we have noticed that intact males are tougher. Do you have an explanation for that?

*Lister - No, I am afraid I haven't. If you like, you can think about males and females in terms of the male being slightly more stress-sensitive than the female. I just wonder if there are contractile effects in the male due to their increased responsiveness to stressors that are not obvious in the female. We don't know the role of connective tissue either under these circumstances?

*Lundström - Do you have any explanation for the effect of calcium on the meat content in the halothane-positive pigs? How could that work?

*Lister - I don't know but would you expect anything?

*Lundström - No, but there is such a high lean content in the halothane-positive pigs, and, if it is the calcium-release channel that is not working properly, then, I suppose you could have an effect of calcium on the lean content, or am I wrong?

*Lister - I would prefer to think of the mechanisms involved as being rather more separate. Muscle metabolism is closely controlled by calcium-dependent mechanism but I don't see a direct link with muscle deposition via the same mechanisms. Calcium, being a universally important ion of biological control could, perhaps, influence the activity of enzymes involved in protein turnover to enhance protein anabolism.

*Essén-Gustavsson - I think it is very interesting that you bring up fat again, and lipid metabolism. Again, we are coming back to the muscle fibers, and what substrates they utilize, and so on, and that is really important. We seem to keep forgetting about fat and fatty acids. You really showed that it is an important substrate, and we should think more about it.

*Lister - I think it almost certainly has a more important role to play in DFD. Since fat metabolism can only occur in a well-oxygenated muscle, then, the sooner there is any physiological event that switches off the use of fat, then the more urgent is the need to rely on glycolysis, and I think this is where the story begins to get interesting.

*Tarrant - Would you agree that the use of β-agonist is simply an artificial and controlled way of doing to pigs what the Pietrain was able to do naturally? You mentioned the lipid mobilizing physiology of the Pietrain, and this is the basis of the nutrient partitioning effect, so, it seems that the β-agonists are an artificial way of doing what the Pietrain was able to do naturally.

*Lister - I suspect that may well be the case, and that might be why β-agonists present so many problems in their use.

*Tarrant - Is it possible to link the halothane molecule with the calcium release channel in the sarcoplasmic reticulum, namely the ryanodine receptor?

Or succinylmethoniumchloride? Is there evidence yet that these compounds act directly on the receptor, or are there intermediate steps?

*Lister - I don't know. The only evidence yet is that the receptor can be blocked with dantrolene.

*Mickelson - Halothane can activate the ryanodine receptor, but it can do so with normal pigs just about as well as it can with the stress-susceptible pigs. As far as dantrolene goes, I don't think anyone really knows how it works, at the subcellular level.

*Lister - It is accepted though that it has to work on calcium fluxes, and it has been argued that it is a useful block for the ryanodine channel.

*Mickelson - It depends on what system you happen to be looking at. At the whole cellular level, dantrolene will block calcium fluxes. They are trying to identify which molecule the dantrolene interacts with, at the subcellular level. It's pretty much unknown.

*Jorgenson - To come back to your graph about the influence of magnesium sulfate, and to try to look at it from a more modern aspect, with respect to the ryanodine receptor. It seems that if, in the intact animal, it is possible to raise the magnesium status, succinylmethonium will not produce as significant of a decrease, a post-mortem decrease, of phosphocreatine in the muscles. And, it seems that one reason may be that magnesium binds to the ryanodine receptor, and inhibits the binding of ryanodine itself, although these investigations are very preliminary. If ryanodine binds there, it opens the channel and increases calcium influx, and it seems that if magnesium competes on that level, ryanodine releases much faster, and actually this might be one explanation for the beneficial influence of magnesium in the intact animal, but the results are preliminary.

Growth Physiology and Post-mortem Metabolism in Porcine Muscle

H.J. Swatland

Department of Animal and Poultry Science
University of Guelph, Guelph, Ontario N1G 2W1, Canada

Introduction

Extensive bibliographies on most of the topics covered here are available (Bendall, 1973; Swatland, 1984a; Bendall and Swatland, 1988), and only sources of data and a few recent studies are referenced here. The possibility that selection for rapid muscle growth might lead slowly to a deterioration in meat quality is of concern to many of those involved in animal production and food science. However, it is difficult to prove that such a change is in progress.

The study of how cellular growth patterns might influence pork quality has been greatly complicated by the occurrence of the porcine stress syndrome (PSS). By 1973, it was evident that accelerated post-mortem glycolysis in PSS pigs is caused by a post-synaptic abnormality in the calcium ion movements that regulate muscle contraction (Bendall, 1973). It is now known that genetic substitution of a nucleotide causes replacement of a single amino acid in the calcium-release channel of the sarcoplasmic reticulum. This channel is identified by its binding of the plant alkaloid ryanodine (Fujii *et al.*, 1991) and the defect affects all histochemical types of skeletal muscle fibers (Ervasti *et al.*, 1991). Thus, in addition to halothane testing for homozygotes, it is now also possible to detect heterozygotes using the polymerase chain reaction, restriction endonuclease, and electrophoretic detection of defective DNA fragments.

We must wait to see whether the elimination of residual heterozygous carriers, if it is commercially feasible, will lead to the extinction of pale, soft, exudative (PSE) pork as a commercial problem. From the work of Eikelenboom and Nanni Costa (1988), there are some doubts. However, it is important to keep advancing on the broader issues involved, whatever the outcome of the PSS problem.

When we increase the rates of lean muscle growth in farm animals using improvements in genetics, nutrition and husbandry, how are we achieving this at the cellular level? Are there certain changes that we must avoid in order to preserve meat quality? The answers to these questions will become increasingly important in global markets if the efficiency of meat production reaches a plateau and competition becomes based on meat quality. These questions are important in all species of food animals - not just pigs.

Myoblasts and myotubes

Very little is known about the factors that determine the numbers of presumptive myoblasts in embryonic pigs, although this factor alone could have an enormous effect on the post-natal potential for meat production. Presumptive myoblasts divide many times by mitosis, and eventually become non-mitotic myoblasts capable of synthesizing muscle proteins. The first myoblasts take quite a long time to form multinucleate fibers because the early fibers or myotubes must establish the architecture of the future muscle.

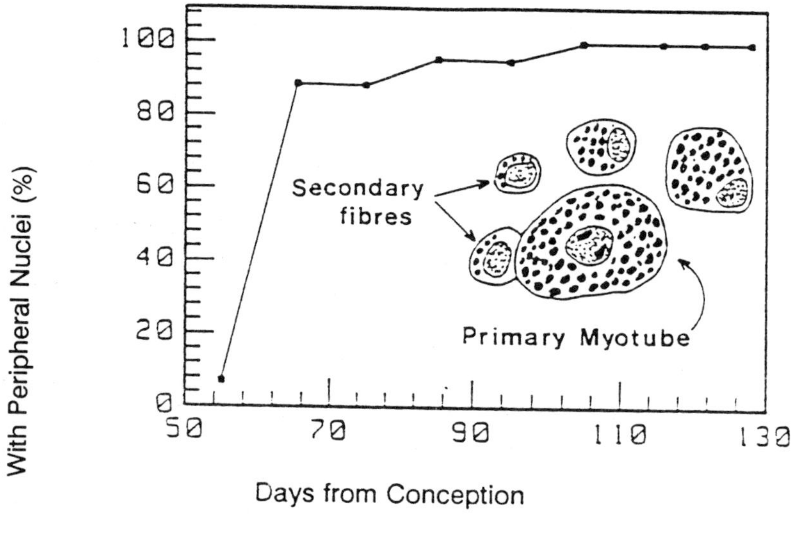

Figure 1. Pre-natal increase in the number of muscle fibers with peripheral nuclei, initially by the formation of secondary fibers and later by the maturation of primary myotubes (from Swatland, 1971).

Approximately 80% of the skeletal muscle fibers of pigs are derived from later myoblasts that use the primary myotubes as a contractile template upon which to establish their orientation. This process occurs quite rapidly, and few of the secondary fibers that are formed develop a hollow tube of myofibrils like the primary myotubes. Thus, the proportion of fibers with peripheral nuclei (secondary fibers) increases very rapidly and may be used to estimate the contribution that secondary fibers make to the total number of muscle fibers (Fig. 1). The rapid early rise to 65 days in the proportion of fibers with axial nuclei is caused by the formation of new secondary fibers, whereas the slower decrease after this time is formed by the movement of nuclei from the axis to the periphery in primary myotubes.

Radial growth and myofibrillar proliferation

The radial growth of fetal muscle fibers is slow at first, but rapidly increases (Fig. 2). Part of the rapid increase is caused by the real growth of secondary fibers which is accompanied by a proliferation of myofibrils; however part is only an illusion created by the maturation (centrifugal movement of nuclei) of primary myotubes, which are larger than secondary fibers in the earlier stages of muscle formation.

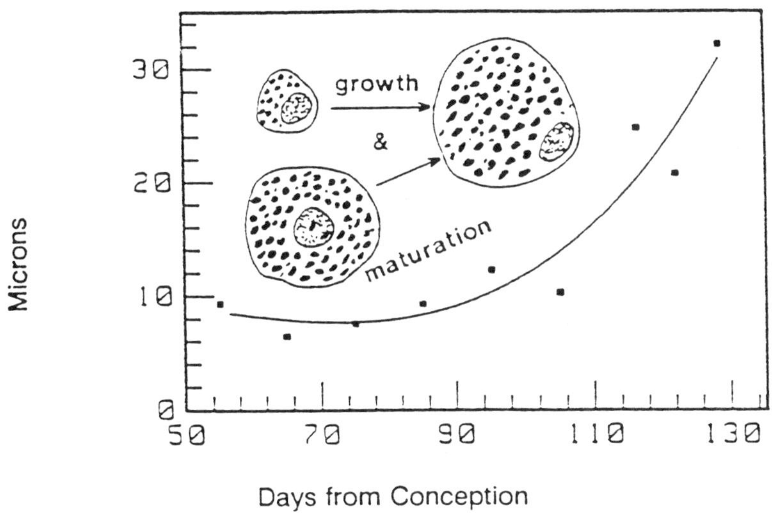

Figure 2. Pre-natal growth in the diameter of the myofibrillar area in fibers with peripheral nuclei (from Swatland, 1971).

The radial growth of muscle fibers is associated with a large increase in the number of myofibrils seen in transversely sectioned fibers (Fig. 3). The myofibrils in porcine muscle fibers tend to form a linked meshwork so that the myofibrils seen in one plane of sectioning may not match exactly to those seen in another plane of sectioning farther along the fiber. When myofibrils reach a certain cross-sectional area by the accumulation of new filaments, they appear to develop a split that often links through a hole in the middle of the myofibril - at least, that is what may be seen when the filament lattice shrinks post-mortem. This may explain why many myofibrils have a periodic-acid Schiff (PAS) core in fibers with a high glycogen content. Perhaps calcium-activated proteases are involved in weakening the Z-lines of myofibrils that have grown too large for the sarcoplasmic reticulum to keep them free of calcium ions during relaxation.

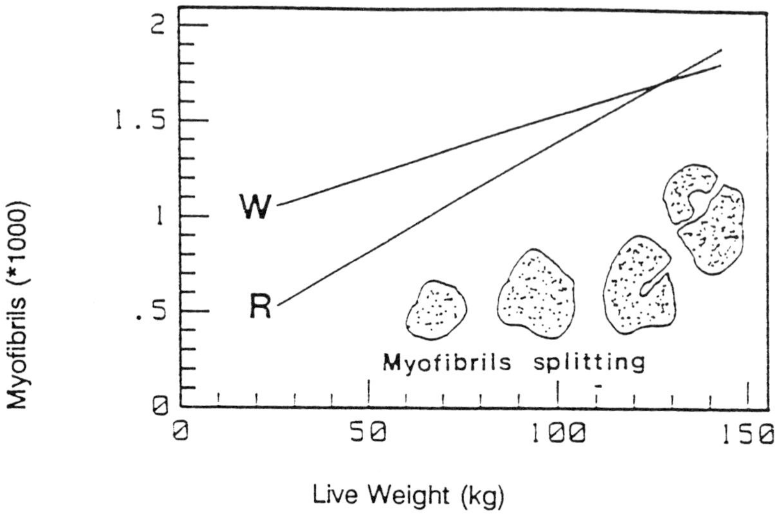

Figure 3. Proliferation of myofibrils per fiber in white (ATPase$^+$ SDH$^-$;W) and red (ATPase SDH$^+$;R) muscle fibers of porcine *longissimus dorsi* muscle (Swatland, 1976a).

As myofibrils proliferate, and muscle fibers grow radially, there is an increase in the number of nuclei per fibre (Fig. 4). It is difficult to prove whether the formation of new myonuclei from satellite cells is regulated by the radius of the fibre, or whether extra myonuclei enable a fibre to grow to a larger radius. But as soon as pigs are placed on a restricted diet that arrests growth, there is almost an instant cessation of the formation of new myonuclei.

Longitudinal growth and sarcomere proliferation

Although radial growth is important, the longitudinal growth of fibers and alterations in fiber arrangement are equally important. For example, Figure 5 shows the patterns of growth for the weight (W), muscle cross-sectional area (A), muscle length (L), and mean muscle fibre cross-sectional area (a) of porcine biceps femoris. One would expect differences in the pattern of growth for one-dimensional (length), two-dimensional (area) and three-dimensional (weight) features of muscle size, but it is surprising to see that muscle area peaks so early in development.

In short, it appears that the arrangement of muscle fibers may change as muscles grow. One very important consequence of this is that the apparent numbers of muscle fibers seen in a cross-sectional area of a muscle (at some standardized point such as its mid-length) may change during post-natal development (Fig. 6), even though the real number may reach a maximum some time before birth. To complicate matters even more, it is very difficult to find the length of individual fibers in most muscles of the pig.

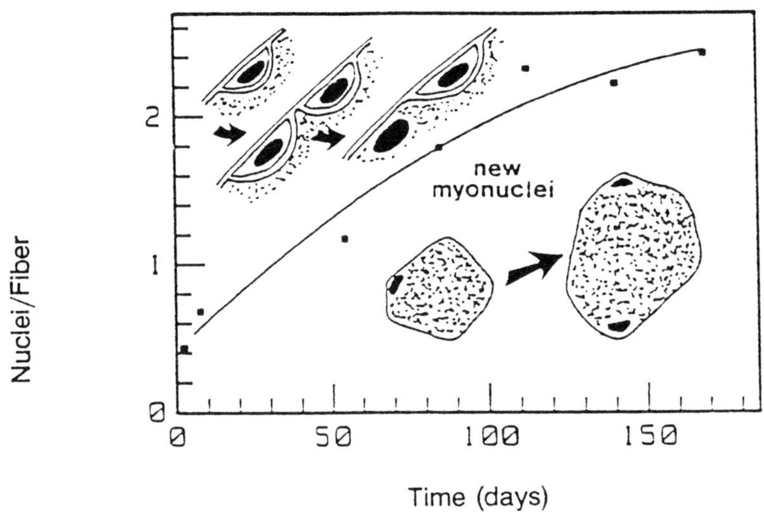

Figure 4. Increase in the number of nuclei seen in 10μm transverse sections of muscle fibers from paraffin-embedded samples of porcine *sartorius* muscle (Swatland, 1977a).

The conclusion that individual muscle fibers may grow along the length of their muscle so that the proportion of fiber length/muscle length increases with age was forced upon us by quantitative histological data, but initially was difficult to understand in microstructural terms. Fortunately, improvements in understanding the microanatomy of the cytoskeleton in skeletal muscle have supplied many of the missing answers. With desmin, vimentin, dystrophin and possibly other proteins involved in linking myofibrils laterally through to the plasma membrane (Granger and Lazarides, 1978; Pierobon-Bormioli, 1981; Wang and Ramirez-Mitchell, 1983; Pardo *et al.*, 1983; Kentish, 1983; Wakayama and Shibuya, 1991), and with corresponding connections between fibers (Shear and Bloch, 1985; Orcutt *et al.*, 1986), the anchorage of intrafascicular terminating muscle fibers is not the problem it was once thought to be. Intrafascicularly terminating fibers now have been described by electron microscopy (Trotter, 1990, 1991), leaving little doubt that they are an important histological feature of the muscles of larger animals.

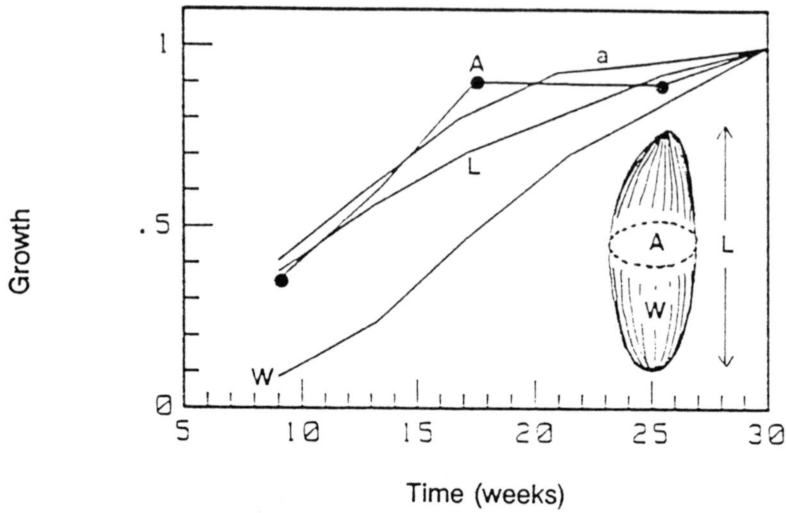

Figure 5. Patterns of growth in porcine *biceps femoris* for weight (W), length (L), area (A), and mean muscle fiber cross-sectional area (a). Data came from Swatland (1982) corrected for sarcomere length and normalized to reach 1 at 30 weeks.

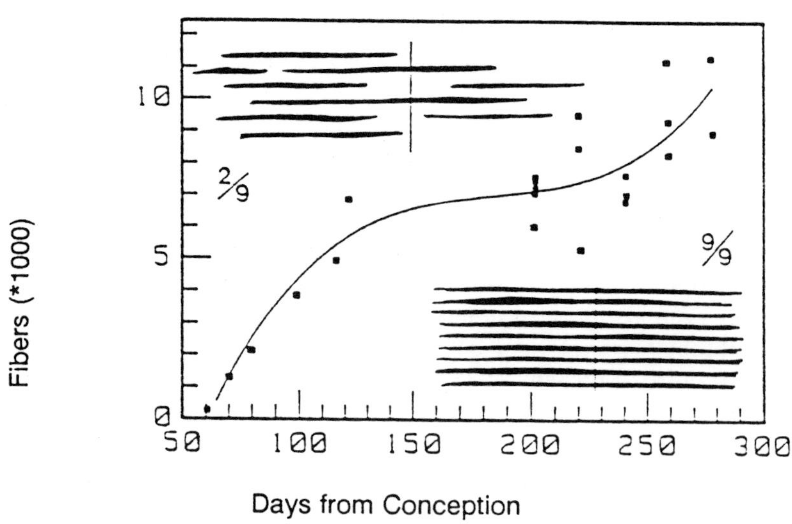

Figure 6. Changes in tne number of muscle fibers at the mid-length of the porcine *sartorius* during pre- and post-natal development (Swatland, 1973, 1976b).

Only muscles with very short fasciculi, such as those of the *peroneus longus*, are lacking intrafascicularly terminating muscle fibers. Thus, fascicular length in such muscles is the same as fiber length, and this enables the rate of sarcomere formation to be estimated (Fig. 7).

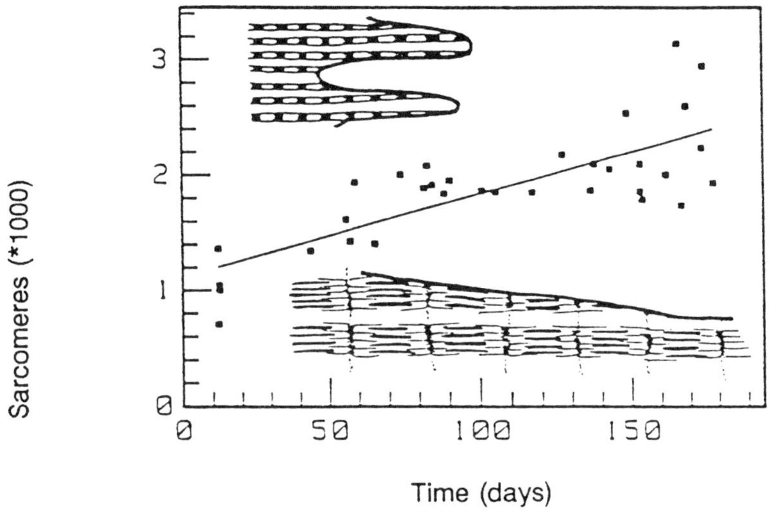

Figure 7. Increase in the number of sarcomeres along the length of fasciculi in the porcine *peroneus longus* muscle (Swatland, 1978).

New sarcomeres are added to each end of the muscle fiber at oblique sites of insertion of the myofibril into the plasma membrane. Very often the membrane at the end of the muscle fiber is folded extensively into finger-like projections to create the surface area for the oblique attachment of myofibrils. Because several sarcomeres are in various stages of formation at any one time, it is difficult to calculate exactly how long each new sarcomere takes to form, but it might be as short as about 40 min.

Longitudinal growth also contributes to muscle areas. In the rib-eye or loin-eye areas created from the *longissimus dorsi* when it is cut into chops perpendicularly to the vertebral column, the muscle fibers run through the chops at an angle. Thus, the length of fasciculi is a major determinant of the depth of the muscle (Fig. 8) and, hence, its cross-sectional area and apparent number of muscle fibers.

Although the involvement of tapered intrafascicularly terminating muscle fibers enables us to solve some otherwise impossible problems in accounting for dramatic post-natal increases and decreases in apparent numbers of muscle fibers, tapered fibers also create some new problems of their own. At worst, they may become a source of error in studies of muscle growth.

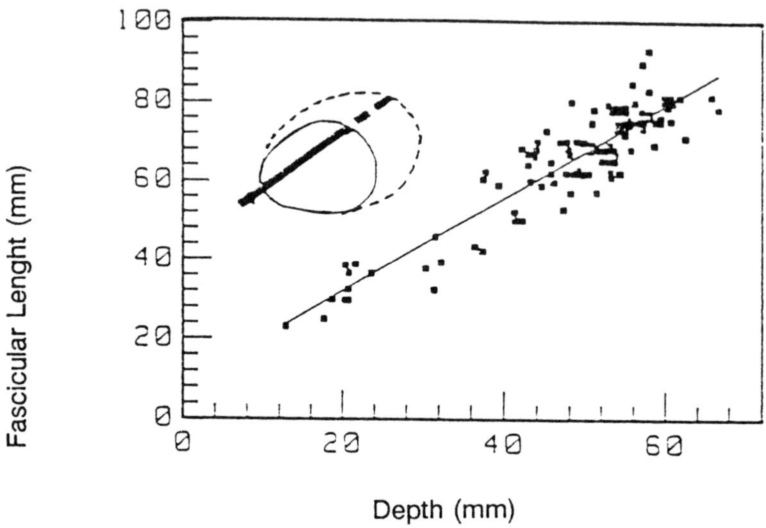

Figure 8. Relationship between the depth of the *longissimus dorsi* muscle seen in loin-eye areas and the fascicular length in that region of the muscle (Swatland, 1976c).

Histochemical fiber types

Before considering the problems created by tapered fibers, we need a brief look at histochemical fiber types themselves. Many researchers agree that porcine skeletal muscles fibers tend to be specialized for either fast or slow contraction (both of which are faster than the extremely slow tonic contractions found in the slow fibers of lower vertebrates). Speed of contraction may be given histochemically by a strong alkaline or formaldehyde-resistant ATPase reaction (slow fibers have ATPase activity that is demonstrated by other reactions that inhibit the reactions of fast fibers). Almost all the slow fibers ($ATPase^-$) have strong SDH activity (SDH^+), whereas fast fibers ($ATPase^+$) may show a range from strong to weak SDH activity (SDH^+ to SDH^-). Beyond this point there is little agreement with regard to either nomenclature or statistical validity.

The older terminology of red, intermediate and white (R, I and W) is used here for the sake of convenience, but microspectrophotometry shows that R, I and W clusters are seldom clearly separate (Fig. 9).

In biochemical spectrophotometry, the optical absorbance of a dilute solution in a cuvette with a fixed light path is generally proportional to the number of absorbing molecules or chromophores, but in microscopy this is seldom the case.

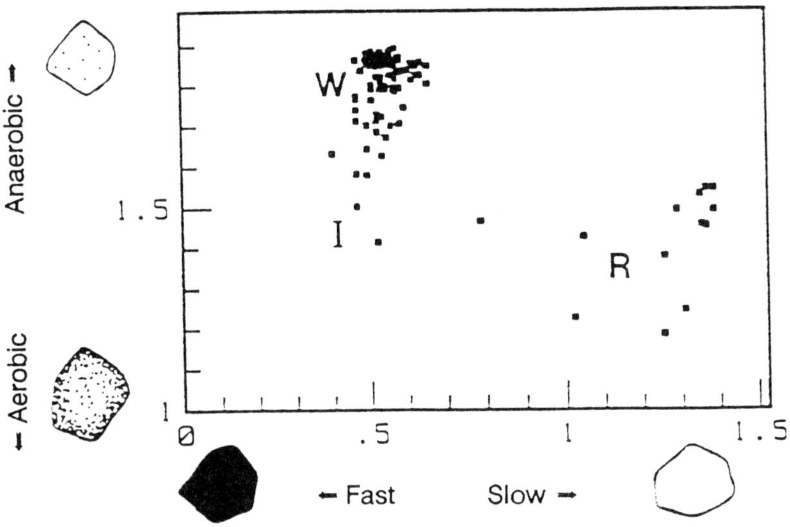

Figure 9. Red (R), intermediate (I), and white (W) muscle fibers defined by their aerobic activity (absorbance of diformazan deposited by the SDH reaction) and contraction speed (absorbance of cobalt sulfide deposited by the alkali-stable, formaldehyde-resistant ATPase reaction). Data came from the *longissimus dorsi* of a pig at 50kg live weight (Swatland, 1977b).

The optical path length through sectioned fibers is extremely difficult to standardize, because of cryostat errors and subsequent shrinkage of the section, and chromophores are often concentrated to the point of crystallization or precipitation while being distributed irregularly in the field. By analogy with biochemical spectroscopy, it is as if the chromophores have clumped to a single point, thus allowing the uninterrupted transmittance of white light - something that would produce a marked change in absorbance. Corrections for this problem (distributional error) are possible, but the cluster boundaries may change. Thus, there are those who believe firmly in the existence of fiber types as separate clusters (Spurway, 1981) while there are others who believe in a continuum of fiber types (Hintz *et al.*, 1980; Billeter *et al.*, 1980; Staron *et al.*, 1983; Dahl and Roald, 1991).

Changes in fiber-type ratios and allometric growth

Muscles change during development as they adapt to the increasing demands of gravity. The grouping of secondary fetal fibers around primary myotubes starts as the basis for the grouping of white fibers around red and intermediate fibers, so that myoglobin is often concentrated in the axis of pork muscle fasciculi. However, in muscles that must adapt to the demands of gravity immediately after birth, particularly muscles that adduct the hind limbs to prevent the splay-leg condition, there is a marked change in the fast/slow ratio of fiber types based on the ATPase reaction (Fig. 10).

In the *longissimus dorsi*, the post-natal changes in fiber types are quite different and are dominated by a decrease in the proportion of fibers with strong aerobic activity and an increase in anaerobic fibers (Fig. 11). In the biceps femoris, adaptation to changes in muscle physiology is associated with the allometric radial growth of muscle fibers, such that fast ATPase$^+$ fibers are positively allometric with respect to slow ATPase$^-$ fibers (Fig. 12). It is quite likely that each muscle of the pig's body responds to growth by developing its own pattern of responses involving changes in fiber type ratios and allometric growth.

How does the neuromuscular junction adapt to growth of the muscle fibers? Synaptic area is difficult to measure (because it has a complex shape and is highly folded) while fiber volume also is difficult to measure (because we need to know fiber length). Hence, the data shown in Figure 13 are far from satisfactory. They suggest that fibers might outgrow their neuromuscular junctions, but further studies are needed to support such a possibility.

Post-natal changes in apparent fiber numbers are probably caused by the differential longitudinal growth of intrafascicularly terminating fibers, but what guarantee do we have that the various fiber types grow in length at the same rate? Obviously, this is something that we must prove and cannot take for granted. If a particular fiber type can grow in length faster than another fiber type, we would expect the former type to increase when counted in transverse sections. Depending on the state of growth, the mean cross-sectional area of this fiber type might either decrease (as tapered endings grow into the plane of sectioning) or increase (as tapered endings are replaced by thicker parts of the same fiber). This would be difficult to distinguish from genuine allometric growth in fiber diameters.

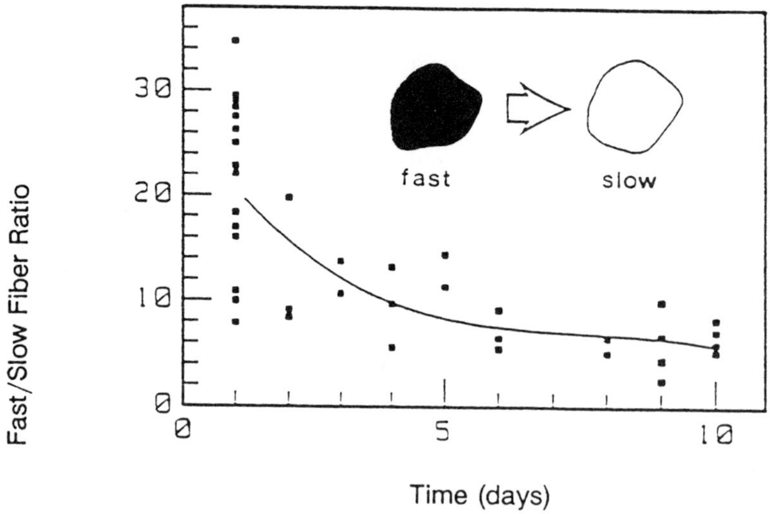

Figure 10. Change in the ratio of fast/slow fiber types based on the ATPase reaction of sartorius muscles in neonatal pigs (Swatland, 1975a).

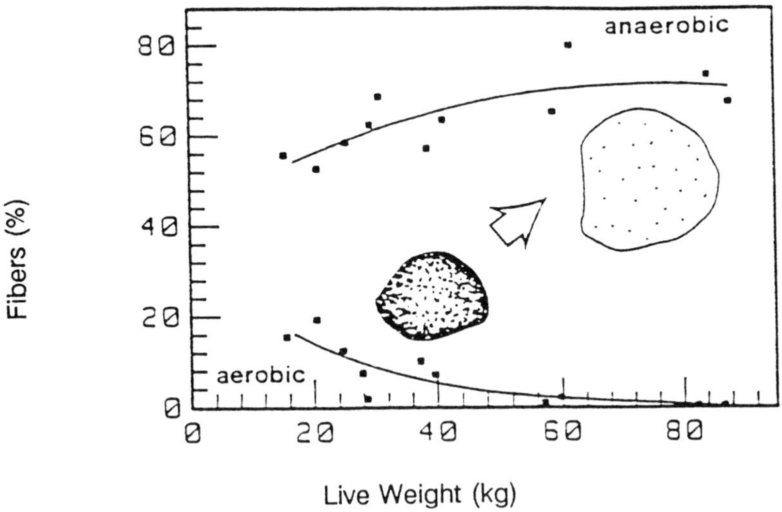

Figure 11. A shift from aerobic to anaerobic activity in fast fibers ($ATPase^{+}$) of the porcine *longissimus dorsi* muscle (Swatland, 1977c).

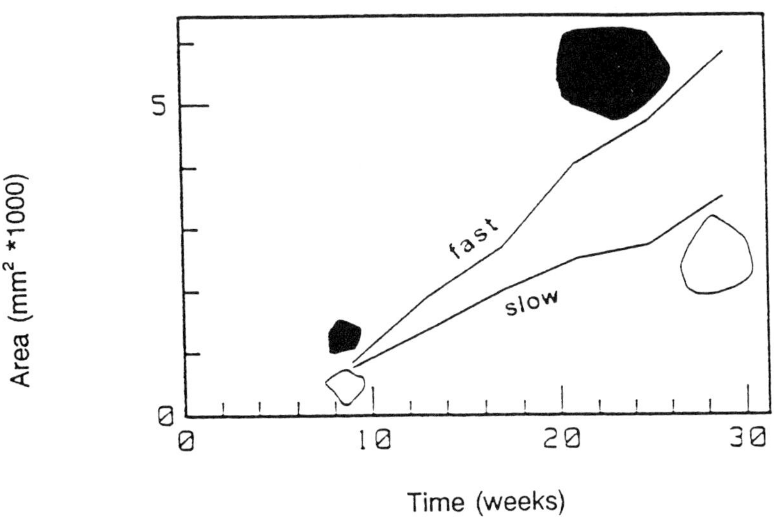

Figure 12. Allometric growth in cross-sectional areas of fast muscle fibers ($ATPase^{+}$) relative to slow fibers ($ATPase^{-}$), from Swatland (1983).

If we can achieve something useful in agriculture by counting and measuring fiber types, we are clearly justified in doing so, but we should remember that muscle growth may be more complicated than it first appears. Who knows, perhaps the differential longitudinal growth of muscle fibers is a fundamental aspect of muscle development, basic to both the training of Olympic athletes and the production of high-quality pork?

Mitochondrial distribution within muscle fibers

Oxygen arrives at the muscle fiber in capillaries that wind over its surface, and many mitochondria (in fibers that have a high content of mitochondria) are concentrated in the periphery of the fiber, under the plasma membrane. The oxygen distribution pattern may be seen in sections reacted for SDH, and measured quantitatively by mapping the distribution of the diformazan reaction product of the SDH reaction using a scanning microscope photometer (Fig. 14). Radial gradients may be averaged through 360° by taking concentric zones as the fiber matrix is eroded computationally. There is a divergence in the growth pattern of radial gradients between white fibers and red plus intermediate fibers (Fig. 15).

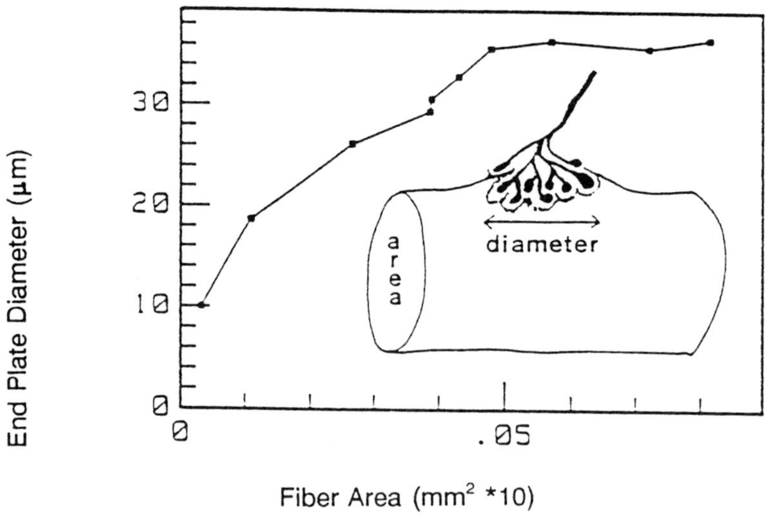

Figure 13. Growth of neuromuscular junctions in porcine *peroneus longus* muscle (Swatland and Cassens, 1972).

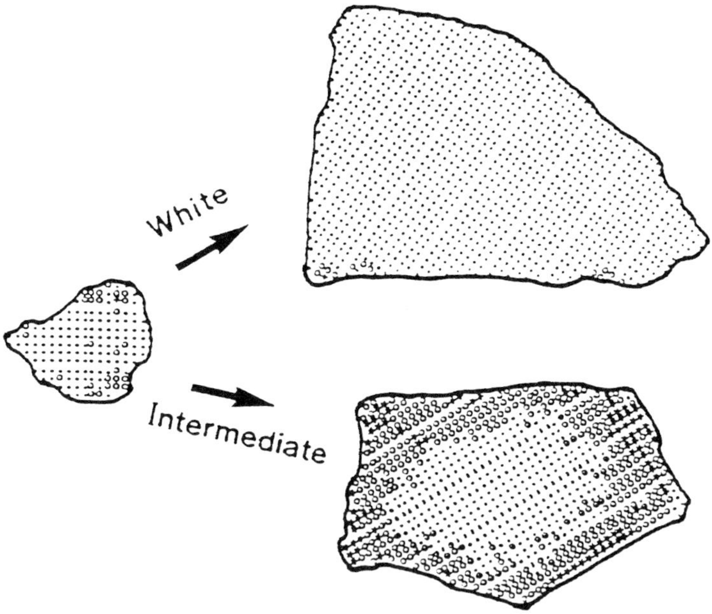

Figure 14. Transverse sections of porcine *longissimus dorsi* reacted for SDH and mapped by scanning photometry, showing different patterns of development in their radial gradient of SDH activity (Swatland, 1984b).

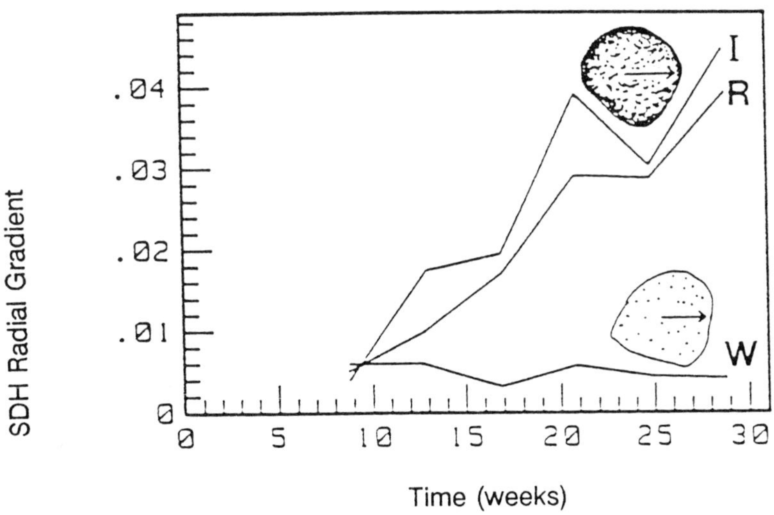

Figure 15. Development of radial gradients of SDH activity in intermediate (I), red (R), and white (W) fibers of porcine *longissimus dorsi* muscle (Swatland, 1984b).

Until recently, the dominant theoretical model of oxygen diffusion into muscle fibers was based on centrifugal diffusion from capillaries between the fibers. In this case, SDH should be concentrated into segments of the fiber defined by the nearest capillary. There is little evidence to support this, and now it appears that the constant contraction of muscle fibers through the meshwork of their capillary bed smooths out the distribution of oxygen so that a more appropriate model is one of centripetal diffusion based on individual fibers. The agricultural significance of this may be that the central cores of large muscle fibers, which may develop massive cores of glycogen in pigs (Fig. 16), may do so because they are anaerobic. As pigs grow heavier, glycogen cores become larger and more numerous. One would expect this to provide a ready substrate for post-mortem glycogenolysis, since the cores are loaded with phosphorylase, but this is only an assumption and has not yet been proven.

Slaughter

The pig has little direct cortical representation in its cerebrum, but responses to touching of the eye, ear and foot are detectable by electrocorticography (ECoG) while a pig is in the trap awaiting stunning (Fig. 17). However, it is difficult to find the source of the reflex muscle activity that occurs as pigs are exsanguinated. The ECoG activity shown in Figure 18 was concurrent with rhythmic muscle activity and this brain activity might have caused the muscle reflexes or, alternatively, it might have been a sensory response to muscle activity initiated in the brain stem or spinal cord.

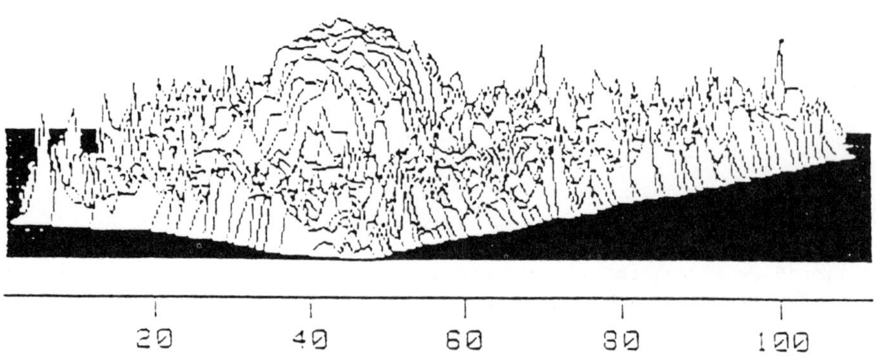

Figure 16. Scanning of a glycogen core in a PAS-stained fiber from a porcine *longissimus dorsi* muscle (Swatland, 1975b,1990).

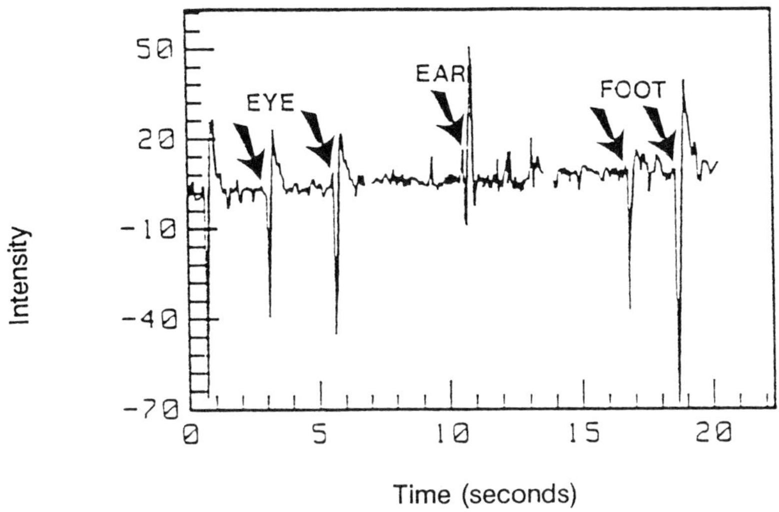

Figure 17. ECoG responses to touching the eye, ear, and hind foot of a pig waiting to be stunned (Swatland *et al.*, 1984).

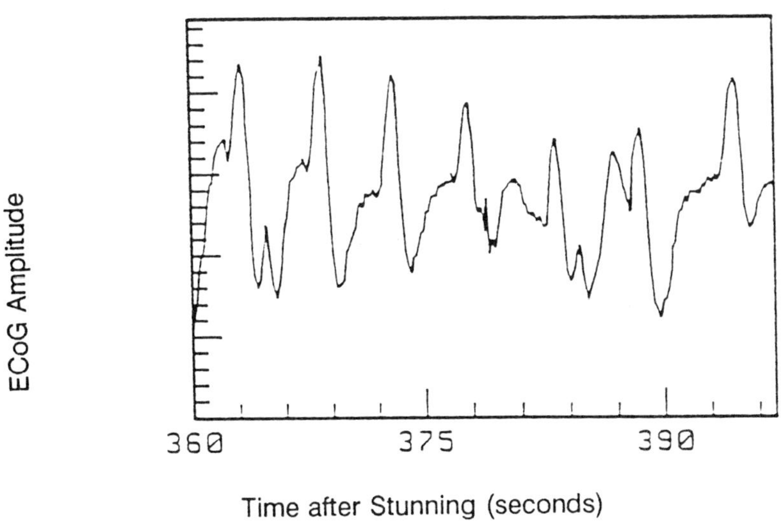

Figure 18. ECoG activity concurrent with rhythmic muscle activity in a stunned pig (Swatland, 1976d).

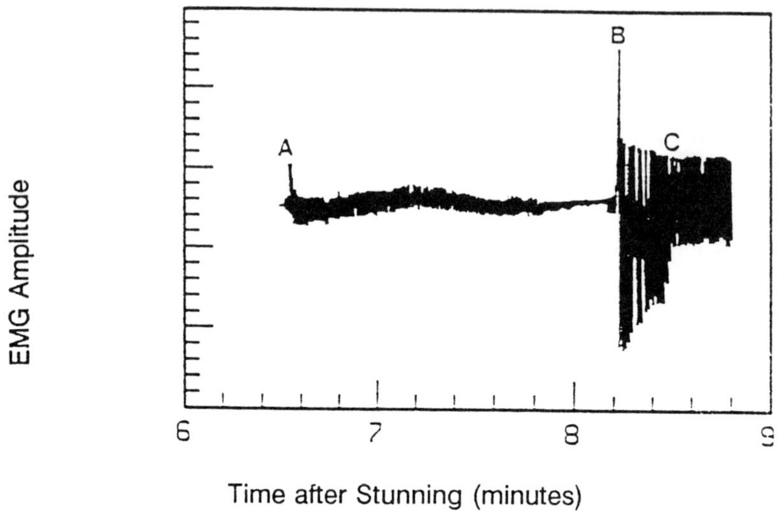

Figure 19. An experiment to show the survival time of a neural pathway from the spinal cord to the ham of in a slaughtered pig (Swatland, 1976e). Activation of a spinal electrode, A; setting maximum gain on the amplifier, B; and reaching a constant end point where the stimulus artifact is stronger than the EMG signal, C.

The lower motor pathway of the pig is viable for at least several minutes after exsanguination, during the time that reflex activity may stimulate glycolysis. In the experiment shown in Figure 19, spontaneous muscle reflex activity had finished after stunning and exsanguination, and a spinal electrode was activated at point A, while electromyographic (EMG) activity was recorded from the ham. The muscle response steadily diminished, then the gain of the amplifier was set to maximum at point B. The muscle response continued to diminish until, at point C, it was reduced to the constant level of a stimulus artifact (the EMG amplifier picking up direct signals from the stimulatory electrode in the spinal cord). Thus, for up to 8.5 min after the start of stunning, this pig had a viable pathway from the spinal cord, through the neuromuscular junction, to the muscle. The agricultural significance of this is that neural activation of muscle contraction may accelerate post-mortem glycolysis. Paradoxically, PSS pigs may have less post-mortem neural activity than normal, so that this mechanism may be a major factor in explaining variations in the rate of pH decline among normal pigs.

Post-mortem glycolysis and development of rigor-mortis

Different histochemical fiber types have a heterogeneous pattern of post-mortem metabolism if they have different levels of glycogen, enzyme activation, and neural thresholds. With nuclear magnetic resonance (NMR) spectroscopy (Fig. 20), two peaks for inorganic phosphate (P_i) may appear in pork post-mortem.

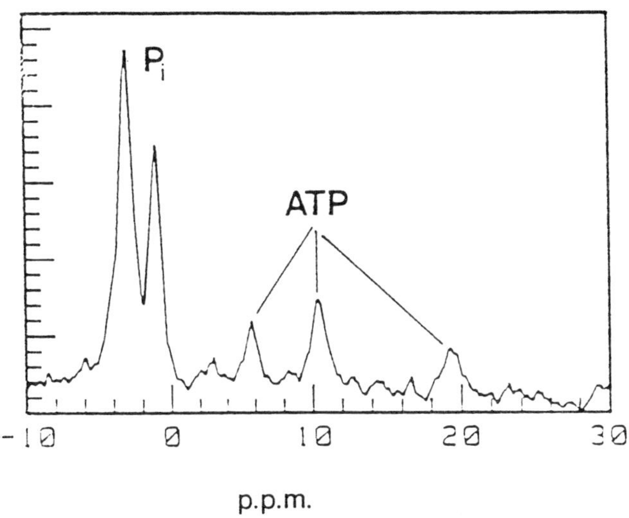

Figure 20. Post-mortem NMR spectroscopy of porcine *longissimus dorsi* showing two peaks for inorganic phosphate (P_i).

Although this is often thought to be a consequence of intra- and extra-cellular compartments differing in pH, it might equally well be caused by two populations of fiber types differing in rate of post-mortem glycolysis and intracellular pH. Stronger evidence of cellular heterogeneity in post-mortem glycolysis comes from quantitative histochemistry of the PAS reaction, where glycogenolysis and lactate production may be followed in individual fibers by a loss of PAS staining. Anaerobic fibers (mostly $ATPase^+$, SDH^-) make the major contribution to the lactate that accumulates post-mortem in pork muscles. When a lot of glycogenolysis can be detected, most of it comes from these fibers (Figure 21).

PAS microphotometry has some underlying problems. Glycogen forms granules whose fine structure is still being discovered (Manners, 1991) and glycogenolysis may involve structural rearrangements that change the accessibility of outer amylose chains to periodate and Schiff's reagent (Ovadia and Stoward, 1971; Puchtler *et al.*, 1974). Thus, when the absorbance of PAS-stained glycogen increases transiently post-mortem (Fig. 22) it might not indicate that glycogen has increased, only that it has altered its structure. Another problem is that the disappearance of glycogen cannot automatically be linked with glycolysis and lactate production, because of a possibility of hydrolysis through other pathways (Calder and Geddes, 1990). Individual muscles of the pork carcass develop rigor mortis sooner or later after slaughter (Fig. 23), and much may be learned at this time about the ultimate properties of the meat.

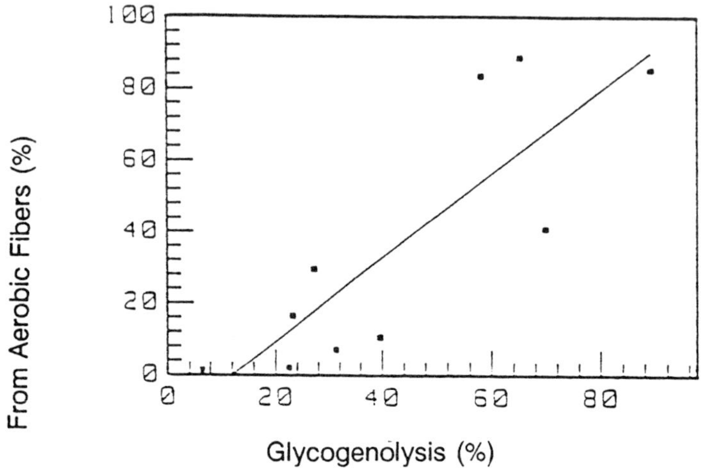

Figure 21. Contribution of anaerobic fibers to total glycogenolysis (Swatland, 1976f).

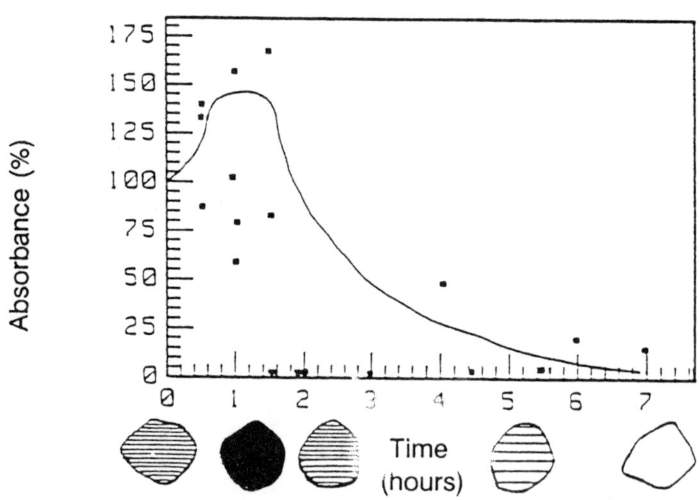

Figure 22. A transient increase post-mortem in the absorbance of PAS-stained glycogen in anaerobic fibers from the porcine *longissimus dorsi* muscle, where pairs of samples were adjusted so that the glycogen absorbance at the start of slaughter was 100% (Swatland, 1977c).

Figure 24 shows how rigor development may also be measured by the decline in the bulk electrical capacitance of the meat, probably caused by ATP-driven pumps in the cell membrane becoming freely permeable to the test current as ATP is depleted. Thus, ^{31}P-NMR, rigorometry and capacitance measurements all tend to give convergent results - with the latter being measured most easily.

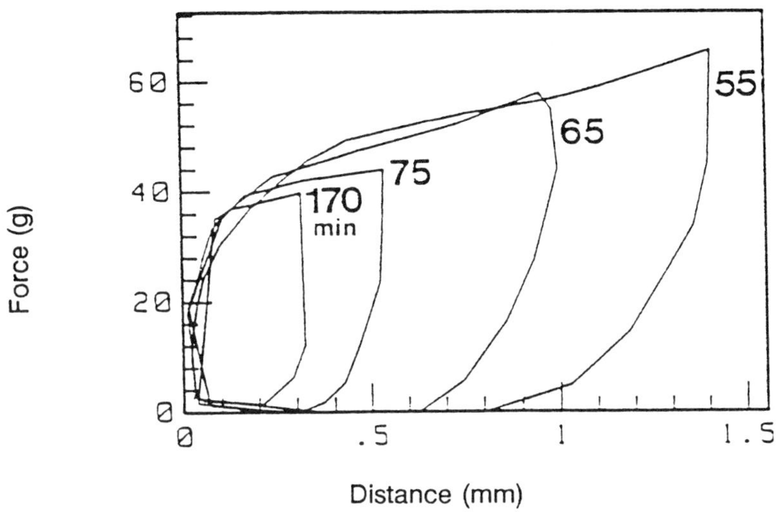

Figure 23. Gradual development of rigor mortis between 55 and 170 min post-mortem in the adductor muscle of a pork carcass hanging in a meat cooler (Swatland, 1985).

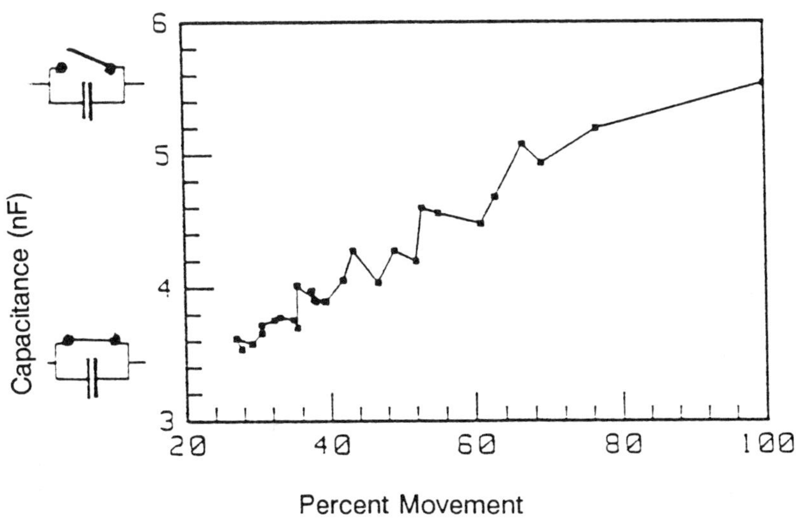

Figure 24. Relationship between membrane capacitance and rigor development in the adductor muscle of a pork carcass hanging in a meat cooler (Swatland, 1985).

Conclusion

We have followed some aspects of muscle growth and meat quality in pigs from the embryo to the abattoir, touching on a few points and ignoring many of the detailed problems. There has not been time to examine how connective tissue, myoglobin and intramuscular fat are involved in the determination of meat quality, and many other topics also have been passed over. However, this integrated approach, encompassing both cell biology and abattoir technology, offers a secure scientific foundation to understand how muscle growth and meat quality are related. The subject is both agriculturally important and scientifically sound, and work in this area is interesting. How may we increase the efficiency of lean meat production without ruining the quality of the meat?

References

Bendall, J.R. 1973. Post-mortem changes in muscle. In: G.H. Bourne (ed.) The Structure and Function of Muscle, Vol. 2, Part 2, 2nd edn., Academic Press, New York. pp. 244-309.

Bendall, J.R. and Swatland, H.J. 1988. A review of the relationships of pH with physical aspects of pork quality. Meat Sci. 24, 85-126.

Billeter, R., Weber, H., Lutz, H., Howald, H., Eppenberger, H.M. and Jenny, E. 1980. Myosin types in human skeletal muscle fibers. Histochemistry 65, 249-59.

Calder, P.C. and Geddes, R. 1990. Post mortem glycogenolysis is a combination of phosphorolysis and hydrolysis. Int. J. Biochem. 22, 847-56.

Dahl, H.A. and Roald, L. 1991. How unequivocal is the muscle fibre type concept? Anat. Embryol. 184, 269-73.

Eikelenboom, G. and Nanni Costa, L. 1988. Fibre optic probe measurements in Landrace pigs of different halothane phenotypes. Meat Sci. 23, 9-19.

Ervasti, J.M., Strand, M.A., Hanson, T.P., Mickelson, J.R. and Louis, C.F. 1991. Ryanodine receptor in different malignant hyperthermia-susceptible porcine muscles. Am. J. Physiol. 260, C58-66.

Fujii, M., Otsu, K., Zorzato, F., De Leon, S., Khanna, V.K., Weiler, J., O'Brien, P.J. and MacLennan, D.H. 1991. Identification of a mutation in the porcine ryanodine receptor that is associated with malignant hyperthermia. Science 253, 448-51.

Granger, B.L. and Lazarides, E. 1978. The existence of an insoluble Z disc scaffold in chicken skeletal muscle. Cell 15, 1253-68.

Hintz, C.S., Lowry, C.V., Kaiser, K.K., McKee, D. and Lowry, O.L. 1980. Enzyme levels in individual rat muscle fibers. Am. J. Physiol. 239, C58-C65.

Kentish, J.C. 1983. Direct attachment of membrane to the Z-band of rabbit skeletal myofibrils. J. Muscle Res. Cell Motil. 4, 663-69.

Manners, D.J. 1991. Recent developments in our understanding of glycogen structure. Carbohydr. Polym. 16, 37-82.

Orcutt, M.W., Dutson, T.R., Wu, F.Y. and Smith, S.B. 1986. The fine structure of the endomysium, perimysium and intermyofibrillar connections in muscle. Food Microstruct. 5, 41-51.

Ovadia, M. and Stoward, P.J. 1971. Some quantitative histochemical studies of the periodic acid oxidation of glycogen *in situ*. Histochem. J. 3, 233-40.

Pardo, J.V., D'Angelo Siliciano, J. and Craig, S.W. 1983. A vinculin-containing cortical lattice in skeletal muscle: transverse lattice elements ("costameres") mark sites of attachment between myofibrils and sarcolemma. Proc. Natl. Acad. Sci. USA 80, 1008-12.

Pierobon-Bormioli, S. 1981. Transverse sarcomere filamentous systems: "Z- and M cables". J. Muscle Res. Cell Motil. 2, 401-13.

Puchtler, H., Meloan, S.N. and Brewton, B.R. 1974. On the bonding of Schiff's reagent in the PAS reaction: effect of molecular configurations in models. Histochemistry 40, 291-9.

Shear, C.R. and Bloch, R.J. 1985. Vinculin in subsarcolemmal densities in chicken skeletal muscle: localization and relationship to intracellular and extracellular structures. J. Cell Biol. 101, 240-56.

Spurway, N.C. 1981. Objective characterization of cells in terms of microscopical parameters: an example from muscle histochemistry. Histochemical. J. 13, 269-317.

Staron, R.S., Hikida, R.S. and Hagerman, F.C. 1983. Re-evaluation of human muscle fast-twitch subtypes: evidence for a continuum. Histochemistry 78, 33-9.

Swatland, H.J. 1971. Development and Innervation of Muscle Subject to Selective Breeding. Ph.D. Thesis, University of Wisconsin, Madison.

Swatland, H.J. 1973. Muscle growth in the fetal and neonatal pig. J. Anim. Sci. 37, 536-45.

Swatland, H.J. 1975a. Histochemical development of myofibers in neonatal piglets. Res. Vet. Sci. 18, 253-7.

Swatland, H.J. 1975b. Relationships between mitochondrial content and glycogen distribution in porcine muscle fibers. Histochem. J. 7, 459-69.

Swatland, H.J. 1976a. Proliferation of myofibrils in the porcine longissimus muscle. J. Anim. Sci. 42, 1434-9.

Swatland, H.J. 1976b. Effect of growth and plane of nutrition on apparent muscle fiber numbers in the pig. Growth 40, 285-92.

Swatland, H.J. 1976c. Longitudinal fascicular growth in the porcine longissimus muscle. J. Anim. Sci. 42, 63-66.

Swatland, H.J. 1976d. An electrocorticographic study of necrobiosis in the brains of electrically stunned and exsanguinated pigs. J. Anim. Sci. 43, 577-82.

Swatland, H.J. 1976e. Electromyography of electrically stunned pigs and excised bovine muscle. J. Anim. Sci. 42, 838-44.

Swatland, H.J. 1976f. Cellular patterns of post-mortem glycogenolysis in *longissimus dorsi* muscles of pigs at different live weights. Histochem. J. 8, 455-61.

Swatland, H.J. 1977a. Accumulation of myofiber nuclei in pigs with normal and arrested development. J. Anim. Sci. 44, 759-64.

Swatland, H.J. 1977b. Transitional stages in the histochemical development of muscle fibers during post-natal growth. Histochem. J. 9, 751-57.

Swatland, H.J. 1977c. Cytophotometry of post mortem glycogenolysis in different histochemical types of muscle fibres of the pig Histochem. J. 9, 163-70.

Swatland, H.J. 1978. Longitudinal growth and rate of new sarcomere formation in porcine muscle. Anim. Sci. 46, 118-24.

Swatland, H.J. 1982. Histology of allometric growth in hindlimb muscles of pigs. J. Agric. Sci. Camb. 98, 629-32.

Swatland, H.J. 1983. Aerobic activity in the axis of growing myofibers in the porcine biceps femoris. J. Anim. Sci. 56, 1324-28.

Swatland, H.J. 1984a. Structure and Development of Meat Animals. Prentice-Hall, New Jersey.

Swatland, H.J. 1984b. The radial distribution of succinate dehydrogenase activity in porcine muscle fibres. Histochem. J. 16, 321-29.

Swatland, H.J. 1985. Development of rigor mortis in intact sides of pork measured with a portable rigorometer. J. Anim. Sci. 61, 882-86.

Swatland, H.J. 1990. Intracellular glycogen distribution examined interactively with a light microscope scanning stage. J. Comput. Assist. Microsc. 2, 233-7.

Swatland, H.J and Cassens, R.G. 1972. Peripheral innervation of muscle from stress susceptible pigs. J. Comp. Pathol. 82, 229-36.

Swatland, H.J., Brogna, R.J. and Lutte, G.H. 1984. Electrical activity in the cerebral hemispheres of electrically stunned pigs. J Anim. Sci. 58, 68-74.

Trotter, J.A. 1990. Interfiber tension transmission in series-fibered muscles of the cat hindlimb. J. Morphol. 206, 351-61.

Trotter, J.A. 1991. Dynamic shape of tapered skeletal muscle fibers. J. Morphol. 207, 21-23 .

Wakayama, Y. and Shibuya, S. 1991. Antibody-decorated dystrophin molecule of murine skeletal myofiber as seen by freeze-etching electron microscopy. J. Electron. Microsc. 40, 143-45.

Wang, K. and Ramirez-Mitchell, R. 1983. A network of transverse and longitudinal intermediate filaments is associated with sarcomeres of adult vertebrate skeletal muscle. J Cell Biol. 96, 562-70.

Discussion

*Kallweit - You were talking about satellite cells which form the new nuclei. Afterwards, these satellite cells will disappear again, as far as I know. How long will this process, post-partum, go on to make satellite cells, and to form new nuclei? How long will this last?

*Swatland - As far as I know, from research on experimental animals, some satellite cells will survive to provide a source of muscle growth for new nuclei in young animals, and for regeneration in old animals. So, the number gets less and less when you measure them, but as far as I know, there are always satellite cells. We, in agriculture, are only dealing with very young animals, up to the point of sexual maturity. So, there is no shortage of satellite cells for growth. But, as I said in my presentation, we don't know whether the muscle fiber asks for the new satellite cells, or the appearance of the new satellite cells enables the muscle fiber to grow. If you starve an animal to the point of stopped growth, then the number of new nuclei stops dead! And then, when you refeed the animal to get growth again, new nuclei appear. So, you can't see which is causing which. It is a very, very profound question.

*Lister - Would you tell us a little bit more about how you characterized the muscle fiber type in the newborn pig, and the change from white to red, or fast to slow? I ask this because the newborn pig is born without any fat, and so it has no energy reserves, other than carbohydrates. I just wondered whether, if your measuring the fast to slow change, you're simply measuring the disappearance of glycogen, on which the pig has to survive for the first several days?

*Swatland - In the slide I showed, the fibers were typed using the ATPase, so we are looking at the genetically programmed proteins, making no inference about the source of the muscle energy, although I agree with you that this is somewhat of a puzzle in the pig. But even the ATPase is suspicious, because nobody looks at the degree of sarcomere length, and one would expect, from the biochemistry, that if there are short sarcomeres, with a big overlap between actin and myosin, then there will be a strong myosin ATPase activity. And in fact, from some of my research, I could show that I must control the sarcomere length to control the strength of the ATPase reaction. There's also another big worry, because there's a fetal form of actomyosin which has not been properly characterized, and how this would appear, using the adult histochemical reaction. So, I'm not totally certain about that, there could be an artifact in there. But, it does make sense that the pig is getting more slow fibers in an abductor muscle to prevent a spraddle legged condition.

*Lundström - You told us that you had an increase in the PAS-staining, after slaughter. That is quite interesting, for we have found a pH increase when we followed the muscle very carefully after slaughter, so, something has happened, and it could be glycogen. Whether glycogen will increase or not, I don't know, but it's very interesting.

*Swatland - It's always great to hear one more piece of evidence. Yes, it's fascinating, thanks.

*Lundström - Yes, and there is some data too.

*Henckel - We also measured the glycogen content after slaughter, and found a small increase that varies widely from animal to animal. So, they do produce glycogen, or build-up glycogen after slaughter, but only during the first hour.

*Swatland - Yes, right. Also, at that time, the muscle fiber takes up fluid by osmosis, so this is what makes all the early detection of PSE difficult. Because the optical, and the electrical characteristics of the muscle reflect the disappearance of intracellular fluid because of osmosis, which is linked to glycogen synthesis, or glycogenolysis.

*Mickelson - I wonder if you could expand just a little bit on how Dr. Bendall showed lactate could be converted into glycogen.

*Swatland - This is not very reliable, I'll have to go home and find the papers, but, I think his research might have been on amphibian muscle or rabbit muscle. He was always annoyed that it wasn't on pork muscle. I think that he identified the pathways, although he was unable to isolate the enzymes. Somebody else, has identified the enzymes responsible for this. It's a big worry, because there could be lysosomal activity too, taking up glycogen, so there could be an alternative pathway, whereby glycogen could go through lysosomes, and not cause a decrease in pH, which would totally mess up all the data where your lactate was used to make inferences. (Later information: The references are Bendall and Taylor, 1970, Biochem. J., 118, 887, and Connett, 1979, Amer. J. Physiol., 237, C231.)

*Essén-Gustavsson - Just from the human work, it's clear that you can make glycogen out of lactate in the muscle. That is absolutely clear today, because with intermittent exercise, they are down to very short periods, they find high levels of glycogen after a few minutes. And, it doesn't go via the liver, it's local. It's in the type IIb fibers. So, you can make glycogen out of lactate, that's clear.

*Mickelson - Has it been shown by radioactive carbon tracing?

*Essén-Gustavsson - No, this is just quantitative biochemical work. But, there has been a debate about that in the human side also. There is convincing evidence today that you can make glycogen out of lactate.

*Puolanne - You said that you are not sure that there are different fiber types. But, the light chains of myosin are different in fast and slow muscles. Could you comment on that?

*Swatland - Yes, you are right. But, the fiber type concept is something that we get from histochemistry. We do ATPase reactions, SDH , and so on. And, when this is done, there isn't a perfect segregation, the clusters are smeared, and joined together. And, even if there is basic biochemical difference, what happens to a fiber as it changes from one type to another? Because of neurotropic factor, we would expect, at some intermediate point, to have both types there. So, if you wish to claim that the clusters are totally separate, then one fiber type cannot change into another. You are forced into thinking that changes in ratio are due to longitudinal growth, which could be possible. The worrying thing is that we are standing here speculating, when after 30 - 40 years of research, I should be able to tell you, but I don't know.

*Henckel - If we do electrophoretic analysis, on myosins, we can get three distinct bands of myosins in adult animals. And, there are also reports that co-expression of myosin isoforms can exist within a fiber. So, you have both slow myosin and fast myosins within the same fiber. And, myosins from fast twitch oxidative, and fast twitch glycolytic fibers within the same fiber. The possibility of changing between these types seems to appear at the expression level.

*Swatland - So, to summarize your position, you believe in a continuum of fiber types, rather than distinct fiber types.

*Henckel - I think that there is a continuum also concerning the myosin types.

*Swatland - Yes, I agree.

Muscle-fiber Characteristics in Pigs and Relationships to Meat-quality Parameters - Review

B. Essén-Gustavsson

Department of Medicine and Surgery, Faculty of Veterinary Medicine, Swedish University of Agricultural Sciences, S-750 07 Uppsala, Sweden

Introduction

During recent decades, pigs have been selectively bred for the rapid production of lean meat. Meat comes from skeletal muscle, which is a unique tissue composed of fibers with various contractile and metabolic characteristics. The muscle fibers are innervated, and each fiber belongs to a specific motor unit. A specialized function of the fiber is to convert chemical energy into mechanical energy. The protein systems involved in contraction are activated when calcium is released from the sarcoplasmic reticulum and binds to troponin c. When a muscle fiber contracts, it needs energy in the form of ATP, which can be regenerated via different metabolic pathways. Important substrate sources for ATP regeneration by oxidative processes are blood-borne substrates such as glucose and free fatty acids, and intramuscular substrates such as glycogen and triglycerides. Glycogen is the most important substrate for ATP regeneration by anaerobic processes, which are often associated with high production of lactate. Capillarization in a muscle is therefore of great importance, both for delivery of oxygen and substrates to each fiber and for the elimination of end products, such as lactate.

The proportions of different fiber types can vary markedly in different species, and among muscles, depending on function (Essén-Gustavsson, 1986). The variations in fiber types among species are caused by both genetic and environmental factors. Age, nutrition, physical activity and hormones are factors that are known to have an influence on fiber-type composition and metabolic profiles. Different demands are placed on the muscle fibers and their energy turnover in connection with postural work and different types of activity. The present-day domestic pig is almost totally inactive, however, thanks to selection, it has a fantastic capacity for rapid muscle growth. An important question is therefore how the muscle fibers, and their metabolic profiles, adapt to this rapid muscle growth, and how this affects meat quality. It has been possible, using histochemical and biochemical techniques on muscle samples obtained both from the living animal and in connection with slaughter, to increase our knowledge of muscle fiber characteristics and their relation to meat quality.

Muscle-fiber characteristics

Histochemical staining techniques can be used to describe the complex characteristics of a muscle and to identify different fiber types. Staining for myosin ATPase is used to distinguish between slow-contracting type I fibers, and fast contracting type II fibers. The myosin ATPase activity is known to be related to the contractile properties of muscle fibers (Barany, 1967). Type II fibers can be further subgrouped into type IIA, IIB and IIC fibers after acid pre-incubations (Brooke and Kaiser, 1970). Metabolic properties in a fiber can be evaluated by staining for NADH-tetrazolium reductase (NADH) and succinate dehydrogenase (SDH), both of which indicate the oxidative capacity of the muscle fibers. Staining for lactate dehydrogenase (LDH) and phosphorylase has been used to estimate glycolytic potential in a fiber. Information about the lipid content within the different fibers has been obtained by staining with Oil red O, and Sudan black B. The periodic acid Schiff (PAS)-stain indicates the glycogen content of the fibers. Histochemical staining can also be done for the capillaries around the fibers and for the oxygen binding haemeprotein myoglobin, which contributes to the differences in colour between muscle fibers. Section thickness, incubation time, temperature and other factors, which are not always easy to control, may influence the staining intensity of the fibers.

Fibers in a muscle show similarities and differences in contractile and metabolic properties and they can therefore be grouped into types. However, it is important to remember that a marked variation in contractile properties, oxidative and glycolytic capacity, and glycogen and lipid content may occur between fibers, even if they demonstrate similar staining intensity for myosin-ATPase, NADH, SDH, LDH, PAS and Oil red O. The reason for this is that histochemical stains have certain limitations and are only semi-quantitative, but, they can still give valuable information about the nature of the fibers in a muscle.

Different classification systems based on either a single histochemical stain or a combination of stains are often used to identify the fibers. The ATPase stain is related to the contractile properties of the fibers and identifies type I, IIA, IIB and IIC fibers. This system has been used for pig muscle (Szentkuti *et al.*, 1981; Bader, 1982b; Essén-Gustavsson and Lindholm, 1984; Karlsson *et al.*, 1992). When fibers have been classified as red, intermediate or white, this is usually based on stainings for NADH or SDH which are related to the metabolic profile of fibers (Rahelic and Puac, 1981; Rede *et al.*, 1986).

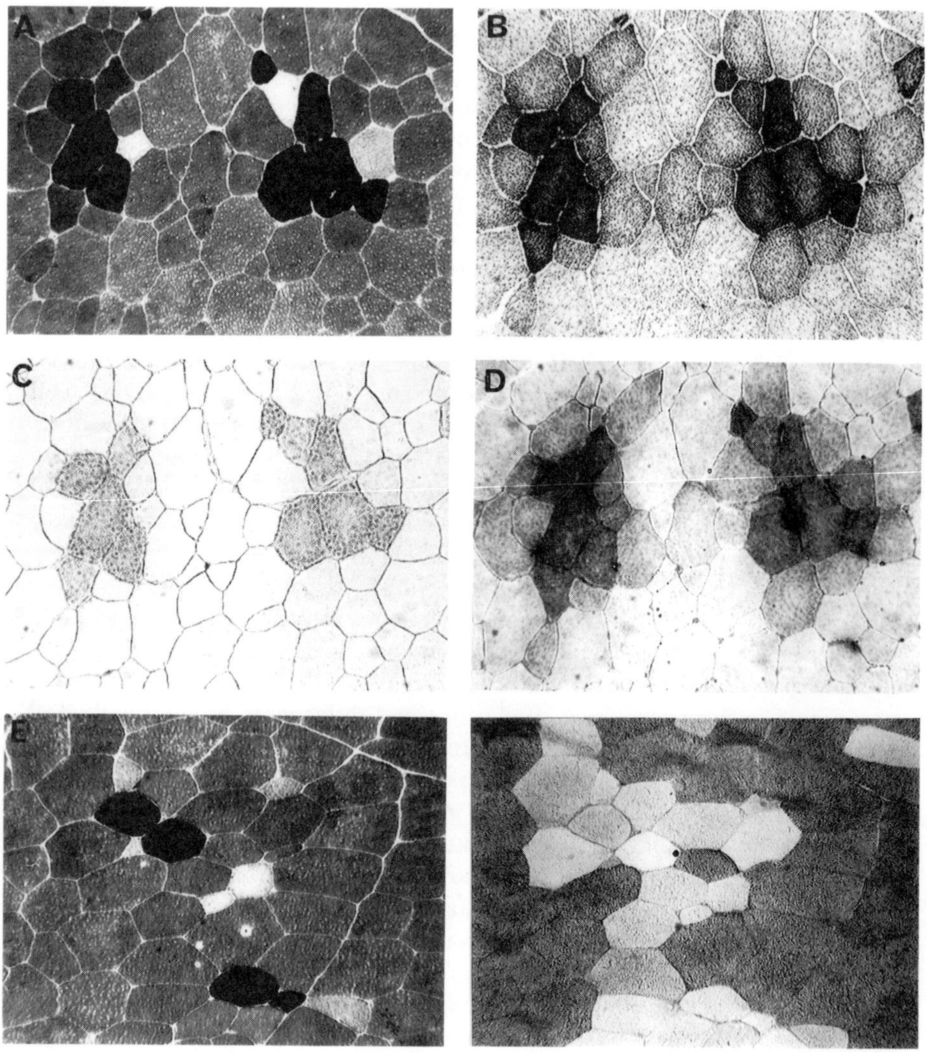

Figure 1. Cross-sections from *M. longissimus* of a domestic pig (Yorkshire) stained for myosin ATPase after acid pre-incubation (pH 4.6) = A and E, NADH-tetrazolium reductase = B, Oil red O = C, Sudan black = D and Periodic acid Shiff = F.

The ATPase stain, in combination with a stain for the oxidative and/or glycolytic capacity of the fibers, has been used to classify fibers as **β**R, **α**R or **α**W (Ashmore and Doerr, 1971; Kiessling and Hansson, 1983; Solomon and West, 1985) and as slow-twitch-oxidative (STO), fast-twitch-oxidative (FTO) or fast-twitch-glycolytic (FTG) (Salomon *et al.*, 1983). Histochemical stainings of *M. longissimus dorsi* are shown in Figure 1.

Muscle-fiber characteristics in different muscles and breeds

There are marked differences in fiber-type composition, fiber areas and metabolic profiles among muscles both within and between pigs (Rahelic and Puac, 1981; Essén *et al.*, 1980; Bader, 1982a; Kiessling and Hansson, 1983; Monin *et al.*, 1987; Jensen *et al.*, 1988). *M. longissimus dorsi* has a large proportion of type IIB fibers (80-90%), while *M. vastus intermedius* has a large proportion of type I fibers (70-80%). The deeper parts of the muscles have a greater proportion of type I fibers, and a higher oxidative capacity than the more superficial parts. Many muscles in the pig have a unique distribution of fibers with type I fibers grouped in clusters, type IIA fibers in their close vicinity, and these are surrounded by type IIB fibers. It is noteworthy that oxidative capacity can vary markedly within all fiber types in a muscle (Fig. 1). Type IIB fibers close to type I and IIA fibers usually display a higher oxidative capacity compared with type IIB fibers in the vicinity of the fascicular border. In domestic pig breeds, type IIA and IIB fiber areas are larger than type I fibers. The metabolic profile can differ in *M. gluteus* and *M. longissimus* of different breeds of pigs (Hampshire, Yorkshire, Swedish Landrace), which nevertheless have similar fiber type compositions (Essén-Gustavsson and Fjelkner-Modig, 1985).

Compared with domestic pigs, the oxidative capacity and the type IIA/IIB fiber ratio are higher in wild pigs (Szentkuti *et al.*, 1981; Bader, 1983; Essén-Gustavsson and Lindholm, 1984; Solomon and West, 1985). The oxidative capacity within the different fibers, especially type I and IIA fibers, is higher in wild pigs and in cross-bred pigs (wild pigs x Duroc x Swedish Landrace x Hampshire) compared with domestic pigs (Fig. 2). The percentage of red (oxidative) fibers decreases, and mean fiber diameter increases with the degree of selection of domestic pigs, i.e. from Manguliza and Black Slavonian to Large White and Swedish Landrace (Rahelic and Puac, 1981). Fiber type composition, and capillary-to-fiber ratio are similar in domestic stress-susceptible and stress-resistant pigs, but fiber areas were shown to be larger and capillarization (cap/mm^2) lower in the former (Cooper *et al.*, 1969; Essén-Gustavsson *et al.*, 1992).

Muscle-fiber characteristics and relation to age and growth

It is said that the number of fibers in a muscle is already determined at birth, and therefore, growth of a muscle is related to an increase in both the length and cross-sectional area of the fibers. Based on histochemical stainings for ATPase and phosphorylase, a differentiation of fibers into types is shown to occur in the muscles during the first couple of days of life, but most of the muscle fibers still have a high oxidative capacity (Cooper *et al.*, 1970; Davies, 1972; Swatland, 1975; Szentkuti and Cassens, 1978). Some of these studies have shown an increase in type I fibers, with age, up to sexual maturation and, in older pigs, a decrease in the proportion of type I fibers has been observed. A differentiation into fiber types, based on oxidative capacity, occurs in the first couple of weeks after birth (Cooper *et al.*, 1970; Van Den Hende *et al.*, 1972; Salomon *et al.*, 1983). Throughout development there is a gradual increase in the cross-sectional area composed of non-oxidative fibers and a decline in the cross-sectional area of oxidative fibers.

The fibers increase in size with advancing age in both domestic and wild pigs, but at older age there seems to be a levelling off (Bader, 1983; Salomon *et al.*, 1983; Klosowska *et al.*, 1985). The gradual increase seen in fiber area with age is more pronounced in non-oxidative fibers (Cooper *et al.*, 1970; Kiessling *et al.*, 1982; Klosowska *et al.*, 1985; Wegner and Ender, 1990). In *M. longissimus dorsi*, the proportion of type IIB fibers increases with the size of the transverse section of the muscle (Bader, 1982a). The numbers of capillaries around type I and IIA fibers are greater than those around type IIB fibers (Oksbjerg and Henckel, 1989; Essén-Gustavsson *et al.*, 1992). This indicates that capillary supply is low, and thus the diffusion distance for oxygen and blood-borne substrates will be long in type IIB fibers, in comparison with type I and IIA fibers. With age and growth, there is a small increase in the number of capillaries in contact with type I fibers (Oksbjerg and Henckel, 1989).

However, at slaughter weight there is a predominance of large, type IIB fibers in *M. longissimus dorsi* of the domestic pig and the capillary density (cap/mm^2) will therefore be very low (Essén-Gustavsson, 1986). Enzyme activities measured on whole-muscle samples from the same individual confirm an age-related decrease in oxidative capacity and an increase in glycolytic capacity of muscle (Essén-Gustavsson *et al.*, 1988).

Dietary inclusion of a beta-adrenergic agonist, salbutamol to growing pigs caused a reduction in the relative area of type IIA fibers and an increase in type IIB fibers of *M. longissimus dorsi* (Oksbjerg *et al.*, 1990). This could indicate that a conversion of type IIA to IIB fibers might have occurred.

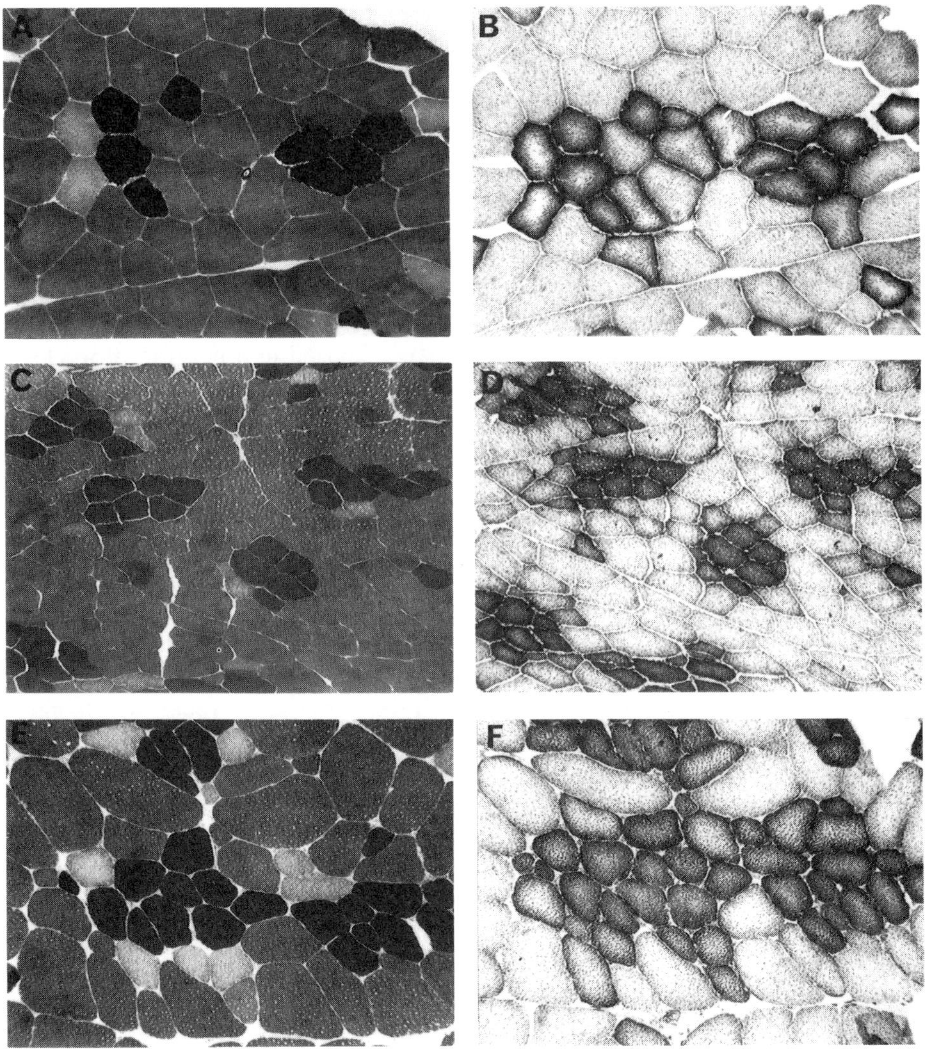

Figure 2. Cross-sections from *M. longissimus* of a wild pig (A, B) and from *M. longissimus* and *M. semitendinosus* (E, F) from a crossbred (wild pig X Duroc X Hampshire X Landrace) stained for myosin ATPase after acid pre-incubation (pH 4.6) = A, C, E and NADH-tetrazolium reductase = B, D and F.

There is an increase in fiber size of all fiber types, but no change in fiber type composition with the administration of porcine somatotropin to growing pigs (Solomon *et al.*, 1990). The difference in metabolic and contractile profiles between fibers is not so pronounced at birth as during post-natal development, which shows the importance of factors such as age and growth for adaptations in muscle characteristics.

Muscle-fiber characteristics and relation to physical activity

Muscles that are involved in maintaining posture, or are in frequent use, have a higher proportion of type I fibers than muscles that are seldom in use (Essén *et al.*, 1980; Bader, 1982b; Kiessling and Hansson, 1983). Type I fiber areas are larger than type IIA and IIB fibers in wild pigs, compared with domestic pigs at similar body weight and thus comparable load (Szentkuti *et al.*, 1981; Bader, 1983). When wild pigs were housed in a pen for 10 weeks, and when domestic pigs were exposed to daily activity for 8 weeks, the proportions of fiber types were not altered, but changes were seen in the fiber areas (Szentkuti and Schlegel, 1985). The fiber area ratio of type II/I was reduced by exercise and increased by inactivity of the pigs. In *M. semitendinosus* of the domestic pigs exposed to exercise, type I fibers were larger than type II fibers, whereas in wild pigs that had been subjected to restricted movement, type II fibers were larger than type I fibers. Cross-bred pigs (wild pigs x Duroc x Swedish Landrace x Hampshire) that live outdoors in large enclosures have a higher type IIA/IIB ratio and a higher oxidative capacity within muscle fibers, compared with domestic pigs. In pigs that are trained on a treadmill for 10-15 min each day for several weeks, no change is seen in fiber type composition, but oxidative capacity increases, and glycolytic capacity decreases in the muscles involved in exercise (Essén-Gustavsson and Lindholm, 1983; Essén-Gustavsson *et al.*, 1988) Alterations in fiber areas, and in the metabolic profiles, especially oxidative capacity, seem to be more easily induced with physical activity than changes in contractile properties. In an active muscle (*M. semitendinosus*), after 3 months of an endurance type of training, fiber sizes were similar but the fraction of type I fibers had increased somewhat (Salomon *et al.*, 1984). In contrast, after interval-training at a higher work intensity, the proportion of type IIB fibers had increased. The adaptation in muscle that takes place as a result of physical activity is only seen in those muscles that are involved during work, and in those fibers that are recruited. Type I and IIA fibers are recruited at lower work intensities, whereas type IIB fibers are recruited at high work intensities. Physical activity, and inactivity, are both important factors that influence adaptations in muscle and especially metabolic profiles and fiber areas.

Glycogen depletion in muscle fibers in connection with slaughter

Glycogen depletion patterns have been studied in *M. longissimus dorsi* and in *M. biceps femoris* of halothane sensitive pigs during exposure to halothane anaesthesia (Essén and Lindholm, 1981). A marked glycogenolysis occurred in these muscles, and to a greater extent in type IIB than in type I and IIA fibers. This is also seen in pigs with the halothane genotype nn, which have a lower glycogen content and twice as many glycogen depleted type IIA and IIB fibers in *M. longissimus dorsi* at slaughter, compared with pigs with NN-genotype (Essén-Gustavsson *et al.*, 1992). A marked glycogen depletion pattern is also observed in *M. longissimus dorsi* of halothane-gene-free pigs immediately after slaughter (Karlsson *et al.*, 1992). Low glycogen levels were seen in most type I and IIA fibers, whereas a great variation in glycogen was seen within type IIB fibers (range 0 - 60% depleted fibers). These halothane-gene-free pigs included entire males, which fought during lairage.

With exercise and stress, there is an enhanced catecholamine response that activates glycogenolysis. The rapid glycogenolysis seen in stress-susceptible pigs is said to be related to abnormalities in the calcium release channel of sarcoplasmatic reticulum (the ryanodine receptor). This defect in the ryanodine receptor is found in both slow- and fast-twitch muscle fibers (Ervasti *et al.*, 1991). Glycogen stored in muscle fibers is a major source of fuel when muscle fibers contract. The stress and movement in connection with transport and handling before slaughter may recruit type I, IIA and IIB fibers to a different extent, giving rise to different glycogen depletion patterns.

Giant fibers

Giant fibers are found predominantly in the border of a fascicle among type II fibers, and are often seen in pigs that are stress-susceptible (Cooper *et al.*, 1969; Szentkuti *et al.*, 1981; Bader, 1982a; Klosowska and Klosowski, 1985). These fibers have a circular cross-section, and staining intensities differ from other fibers (Fig. 3). An impaired energy release and a rapid glycogenolysis with lactate production due to hypoxia are said to cause lesions of membranes and swelling of both the mitochondria and sarcoplasmatic reticulum (Bader, 1982b). More glycogen-depleted fibers, and many giant fibers were found in *M. longissimus dorsi* of pigs that had been transported over a distance of 25 or 70km before slaughter, when compared with pigs that had been housed at the abattoir for 5 days (Klozowska and Klosowski, 1985). Administration of porcine somatotropin (pST) has been shown to result in an increase in fiber area for all fiber types in *M. longissimus dorsi* (Solomon *et al.*, 1990). In some of these pST-treated pigs, giant fibers were observed. Vitamin E deficiency may be associated with the occurrence of giant fibers (Jensen *et al.*, 1988). Giant fibers due to vitamin E deficiency may be caused by peroxidation and free-radical formation, giving rise to membrane damage, which may disturb

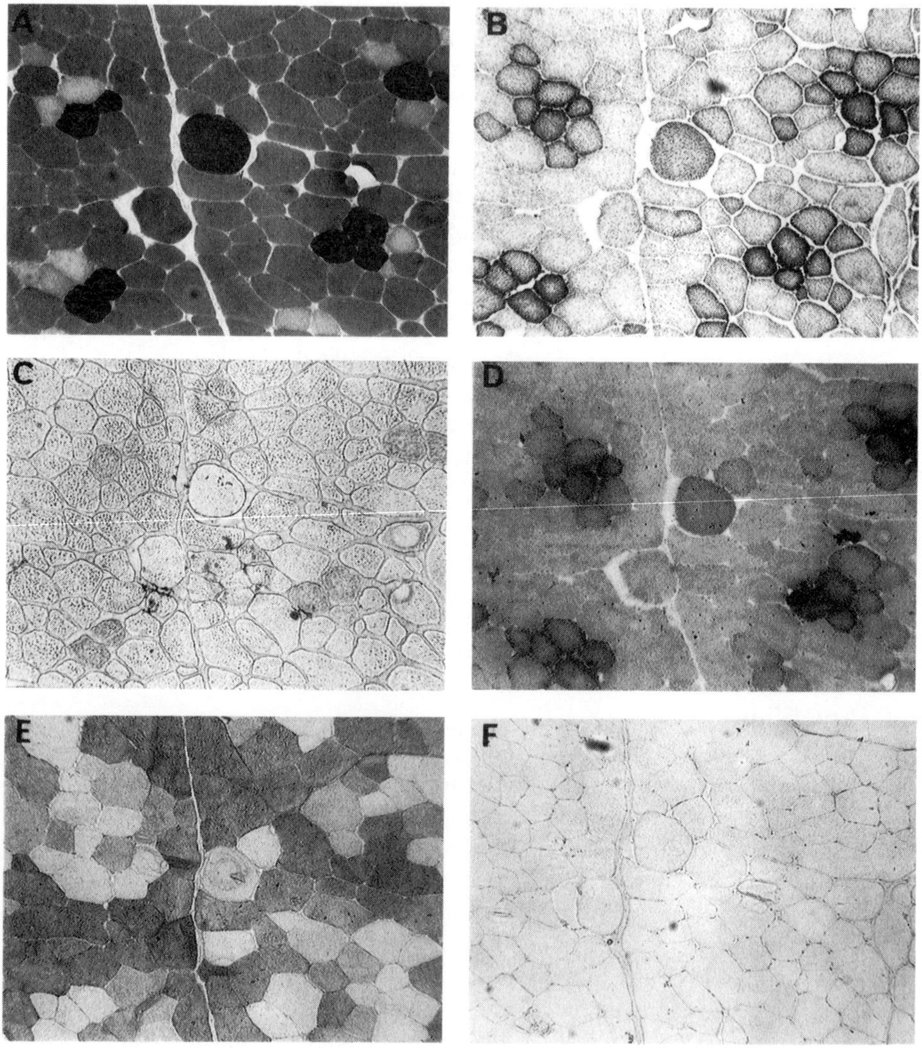

Figure 3. Cross-sections from. *M. longissimus* from a cross-bred pig (Yorkshire * wild pig) stained from myosin ATPase after acid pre-incubation (pH 4,6) = A, NADH-tetrazolium reductase = B, Oil red O = C, Sudan black = D, Periodic acid Schiff = E and capillares = F. Note the two giant fibers in the section.

calcium homeostasis in the cell, and give rise to tissue destruction. The content and storage capacity of vitamin E is related to the oxidative capacity of the muscle and the amount of lipid containing type I fibers (Jensen *et al.*, 1988).

Giant fibers have also been observed in muscles from wild pigs, but their properties were generally similar to αR, and βR fibers, and usually they lie in the middle of a fascicle (Solomon and Eastridge, 1987).

Meat-quality parameters and muscle-fiber characteristics

Aside from fiber-type composition and metabolic profiles, meat-quality traits also differ among muscles (Monin *et al.*, 1987; Karlsson *et al.*, 1992). When *M. longissimus dorsi* is compared with *M. biceps femoris*, it has a smaller proportion of type I and a larger proportion of type IIB fibers and a smaller oxidative but similar glycolytic capacity. Glycogen content is similar in the two muscles, but in *M. longissimus dorsi*, ultimate pH is lower and technological meat quality is poorer, measured as a higher reflectance value and drip loss compared with *M. biceps femoris.*

If there are many glycogen depleted fibers at slaughter, especially of type IIB, technological meat quality will be influenced. In halothane-gene-free pigs, the proportion of depleted type IIB fibers in *M. longissimus dorsi* was positively correlated to ultimate pH, and pH at exsanguination, and negatively correlated to drip loss and reflectance values (Karlsson *et al.*, 1992). When 30% or more of type IIB fibers were depleted of glycogen, the muscles showed a tendency to become DFD (dry, firm and dark), as the meat had high ultimate pH, high water-holding capacity and low reflectance value. It has been clearly documented in pigs with similar fiber type composition and enzyme activities in the muscles that those animals having the genotype nn also have poorer meat quality and a greater content of PSE-meat compared with those with genotype NN (Lundström *et al.*, 1989). The nn-pigs have a larger proportion of glycogen-depleted type IIB fibers at slaughter, even though all pigs were raised, transported and handled in a similar manner. In this case, the proportion of depleted type IIB fibers was correlated negatively with pH at exsanguination, and positively with drip loss and reflectance values. The metabolic response in the nn-genotype is related not only to stress situations prior to slaughter but also to the fact that these pigs have large fiber areas and lower capillarization of the muscle (Essén-Gustavsson *et al.*, 1992). In addition to glycogen depletion patterns, concentrations of adenine nucleotides and lactate, IMP and ammonia levels also differ at exsanguination in pigs with different halothane genotypes (Essén-Gustavsson *et al.*, 1991). These metabolic changes may induce changes of cell membranes that are said to be related to PSE-meat.

Ultimate meat quality has been predicted from glycogen depletion patterns analysed on samples (*M. longissimus dorsi*) taken 2h after slaughter (Severini *et al.*, 1991). Muscle samples having a high initial pH, and numerous

PAS-positive fibers were predicted to give meat of good ultimate quality. Muscle samples that had a large proportion of PAS-negative fibers were expected to give DFD-meat with high initial pH or PSE-meat with low initial pH. The classification of the muscle samples at 24h post-mortem was in good agreement with the prediction made 2h after slaughter, based on initial pH and staining intensity for glycogen. Meat quality is thus related to glycogen depletion patterns and metabolic responses at slaughter.

Comparisons within, and between breeds also indicate that the size and the metabolic profile within the muscle fibers may be important for different parameters of meat quality. Repeated muscle sampling from the same animal (Landrace) at different ages (70 - 220 days) showed that pigs developing PSE-meat have about 10% larger fiber area than pigs developing normal meat (Wegner and Ender, 1990). As a consequence of selection (three generations) for larger or smaller fiber diameters of *M. longissimus dorsi*, the frequency of halothane-positives decreased in the pigs with smaller diameters and increased in the pigs with larger diameters (Wicke *et al.*, 1991). Meat from muscles in which giant fibers have been observed often has PSE (pale,soft,exudative) properties (Klosowska and Klosowski, 1985; Solomon *et al.*, 1990). Several studies indicate that the size of the fibers may be of importance for meat quality. The more rapid decline in pH and reduction in water binding and the lighter colour of the meat of highly selected breeds such as Swedish Landrace and Large White are said to be partly related to the larger proportion of white (non-oxidative) fibers in these breeds, compared with breeds such as Manguliza, wild pigs and Black Slavonic pigs (Rede *et al.*, 1986). With an increase in fiber diameter in *M. longissimus dorsi*, a decrease was found in taste-panel scores for tenderness (Carpenter *et al.*, 1963). Pigs that were raised in pens under the most restricted movement regimen had larger fiber areas and less-tender meat than pigs that were allowed a moderate degree of movement in their daily regimen (Petrov and Tomov, 1985). Taste panels have rated meat from pigs with restricted movement to be less tender than meat from trained pigs that had a higher oxidative and lower glycolytic capacity in *M. longissimus dorsi* (Essén-Gustavsson *et al.*, 1988).

Fat in muscle is also a factor that may be of importance in meat quality. In sensory tests, meat of fat-type pigs obtained high tenderness scores, which may have been due to the large fat content (Rede *et al.*, 1986). Meat from half-Chinese cross-bred pigs has been judged to be more tender, juicy and tasty than meat from purebred European pigs (Touraille *et al.*, 1989). Chinese pure-bred pigs have received higher sensory scores than did Landrace and Duroc (Suzuki *et al.*, 1991). The pigs in both these studies with high sensory scores had more intramuscular fat than the other pigs. Intramuscular lipid content varies little with respect to metabolic type of muscle (Leseigneur-Meyneir and Gandemer, 1991). However, intramuscular fat represents the lipid that is stored in both intracellular and intercellular areas, and between fascicles. Domestic pigs fed a low-protein diet had a higher intramuscular fat content (IMF) and lower shear force values in *M. longissimus dorsi* than pigs fed a high-protein diet (Essén-Gustavsson *et al.*, 1992). Fiber type composition, and oxidative and glycolytic capacity, did not differ between the pigs. IMF values correlated to triglyceride content in the muscle, which in turn was negatively related to mean fiber area. The histochemical stainings showed that all type I fibers and about 26% of type IIA fibers contained neutral lipids. Neutral lipids were seen in only 1% of type IIB fibers. High triglyceride content in muscle fibers may be one factor of importance for meat quality. Tenderness of meat was lower in Yorkshire pigs, compared with Swedish Landrace and Hampshire pigs. These pigs had muscles with higher oxidative and lower glycolytic capacity, and a higher triglyceride content, than muscles from the Yorkshire pigs (Essén-Gustavsson and Fjelkner-Modig, 1985). Positive ratings for sensory properties and especially tenderness may thus be related to the oxidative capacity and fat content of muscle fibers.

Conclusion

Histochemical and biochemical properties of muscle fibers (fiber type, fiber area, oxidative and glycolytic capacity and glycogen and lipid content) are factors that may influence meat quality. An important factor for post-mortem changes and meat quality is the metabolic response that takes place in the different fiber types before slaughter. Selection for leaner pigs and for a high proportion of large fibers, especially of type IIB, can result in poor capillarization and, consequently, an insufficient delivery of oxygen and substrates, and elimination of end products.

In the future, fiber types in pig muscle will probably be investigated with more advanced, and sensitive techniques, where it is possible to look at adaptations in different contractile, sarcoplasmic and other proteins that are of importance for cell differentiation and metabolic profiles. Furthermore, single-fiber dissection and quantitative biochemical analyses may increase the knowledge about metabolic and contractile profiles in a fiber.

References

Ashmore, C.R. and Doerr, L. 1971. Comparative aspects of muscle fiber types in different species. Exper. Neurol. 31, 408-18.

Bader, R. 1982a. Enzymhistochemische und histometrische Untersuchungen an Skelettmuskeln von ausgemästeten, gesunden Schweinen der deutschen Landrasse. Zbl. Vet. Med. A, 29, 443-57.

Bader, R. 1982b. Histologische Befunde aus licht- und elektronenmikroskopischen Untersuchungen an der Skelettmuskulatur von gesunden, ausgemästeten Schweinen der Deutschen Landrasse. Zbl. Vet. Med. A, 29, 458-76.

Bader, R. 1983. Vergleichende histometrische und histologische Untersuchungen an der Skelettmuskulatur von Wild- und Hausschweinen. Berl. Münch. Tierärztl. Wschr. 96, 89-97.

Barany, M. 1967. ATPase activity of myosin correlated with speed of muscle shortening. J. Gen. Physiol. 50, 197-218.

Brooke, M.H. and Kaiser, K.K. 1970. Muscle fiber types: How many and what kind? Arch. Neurol. 23, 369-79.

Carpenter, Z.L., Kauffmann, R.G., Bray, R.W., Briskey, E.J. and Weckel, K.G. 1963. Factors influencing quality in pork. A. Histological observations. J. Food Sci. 28, 467-71.

Cooper, C.C., Cassens, R.G. and Briskey, E.J. 1969. Capillary distribution and fiber characteristics in skeletal muscle of stress-susceptible animals. J. Food Sci. 34, 299-302.

Cooper, C.C., Cassens, R.G., Kastenschmidt, L.L. and Briskey, E.J. 1970. Histochemical Characterization of Muscle Differentiation. Development. Biol. 2, 169-84.

Davies, A.S. 1972. Post-natal changes in the histochemical fiber types of porcine skeletal muscle. J. Anat. 113 (2), 213-40.

Ervasti J.M., Strand M.A., Hanson T.P., Mickelson J.R. and Louis C.F. 1991. Ryanodine receptor in different malignant hyperthermia susceptible porcine muscles. Am. J. Physiol. 260, C58-C66.

Essén, B., and Lindholm, A. 1981. Muscle fiber types and glycogen depletion pattern in halothane-sensitive pigs. In: Froysten, T., Slinte, E. and Standal, N. (eds). Porcine Stress and Meat Quality - causes and possible solutions to the problems. Agricultural Food Research Society, Ås, Norway. pp. 42-52.

Essén, B., Lindholm, A and Persson, S. 1980. The effect of exercise and anabolic steroids on enzyme activities and fiber composition in skeletal muscle of pigs. Proceedings of the Fifth meeting of the Academic Society for Large Animal Veterinary Medicine, Glasgow, Scotland.

Essén-Gustavsson, B. and Lindholm, A. 1983. Influence of exercise on muscle metabolic characteristics in pigs and on backfat thickness at slaughter. Fifth International Conference on Production Disease in Farm Animals. Uppsala, Sweden, pp. 356-61.

Essén-Gustavsson, B. 1986. Activity- and inactivity-related muscle adaptation in the animal kingdom. In: Saltin, B. (ed.), Biochemistry of Exercise IV. Intl. Series on Sport Medicine. Vol. 16. Human Kinetics Publishers, Campaigne, Illinois. pp. 435-43.

Essén-Gustavsson, B. and Lindholm, A. 1984. Fiber types and metabolic characteristics in muscles of wild boars, normal and halothane sensitive Swedish Landrace pigs. Comp. Biochem. Physiol. 78A (1), 67-71.

Essén-Gustavsson, B. and Fjelkner-Modig, S. 1985. Skeletal muscle characteristics in different breeds of pigs in relation to sensory properties of meat. Meat Sci. 13, 33-47.

Essén-Gustavsson, B., Lundström, K., Larsson, G., Lindholm, A., Nordin, A-C., Hansson, I and Tornberg, E. 1988. The Effect during Growth of Moderate Exercise on Muscle Metabolic Characteristics *In Vivo* and Relation to Meat Quality and Sensory Properties. 34th International Congress of Meat Science and Technology Brisbane, Australia, pp. 27-30.

Essén-Gustavsson, B., Karlström, K., Lundström, K. and Pöso, R. 1991. Adenine nucleotide breakdown products in muscle at slaughter and their relation to meat quality in pigs with different halothane genotypes. 37th International Congress of Meat Science and Technology. Kulmbach, Germany, pp. 348-50.

Essén-Gustavsson, B., Karlström, K. and Lundström, K. 1992. Muscle fiber characteristics and metabolic response at slaughter in pigs of different halothane genotypes and their relation to meat quality. Meat Sci. 31, pp. 1-11.

Essén-Gustavsson,.B., Karlsson, A, Lundström K and Enfält A.C. 1992. Intramuscular Fat Content and Lipid in Muscle Fibers of Pigs Fed High and Low Protein Diet and Relation to Meat Quality. 38th International Congress of Meat Science and Technology, Clermont Ferrand, France, pp. 41-4.

Jensen, M., Essén-Gustavsson, B and Hakkarainen, J. 1988. The effect of a diet with a high or low content of vitamin E on different skeletal muscles and myocardium in pigs. J. Vet. Med. A35, 487-97.

Karlsson, A., Essén-Gustavsson, B. and Lundström, K. 1992. Muscle Glycogen Depletion Pattern in *Longissimus dorsi* Muscle of Pigs Fed High- and Low-Protein Diet. 38th International Congress of Meat Science and Technology. Clermont Ferrand, France. 375-8.

Karlsson, A., Enfält, A-C., Essén-Gustavsson, B. Lundström, K., Rydhmer, L. and Stern, S. Muscle histochemical and biochemical properties in relation to meat quality during selection for increased lean tissue growth rate in pigs. (submitted).

Kiessling, K-H. and Hansson, I. 1983. Fiber composition and enzyme activities in pig muscles. Swedish J. Agric. Res. 13, 257-61.

Kiessling, K-H., Lundström K., Petersson, H. and Stalhammar, H. 1982. Age and feed related changes of fiber composition in pig muscle. Swedish J. Agric. Res. 12, 69-75.

Klosowska, D. and Klosowski, B. 1985. Histological and histochemical investigations on heterogeneity of muscle in pigs. Tag.-Ber., Akad. Landwirtsch.-Wiss. DDR, (Berlin) 236, 191-97.

Klosowska, D., Klosowski, B., Fiedler, I.und Wegner, J. 1985. Veränderungen in der Fasertypenverteilung und Muskelfasergröβe im *M. longissimus dorsi* der Schweine während des Wachstums und Beziehungen zwishen histologischen Merkmalen und Schlachtkörperkriterien. Arch. Tierzucht, (Berlin) 28(2), 171-80.

Leseigneur-Meynier, A. and Gandemer, G. 1991. Lipid composition of pork muscle in relation to the metabolic type of the fibers. Meat Sci. 29, 229-41.

Lundström, K., Essén-Gustavsson, B., Rundgren, M., Edfors-Lilja, I. and Malmfors, G. 1989. Effect of halothane genotype on muscle metabolism at slaughter and it's relationship with meat quality: A within-litter comparison. Meat Sci. 25, 251-63.

Monin, G., Mejenes-Quijano, A., Talmant, A. and Sellier, P. 1987. Influence of breed and muscle metabolic type on muscle glycolytic potential and meat pH in pigs. Meat Sci. 20, 149-58.

Oksbjerg, N.and Henckel, P. 1989. Capillary Supply in *M. longissimus dorsi* of Pigs in Relation to Live Weight. 40th Annual Meeting of the European Association for Animal Production, Dublin, Ireland.

Oksbjerg, N., Blackshaw, A., Henckel, P., Fernandez, J.A. and Agergaard, N. 1990. Alterations in protein accretion and histochemical characteristics of the *M. longissimus dorsi* in pigs caused by Salbutamol. Acta Agric. Scand. 40, 397-401.

Petrov, J. and Tomov, V. 1985. A study on skeletal muscle structure and meat tenderness of pigs raised at different regimens of movement. Zhivetnovnd. Nanki. 22(1), 63-69.

Rahelic, S. und Puac, S. 1980. Durchmesser und Fasertyp von drei Muskeln des Schweineshinkens. Fleischwirtsch. 60(3), 486-88.

Rahelic, S. and Puac S. 1981. Fiber types in *longissimus dorsi* from wild and highly selected pig breeds. Meat Sci. 5, 439-50.

Rede, R., Pribisch, V. und Rahelic, S. 1986. Untersuchungen über die Beschaffenheit von Schlachttierkörpern und Fleisch primitiver und hochselektierter Schweinerassen. Fleischwirtsch. 66(5), 898-907.

Salomon, F-V., Günther, M. und Gruschwitz, F. 1983. Zur Entwicklung von Fasertypenkomposition und Faserdurchmesser im *M. longissimus* des Hausschweines. Anat. Anz., Jena 154, 69-79.

Salomon, F-V., Thüring, M. und Salomon, B. 1984. Über den Einfluβ von Laufbandtraining auf die Fasertypenverteilung und den Faserdurchmesser im *M. semitendinosus* des Schweines. Arch. exper. Vet. Med., Leipzig 38, 6, 800-6.

Severini, M., Trevisani, M. and Loschi, A.R. 1991. A Trial for Early Prediction of PSE and DFD Pig Meat by Measuring pH and the Percentage of PAS-positive Fibers. 37th International Congress of Meat Science and Technology, Kulmbach, Germany, pp. 485-8.

Solomon, M.B. and West, R.L. 1985. Profile of fiber types in muscles from wild pigs native to the United States. Meat Sci. 13, 247-54.

Solomon, M.B. and Eastridge, J.S. 1987. Occurrence of giant fibers in muscles from wild pigs native to the United States. Meat Sci. 20, 75-81.

Solomon, M.B., Campbell, R.G. and Steele, N.C. 1990. Effect of sex and exogenous porcine somatotropin on *longissimus* muscle fiber characteristics of growing pigs. J. Anim. Sci. 68, 1176-81.

Suzuki, A., Kojima, N., Ikeuchi, Y., Ikarashi, S., Moriyama, N., Ishizuka, T. and Tokushige, H. 1991. Carcass composition and meat quality of chinese purebred and European x Chinese crossbred pigs. Meat Sci. 29, 31-41.

Swatland, H.J. 1975. Histochemical development of myofibers in neonatal piglets. Res. Vet. Sci.18, 253-7.

Szentkuti, L. und Schlegel, O. 1985. Genetische und funktionelle Einflüsse auf Fasertypenanteile und Faserdurchmesser im *M. longissimus dorsi* und *M. semitendinosus* von Schweinen. Untersuchungen an trainierten Haus- und immobil gehaltenen Wildschweinen. Dtsch. tierärztl. Wschr. 92, 93-7.

Szentkuti, L. und Cassens, R.G. 1978. Die Verteilung der Fasentypen I, IIA und IIB im *M. longissimus dorsi* und *M. semitendinosus* von Schweinen verschiedenen Alters. Dtsch. Tierärztl. Wschr. 5, 23-7.

Szentkuti, L., Niemeyer, B. und Schlegel, O. 1981. Vergleichende Untersuchung von Muskelfasertypen mit der Myosin-ATPase-Reaktion im *M. longissimus dorsi* von Haus- und Wildschweinen. Dtsch. tierärztl. Wschr. 88, 407-11.

Touraille, C., Monin, G. and Legault, C. 1989. Eating quality of meat from European x Chinese crossbred pigs. Meat Sci. 25, 177-86.

Van Den Hende, C., Muylle, E. Oyaert, W. and De Roose, P. 1972. Changes in muscle characteristics in growing pigs: Histochemical and electron Microscopic Study. Zbl. Vet. Med. A19, 102-10.

Wegner, J. und Ender, K. 1990. Mikrostrukturelle Grundlagen des Wachstums von Muskel- und Fettgewebe und die Beziehung zu Fleischansatz und Fleischbeschaffenheit. Fleischwirtsch. 70(3), 337-40.

Wicke, M., von Lengerken, G., Fiedler, I., Altmann, M. and Ender, K. 1991. Einfluss der Selektion nach Merkmalen der Muskelstruktur des *M. longissimus* auf Belastungsempfindlichkeit und Fleischbeschaffenheit beim Schwein. Fleischwirtsch 71(4), 437-42.

Discussion

*Puolanne - Is there a method to measure the capillary capacity, instead of capillary density, and whether it is different if it is hormone induced, or work induced?

*Essén-Gustavsson - That is very interesting. No, I'm not aware of any methods. The way we measure, to get some kind of idea about capillaries today, is of course only with histochemical staining techniques, to count the capillaries around the fibers. But, you mean how to measure capillary capacity and how to express the network. I don't know any technique to do that.

*Puolanne - Because we know that the speed of blood in muscle is about 2mm per second, that may have an influence on these aspects, too.

*Essén-Gustavsson - Yes, definitely.

*Henckel - Just to get an idea of the capillary network, we have used ink dress, injection of the inks, but the animal must be killed in order to do it, so, that's the problem. If you want to see the network, how it's dispersed, it is possible to make mathematical models of capillary distributions and growth. When we do the cuttings on the samples we get, sometimes longitudinal sections of capillaries are cut as well, and if you include this number, that's actually capillaries that are cut perpendicularly, and capillaries that are cut longitudinally, you can make mathematical models of the network of capillaries, but it's quite hard work to do.

*Mickelson - I have a question related to that area. You mentioned that there are two sources of fatty acids for a type I fiber to use, those coming through the capillaries, and the triglyceride. What is the relative rate at which a type I fiber uses those sources? What is the turnover rate of the intrafiber triglycerides?

*Essén-Gustavsson - I don't know the rate. What we know today is that lipid is stored in fibers. What is the importance of the lipid stored there? Some think that when fatty acids enter the fiber, they first form a triglyceride, whereas the triglyceride itself releases another fatty acid. So, there's sort of a triglyceride pool where the fatty acids go to before they get released and are oxidized in the mitochondria. Now the thing that is interesting is that in order to make new triglycerides, you need α-glycero phosphate, and the only way to get that is from glycogenolysis. What we know, not from pigs, because no one has measured it on pig muscle, is that triglycerides are utilized in long-distance work. So, it is definitely important, even though the pig is not an endurance runner.

*Mickelson - One of the reasons I asked the question was that there's an alternative candidate gene for porcine malignant hyperthermia defect, more likely human MH. It is a hormone-sensitive lipase that is supposedly active in skeletal muscle to some limited extent. This theory is that it generates free fatty acids that supposedly damage membranes. So, I wondered if you could shed any light on whether that enzyme is really there or not.

*Essén-Gustavsson - I don't know anything about that.

*Tornberg - I would like to comment on meat quality, in relation to the fiber-type composition in muscle. When we talk about meat quality, what do we mean? Is it the occurrence of PSE? Or, is it the tenderness, the flavor and the juiciness of the meat? With the occurrence of PSE, I can agree that the higher the percentage of IIB fibers, the higher the propability for PSE to occur. But, with regard to tenderness, I think it's the other way around. We have measured, in our studies of rigor development, the amount of intermediate fibers and red fibers in three beef muscles. The degree of contraction, during rigor is highly correlated to the toughness of the meat evaluated by sensory taste panel after 14 days ageing. So, the more red fibers there are, the more contracted the meat and consecuently the tougher the meat.

*Essén-Gustavsson - When you say red fibers, what have you analysed, NADH, or SDH, or what ?

*Tornberg - ATPase.

*Essén-Gustavsson - Then it's not oxidative capacity, you have measured but a marker for the contractile properties. Your red and intermediate fibres correspond to type I and II fibres.

*Tornberg - Well, anyhow, if you summarize I and IIA, you got a very good correlation between the degree of contraction, and the sum of these fiber types, for three different muscles. Moreover, we have a high correlation between thoughness, evaluated after 14 days ageing, and the degree of contraction. So, it's the other way around, the more type I and IIA fibers, the less tender the meat.

*Essén-Gustavsson - Well, maybe there are several factors we need to cope with to get good meat quality. Fibers can be recruited in different ways just before slaughter, or at slaughter. The metabolism in the fibers then, will affect meat quality. I think that's very important. It's important to remember that the fibers have both differrent contractile and metabolic properties. How should we cope with them to get the perfect meat quality.

*Vögeli - I want to make a remark about what you said about vitamin E. Vitamin E protects meat from oxidation, as you have said. I have analysed the genetic background of lipoproteins, these are important transport molecules for vitamin E, and, I have detected some genetic differences by producing antibodies that detect different allotypes. I have seen that some genotypes of these lipoprotein systems are positively correlated with oxidation stability, and some are negatively correlated, as measured by malon-aldehyde, and oxidation products. And, I could see that besides the feed effect, the genetic effect of these lipoproteins are very important. I have analysed the α-lipoprotein-B types, and I could see that, besides the feed, this genetic basis is also very important.

*Jensen - I'd like to make a comment on that because it agrees very well with our studies in the live animal. We have fed different litters, on the same amount of vitamin E, but different levels of α-tocopherol has been observed in liver fat and muscle. So, I do agree with you that it must be a genetic background as well.

*Vögeli - For instance, we have produced patties, and we could see it after 1 day, we produced these patties and put them in the refrigerator, but also after 1 month deep freezing, and 3 months, we had the same result. If you have some certain lipoprotein genotypes, it increases this malon-aldehyde production, with time. And this is important because more and more products are produced and frozen. It will be very important to have good oxidation stability in the future.

*Jensen - You have made all your determinations on meat, and we have done this in the live animal, and that just stresses the importance of having a good defense system against free radicals, both in the live animal, and after slaughter, when you talk about meat quality.

*Essén-Gustavsson - Well, we discuss oxidative capacity, peroxidation, free radicals, etc. Maybe we have to make pig muscles more oxidative somehow!

*Henckel - Why don't we just select for capillary density? Capillary density is related to growth rate, I think it's 0.6, and you'll have less drip loss, so it's rather simple.

*Essén-Gustavsson - After all these years, we understand that marked changes in glycogenolysis and pH are bad, that is, either you get PSE or DFD at slaughter. Now we have to try to avoid this and the question is how.

*Tarrant - The slides that you showed with the glycogen negative fibers, some of them were *in vivo*, and some of them were post-mortem. Now, *in vivo*, I think the glycogen negative fiber can recover. But, post-mortem, is it your opinion that those fibers are in rigor mortis? And, if they are in rigor, do you think that it's permanent, or is it possible for them to perhaps recover some ATP?

*Essén-Gustavsson - Yes, I think they are in rigor. As far as post-mortem recovery, I think, it's not the glycogen depletion that's important itself, it is whether or not there's ATP, or CP there. It's the phosphate stores that are the most important. If there's no ATP, no CP, no glycogen, that fiber has to be in rigor, doesn't it?

*Tarrant - As far as I know, there is no histochemical test to demonstrate the rigor state. Maybe this is because most of histochemistry comes from medical studies, and biological studies where there's no interest in rigor. This is a post-mortem phenomenon, it's something that perhaps meat scientists are mainly interested in. Is there a possibility for a new histochemical test for the rigor state?

*Essén-Gustavsson - That would be a method of interest if someone could find that. But, when ATP has produced IMP, the question is, can IMP form AMP again? It can, if you have proteins there. There must be aspartate, and the proper enzymes. I also do some research with horses. After a horse race, we have found, *in vivo*, some fibers with no ATP. So, the question is, can they work, or not? Exactly what you said. I mean, for pigs, having these marked glycogenolysis, they have to work with ATP. When these pigs have to race to slaughter, ATP is probably zero.

*Tarrant - It also has implications post-mortem, though. Getting back to post-mortem, in relation to Gerald Offer's hypothesis about drip loss, and the binding of the heads of myosin to actin, and resistance to further denaturation. I think it's very important for meat scientists to know when the individual fibers go into rigor, as distinct from knowing when the average mass of muscle goes into rigor.

Calcium (CA^{2+}) Regulation in Porcine Skeletal Muscle - Review

J.R. Mickelson and C.F. Louis

Department of Veterinary Pathobiology, University of Minnesota, 295 Animal Science Veterinary Medicine, St. Paul, MN 55108, USA

Introduction

The function and activity of a wide variety of tissues is regulated by the concentration of ionized calcium in the cytoplasm. In skeletal muscle, a partial list of the major cellular processes regulated by Ca^{2+} would include force generation resulting from thick and thin filament interaction, glycogenolysis, mitochondrial oxidative metabolism, and intracellular protein turnover. The sarcoplasmic Ca^{2+} concentration in resting skeletal muscle is estimated as being in the range 50-500nM. During a muscle twitch, the sarcoplasmic Ca^{2+} concentration rapidly rises to a level of 1-10μM before returning again to the resting level within 100ms. During a sustained tetanus, however, sarcoplasmic Ca^{2+} stays elevated throughout the entire time course of stimulation. To govern the range of its activities, skeletal muscle thus has a requirement for the regulation of the sarcoplasmic Ca^{2+} concentration over the millisecond, as well as second to minute, time scale.

The ability of muscle to regulate its sarcoplasmic Ca^{2+} concentration derives from the activity of specific Ca^{2+} pumps, transporters, and channels located in the surface membranes (sarcolemma and transverse tubule system), sarcoplasmic reticulum (SR), and mitochondria. A 100000-fold Ca^{2+} gradient can be maintained between these membrane systems and the sarcoplasm. At any given instant, the Ca^{2+} concentration in the sarcoplasm is dictated by the balance between two opposing processes; namely, channel-mediated Ca^{2+} influx down its concentration gradient (through surface membrane channels or the Ca^{2+} release channel of the SR), and Ca^{2+} pumps removing Ca^{2+} (into the extracellular space or back into the SR). In this brief review, we will attempt to summarize current knowledge of the mechanism of excitation-contraction coupling, whereby nerve stimulation/surface membrane depolarization results in a rise in sarcoplasmic Ca^{2+}, and muscle contraction. We will then discuss how abnormalities in sarcoplasmic Ca^{2+} regulation are responsible for the porcine stress syndrome (PSS), before concluding with some aspects of Ca^{2+} regulation in post-mortem muscle.

Sarcolemma and transverse tubule membranes

The surface membrane system of skeletal muscle comprises the sarcolemma and transverse tubule system. Although the t-tubules are contiguous invaginations of the sarcolemma, and these two different membranes have a large number of structural and functional similarities, there are also many differences between these two membrane systems, particularly as regards their role in the regulation of sarcoplasmic Ca^{2+} concentrations. Following stimulation by a motorneuron, the controlled movement of Na^+ and K^+ across these surface membranes during depolarization and subsequent repolarization is of course, critical to muscle function. However, with extracellular Ca^{2+} concentration at millimolar levels, it is also the responsibility of the surface membranes to maintain a Ca^{2+} barrier and ensure resting sarcoplasmic Ca^{2+} remains at nanomolar levels. Furthermore, the specialized Ca^{2+} channel located in the t-tubules plays a key role in excitation-contraction coupling by regulating the release of Ca^{2+} from the SR.

Surface membrane ATP-dependent Ca^{2+} transport

Ca^{2+} transport in skeletal muscle surface membranes has been characterized in membrane preparations isolated from rabbit and porcine (*longissimus dorsi*) muscle (Brandt *et al.*, 1980; Ervasti *et al.*, 1989b; Hidalgo *et al.*, 1986; Mickelson *et al.*, 1987). Purified sarcolemma and t-tubule membranes conveniently reseal into vesicles with an inside-out orientation. This places the intracellular (cytoplasmic) membrane surface on the outside surface of the isolated vesicular preparations. The movement of Ca^{2+} across the membrane is then monitored using $^{45}Ca^{2+}$, and the separation of the $^{45}Ca^{2+}$-containing membrane vesicles from the surrounding media by filtration.

In the presence of MgATP, both purified sarcolemma and t-tubule vesicles are capable of accumulating $^{45}Ca^{2+}$ and maintaining a considerable Ca^{2+} gradient. This vectorial Ca^{2+} transport to the interior of the vesicle is the *in vitro* equivalent of transporting Ca^{2+} out of the muscle cell. The rate of surface membrane Ca^{2+} uptake is not rapid (10-20nmol Ca/mg membrane protein/min) compared with the high activity of the SR Ca^{2+} pump (greater than 1 μmol Ca/mg membrane protein/min) (Brandt *et al.*, 1980; Ervasti *et al.*, 1989b; Hidalgo *et al.*, 1986; Mickelson *et al.*, 1987). However, the $Ca^{2+}{}_{0.5}$ for sarcolemmal or t-tubular Ca^{2+} transport (Ca^{2+} concentration at which half-maximal stimulation of Ca^{2+} uptake occurred) was in the range 0.2-0.5 μM, making it possible for this system to participate in the lowering of sarcoplasmic Ca^{2+} to typical resting levels. Interestingly, cAMP-dependent protein kinase and the regulatory protein calmodulin both stimulate the maximal rate of Ca^{2+} transport as well as increase the affinity of the enzyme for Ca^{2+} (Mickelson *et al.*, 1987), indicating that the skeletal muscle surface membrane Ca^{2+} pumps may also be regulated by the metabolic state of the cell.

These observations indicated that the skeletal muscle surface-membrane system has a Ca^{2+}-ATPase similar to that described for other tissues such as the well-characterized erythrocyte Ca^{2+} pump (Wang *et al.*, 1992). It is usually considered that these Ca^{2+} pumps primarily serve a "housekeeping" function, maintaining a low constant rate of activity and thus preventing accumulation of Ca^{2+} within the cell. Ca^{2+} transport by these pumps is at a stoichiometry of 1 Ca^{2+} per ATP (Wang *et al.*, 1992). Molecular cloning techniques have now characterized cDNAs for two forms of a surface membrane Ca^{2+}-ATPase expressed in skeletal muscle (Greeb and Shull, 1989). These proteins of M_r 125-130kD have considerable homology to the surface membrane Ca^{2+} pumps of a wide variety of tissues (Greeb and Shull, 1989, Wang *et al.*, 1992). It is now also clear that there are at least four genes coding for plasmalemmal Ca^{2+} pumps, with even more variants possible due to alternative splicing of the primary transcripts (Greeb and Shull, 1989; Wang *et al.*, 1992).

Surface membrane Na^+/ Ca^{2+} exchange

A Na^+/Ca^{2+} exchange activity has also been described in skeletal muscle sarcolemma and t-tubules (Donoso and Hidalgo, 1989; Ervasti *et al.*, 1989b; Mickelson *et al.*, 1987). This activity is in most regards similar to that observed in other excitable tissues such as nervous tissue and cardiac muscle (Langer, 1992; Nicoll *et al.*, 1990). It has been argued that Na^+/Ca^{2+} exchange is the primary means of Ca^{2+} efflux from cardiac muscle cells (Langer, 1992). Experimentally, Na^+/Ca^{2+} exchange can be observed as $^{45}Ca^{2+}$ transport driven by a Na^+ gradient imposed across the membrane. Other monovalent cations (K^+ or Li^+) do not substitute for Na^+ in this regard. The stoichiometry of Na^+/Ca^{2+} exchange appears to be three Na^+ for each Ca^{2+}, and the affinity for Ca^{2+} is approximately 1μM, slightly less than that of the Ca^{2+}ATPase (Langer, 1992). The Na^+/Ca^{2+} exchange activity of cardiac muscle is very sensitive to the sarcoplasmic Na^+ concentration (Langer, 1992). This may allow the reversal of ion fluxes, resulting in Ca^{2+} movement into and Na^+ movement out of the cell during some situations.

Attempts at isolating and purifying the Na^+/Ca^{2+} exchanger had been met with considerable frustration. However, recently a cDNA coding for a cardiac muscle Na^+/Ca^{2+} exchanger was cloned and expressed in Xenopus oöcytes (Nicoll *et al.*, 1990). The native Na^+/Ca^{2+} exchanger has a M_r of approximately 160 kD and is heavily glycosylated (Langer, 1992, Nicoll *et al.*, 1990). Whether this protein or an isoform is expressed in skeletal muscle and is responsible for the observed Na^+/Ca^{2+} exchange activity of isolated surface membrane vesicles from this tissue remains to be defined.

T-tubule Ca^{2+} channel/dihydropyridine receptor

In contrast to the sarcolemma, the skeletal muscle transverse-tubule system has a high density of voltage-activated Ca^{2+} channels (Sanchez and Stefani, 1978). T-tubule Ca^{2+} channels have been extensively characterized electrophysiologically, pharmacologically, and biochemically. These channels slowly activate during an action potential, and then inactivate, resulting in the net inward flux of Ca^{2+}, which is known as the slow inward current. The channel has a unitary conductance of approximately 10pS and is not activated until the membrane potential reaches approximately -40mV. The pharmacology and kinetics of channel activity distinguish this L-type channel from other classes of surface membrane Ca^{2+} channels. In cardiac muscle the Ca^{2+} current carried in during the action potential plays the major role in activating the further release of Ca^{2+} from the SR (Ca^{2+}-induced Ca^{2+} release) (Fabiato, 1983). However, this is not the case in skeletal muscle, and the role that the Ca^{2+} channel activity itself plays in the regulation of skeletal muscle function is not completely understood.

The L-type Ca^{2+} channels are receptors for several major pharmacologic classes of Ca^{2+} channel blockers, including the dihydropyridines (nitrendipine or PN200-110), phenylalkylamines (verapamil or D600), and benzothiazepines (diltiazem) (Campell *et al*., 1988; Caterall, 1991). T-tubule vesicles isolated from rabbit or pig skeletal muscle typically bind greater than 40pmol nitrendipine/mg membrane protein, with an affinity in the nanomolar range (Ervasti *et al*., 1989a). T-tubules are the richest source of this protein in mammals, so have served as a source of material for the many studies into the function of L-type Ca^{2+} channels.

The multisubunit rabbit skeletal muscle dihydropyridine receptor has been isolated and characterized (Campell *et al*., 1988; Caterall, 1991). The purified complex consists of subunits of M_r 170kD, 150kD, 52kD, and 32kD termed $\alpha 1$, $\alpha 2$, β, and γ, respectively, that exist in a 1:1:1:1 ratio; a fifth subunit, δ, is derived from $\alpha 2$ by proteolytic processing. The $\alpha 1$ subunit is responsible for the Ca^{2+} channel and Ca^{2+} antagonist binding activities, while the other subunits appear to modulate these activities. The $\alpha 2$ and γ subunits are glycosylated while the $\alpha 1$ and, β subunits are substrates for protein kinases. The cDNA-derived amino acid sequences of these proteins have been used to construct models for the structures of the subunits. Of particular interest is the homology between the $\alpha 1$ subunit sequence and that of the voltage-sensitive plasma membrane Na^{+} channel (Rios and Brum, 1987). Both these channels comprise approximately 2000 amino acids that contain four homologous domains (labelled I-IV), each of which in turn contains six transmembrane helical segments (S1-S6). Hypotheses concerning the function of individual domains and segments of these molecules are currently being tested by site-specific mutagenesis. The wide variety of dihydropyridine receptor/Ca^{2+}-channel subtypes found in various tissues appears to be due to several genes coding for each type of subunit, combined with possible alternative splicing of each transcript, resulting in a tissue specific expression of one or more subunit assemblies.

Much of the interest in the skeletal muscle dihydropyridine receptor is not due to its Ca^{2+}-channel activity (which is slow and actually carries little Ca^{2+} current) but rather its ability to act as a plasmalemmal voltage sensor. This molecule is hypothesized to function in excitation-contraction coupling by sensing the change in surface membrane potential and transducing this signal into one for Ca^{2+} release from storage sites in the SR (Rios and Brum, 1987; Tanabe *et al.*, 1987). A structural change in the dihydropyridine receptor, detected as an intramembrane movement of charge, accompanies a change in membrane potential to activating levels (Adams *et al.*, 1990; Rios and Brum, 1987; Tanabe *et al.*, 1988). This charge movement resides in positively charged amino-acid residues in the $\alpha 1$ subunit (likely within the S4 transmembrane segments); it is inhibited by Ca^{2+}-channel antagonists, and appears essential for the eventual release of SR Ca^{2+} (Rios and Brum, 1987). Although the transduction process by which the voltage sensor accomplishes the opening of SR Ca^{2+} channels is not known, it is certainly aided in this task by its juxtaposition to the SR terminal cisternae. Ultrastructural evidence indicates that the dihydropyridine receptors are aligned with the "feet" structures of the SR if not actually touching them (Block *et al.*, 1988). However, physical interaction with the SR feet proteins has not been convincingly demonstrated, and communication between the dihydropyridine receptor and the SR membrane may be effected by a different molecule. This may be a protein or possibly a chemical mediator such as inositol trisphosphate or Ca^{2+} (Caswell and Brandt, 1989). This aspect of t-tubule regulation of skeletal muscle function is a very active field of research.

Sarcoplasmic reticulum

When muscle contracts, the intracellular membrane system of the SR is primarily responsible for the regulation of sarcoplasmic Ca^{2+} concentration. Within a mammalian muscle the SR network surrounds the myofibril at the level of the A-band I-band junction, comprising terminal cisternal sacs which contain the "feet" connections with the t-tubule network. The terminal cisternae branch out, forming an elongated longitudinal system which fuses to reform the terminal cisternae sacs at the next A-I junction. The anatomical distinction between longitudinal tubules and terminal cisternae also correlates with a functional distinction within the SR. Sucrose density gradient centrifugation allows purification of longitudinal and terminal cisternae membrane vesicles derived from rabbit and pig muscle SR (termed light and heavy SR respectively, due to their differing densities) (Meissner *et al.*, 1988). Light SR is highly enriched in the Ca^{2+}-ATPase pump activity while the Ca^{2+}-release and Ca^{2+}-binding activities are enriched in the heavy SR fraction (Ervasti *et al.*, 1991; Meissner *et al.*, 1988; Mickelson *et al.*, 1986). Thus, the longitudinal SR is specialized to remove Ca^{2+} from the sarcoplasm to relax muscle while the terminal cisternae is specialized to release Ca^{2+} to allow contraction.

SR Ca^{2+} uptake

Many aspects of the SR Ca^{2+}-ATPase activity, which couples ATP hydrolysis to the transport of Ca^{2+} from the sarcoplasm to the SR lumen, are now well described. The high density of this enzyme in the longitudinal SR makes possible a rate of removal of Ca^{2+} from the sarcoplasm that would be expected *in vivo* to achieve muscle relaxation; it has also enabled many detailed studies of the properties of this enzyme (Inesi, 1985; Maruyama *et al.*, 1989). Fast kinetic analyses have enabled a detailed understanding of the catalytic activity. During a reaction cycle the enzyme in the E1 state binds MgATP and two Ca^{2+} on the sarcoplasmic side with high affinity (activation at approximately 200nM Ca^{2+}). The ATP is hydrolyzed, with the ADP released and the γ-PO_4 transferred to an aspartyl residue to form an E1-P phosphorylated intermediate. A relatively slow conformational change to an E2-P state then occurs, which is believed to disrupt the Ca^{2+} binding sites, lowering the affinity for Ca^{2+}, and allowing accessibility of the Ca^{2+} to the SR lumen. The phosphoenzyme is hydrolyzed, the P_i released, and the enzyme returned to the E1 state. A similar mechanism is likely for the reaction cycle catalyzed by the surface membrane Ca^{2+}-ATPase.

The SR Ca^{2+}-ATPase can be visualized as comprising a cytoplasmic headpiece, a stalk, and a transmembrane base, with most of the molecular mass residing in the sarcoplasm. With the molecular cloning of the cDNA coding for this 110kD Ca^{2+} pump protein, a model for secondary and tertiary structures of the enzyme has been devised (Brandl *et al.*, 1986; Maruyama *et al.*, 1989). This model proposes that the base consists of 10 transmembrane α-helixes (five helical hairpin loops) which form the transmembrane channel, five α-helical segments that join portions of the headpiece to the base, and three domains in the headpiece (nucleotide binding, phosphorylation, and beta strand domains). The functions of individual amino acid residues within these domains is being tested by mutational analyses (Maruyama *et al.*, 1989).

Different isoforms of the SR Ca^{2+}-ATPase are expressed in Type 1 and Type 2 skeletal muscle fibers. It is now clear that this is due to two different genes for the SR Ca^{2+}-ATPases, namely the fast-twitch isoform and the slow-twitch/cardiac isoform (Brandl *et al.*, 1986). That this is true also for porcine muscle is demonstrated on immunoblots by the ability of a fast-twitch Ca^{2+}-ATPase isoform specific antibody to bind to the pig *longissimus dorsi* Ca^{2+}-ATPase and a slow-twitch/cardiac specific antibody to bind to the pig *vastus intermedius* and cardiac muscle Ca^{2+}-ATPase (Ervasti *et al.*, 1991). Alternative splicing of the slow-twitch cardiac form also gives rise to endoplasmic reticulum Ca^{2+}ATPases in non-muscle tissues (Gunteski-Hamblin *et al.*, 1988).

It is also known that the relaxation rates of slow and fast muscles differ *in vivo,* as do the Ca^{2+} uptake capabilities of SR from these muscles *in vitro* (slow only 20% or so that of fast) (Briggs *et al.*, 1977). This appears due to a slightly lower density of Ca^{2+}-ATPase in the SR membranes of slow-twitch fibers, and a lower cellular volume of SR in slow-twitch fibers, resulting in slow-twitch fibers having approximately 50% of the Ca^{2+}ATPase content of fast-twitch fibers (Fergussen and Franzini-Armstrong, 1988). Thus, there may be other contributing factors to the five to six-fold lower relaxation rate of slow-twitch fibers. Such factors might include the presence of the regulatory protein phospholamban in slow-twitch muscle, which in the non-phosphorylated state lowers the affinity of the Ca^{2+}-ATPase for Ca^{2+}, and intrinsic differences in the catalytic properties of the two Ca^{2+}-ATPase isoforms (Fergussen and Franzini-Armstrong, 1988).

SR Ca^{2+} release

Several lines of evidence converged within the last 4 years to provide much needed information about the nature of the SR Ca^{2+} release process. The SR "feet" structures at the triadic junction between the SR terminal cisternae and the transverse tubules had been identified previously by electron microscopy (Franzini-Armstrong, 1970). Subsequently, these high M_r spanning proteins were isolated from rabbit muscle membrane fractions enriched in triads (Kawamoto *et al.*, 1986). Also, the localization of the high-capacity Ca^{2+}-binding protein, calsequestrin, to the lumen of the SR terminal cisternae was achieved (Jorgensen *et al.*, 1983). However, with the realization that the plant alkaloid ryanodine interacted specifically with some component of the heavy SR Ca^{2+} release mechanism came the ability to use [^{3}H]ryanodine binding as a tool in the isolation of what is now known as the SR Ca^{2+} release channel or ryanodine receptor. We now know that this channel is the primary means by which Ca^{2+} stored in the SR terminal cisternae is released into the sarcoplasm during muscle contraction.

Ca^{2+} release from $^{45}Ca^{2+}$-loaded skeletal muscle heavy SR vesicles is activated by μM Ca^{2+}, mM adenine nucleotide, and inhibited by mM Ca^{2+}, mM Mg^{2+}, and the regulatory protein calmodulin (Meissner *et al.*, 1986; Meissner *et al.*, 1988). Various other factors are known to affect the Ca^{2+} release process, including pH, ruthenium red, caffeine, local and inhalational anesthetics, and ryanodine, which binds to the channel when it is in the open state (induced by μM Ca^{2+} or nucleotides) (Meissner, 1986; Meissner *et al.*, 1986; Meissner *et al.*, 1988; Rousseau *et al.*, 1988). Possible effects of inositol trisphosphate in activating SR Ca^{2+} release and channel activity have been controversial (Valdivia *et al.*, 1990). Investigations into the Ca^{2+} release mechanism were aided greatly by the ability to observe this channel activity following fusion of isolated heavy SR vesicles into planar lipid bilayers and measuring ionic currents through single-channel molecules. In these bilayers the SR Ca^{2+} channel was found to have a very high Ca^{2+} conductance (80-

100pS) with an even higher monovalent cation conductance (400-800pS) in the absence of Ca^{2+} (Lai *et al.*, 1988; Smith *et al.*, 1985). The ability to conduct other ions has often been used experimentally in the study of this channel. The properties of the Ca^{2+} release process are very similar when using $^{45}Ca^{2+}$ flux or single-channel measurements to examine channel activity.

Using [^{3}H]ryanodine binding as a probe, several laboratories simultaneously reported the purification of a high M_r ryanodine binding protein from rabbit muscle heavy SR. This protein appeared to be a tetramer with a subunit size greater than 400kD (Inui *et al.*, 1987; Lai *et al.*, 1988). It was identified as the SR Ca^{2+} release channel by demonstrating that the single-channel activity of the purified ryanodine receptor was identical to the Ca^{2+}-channel activity of native SR vesicles (Lai *et al.*, 1988). Significantly, the ryanodine receptor was also found to be the previously described spanning/foot protein by the similarities in M_r, but more importantly by the quatrefoil "foot-like" structure (27 x 27 x 14nm) observed by electron microscopy (Inui *et al.*, 1987; Lai *et al.*, 1988). This places the SR Ca^{2+}-release channel directly at the triadic junction between the terminal cisternae and the t-tubule. Thus, the SR Ca^{2+}-release channel, ryanodine receptor, spanning protein and foot protein are one and the same.

With the cloning of the rabbit muscle ryanodine receptor (Takeshima *et al.*, 1989), it was demonstrated that the true subunit M_r of this protein was 565kD, with a tetrameric size of 2260kD, making it one of the largest proteins known. Two mammalian ryanodine receptor isoforms have been identified (Otsu *et al.*, 1990). They have approximately 70% homology and are the products of different genes. The skeletal muscle isoform is expressed in both slow- and fast-twitch skeletal muscle, with the second isoform expressed in cardiac muscle and brain (Otsu *et al.*, 1990). Models for the structure of the ryanodine receptor place the transmembrane channel-forming domains in the C-terminal 20% of the molecule, binding sites for nucleotide and calmodulin in the central region, and the major N-terminal portion of this protein comprising the cytosolic foot, which extends from the SR to the t-tubule (Otsu *et al.*, 1990; Takeshima *et al.*, 1989).

Mitochondrial Ca^{2+} transport

Isolated mitochondria are capable of accumulating large amounts of Ca^{2+} into the matrix space, utilizing the respiration- or ATPase-generated proton electrochemical gradient across the inner membrane (Lehninger, 1970). Although most studies have utilized mitochondria isolated from liver, kidney, or cardiac muscle, skeletal muscle mitochondria are also capable of accumulating Ca^{2+} (Mickelson and Marsh, 1980). A second mechanism of transmembrane Ca^{2+} movement is achieved by a Na^{+}-induced Ca^{2+} efflux mechanism in which mitochondria can release Ca^{2+} back into the sarcoplasm (Mickelson and Marsh, 1980).

The Km for mitochondrial Ca^{2+} uptake is in the μM concentration range, making it possible that under certain conditions mitochondria could accumulate Ca^{2+} *in vivo.* Furthermore, the lower SR content and longer relaxation times of slow-twitch fibers, coupled with their higher mitochondrial content, made a mitochondrial contribution to relaxation possible; the situation for cardiac muscle was considered even more likely. Muscle mitochondria also contain large amounts of Ca^{2+} when isolated. However, at present more direct evidence suggests that the role, if any, that mitochondria might play in the Ca^{2+} regulation of skeletal muscle is small. The major difficulty for the mitochondrial hypothesis is that electron-probe microanalysis of the elemental content of flash-frozen cryosectioned muscle detected very little Ca^{2+} in the mitochondrial compartment under any condition (Somlyo and Somlyo, 1986). Several mitochondrial oxidative enzymes are sensitive to Ca^{2+} concentration so it has been proposed that the mitochondrial Ca^{2+} transport was required for the regulation of mitochondrial activity rather than for the regulation of cytoplasmic Ca^{2+} (McCormack and Denton, 1981). There is little doubt, however, that in situations in which there is a prolonged elevation in sarcoplasmic Ca^{2+}, continued mitochondrial Ca^{2+} uptake and resultant overload contributes to mitochondrial destruction and cell death (Duncan, 1978).

Nature of the PSS defect in muscle Ca^{2+} regulation

PSS is characterized by sustained muscle rigidity, increased aerobic and anaerobic metabolism and increased heat production in response to halogenated anesthetics and various stressors. It is also known that a small number of humans also react abnormally to this agent, developing a response known as malignant hyperthermia (MH). For some time it has been reasoned that a defect in Ca^{2+} regulation, resulting in a sustained elevation in sarcoplasmic Ca^{2+} concentration, could be responsible for both PSS and MH. This elevated Ca^{2+} would continue to stimulate contractile activity and glycogen breakdown, resulting in even further stimulation of metabolism and heat production. That sarcoplasmic Ca^{2+} concentration is indeed elevated during exposure of PSS and human MHS muscles to halothane has been experimentally confirmed (Iaizzo *et al.*, 1989; Lopez *et al.*, 1988); however there is disagreement over whether the resting Ca^{2+} concentration in the absence of volatile anesthetics is also elevated in these muscles. The abnormal sensitivity of PSS muscle biopsies to anesthetic and caffeine-induced contractures (Gronert, 1986) gave initial clues as to the nature of the primary defect in Ca^{2+} regulation in susceptible muscles. Over the years a number of laboratories have sought the identity of the PSS and human MH genes, and here we will attempt to summarize some of the more recent findings that shed light on the biochemical basis of this disorder.

Surface membrane structure and function in PSS

The possibility that there are differences in the properties of the skeletal muscle surface membrane system of PSS and normal pigs has been addressed using purified sarcolemma and transverse-tubule membrane fractions. In comparing purified membrane fractions from PSS Pietrain and normal Yorkshire pigs, studies at the University of Minnesota have found that many of the characteristics of the purified PSS and normal surface membranes were not different; this included membrane yield, protein and lipid content, maximal $(Na^{+}+K^{+})$-ATPase, and Na^{+} channel content estimated as $[^{3}H]$saxitoxin binding (Ervasti *et al.*, 1989b; Mickelson *et al.*, 1987). Maximal adenylate cyclase activity and the sensitivity of this system to stimulation by the β-agonist isoproterenol also did not differ between the PSS and normal breeds (Mickelson *et al.*, 1987), indicating that the cAMP second-messenger system was unaltered. The possibility that there are small differences in the fatty acid side-chain distribution among the membrane phospholipid classes can not yet be excluded (Seewald *et al.*, 1991). However, the general similarity between the PSS and normal muscle surface-membrane preparations also indicated that these membranes had not been damaged by post-mortem degradation or preparation-induced artifacts, in addition to ruling out several possible sites of a defect that had been suggested in previous reports.

While examining aspects of skeletal muscle surface membrane Ca^{2+} regulation in PSS, it was found that the Km for Ca^{2+} of the ATP-dependent Ca^{2+} pump, and the V_{max} for Ca^{2+} uptake were similar for PSS and normal sarcolemma and t-tubules (Ervasti *et al.*, 1989b; Mickelson *et al.*, 1987). However, for as yet unknown reasons, the ability of the PSS membrane vesicles to accumulate Ca^{2+} over a longer time period (15-20 min) was approximately 70% that of the normal membranes. The passive Ca^{2+} permeability of PSS and normal surface membrane vesicles to Ca^{2+}, as well as the Na^{+}/Ca^{2+} exchange activity (measured as a Na^{+}-induced $^{45}Ca^{2+}$ efflux) were also not different for the PSS and normal membranes. Halothane concentrations 4-5 times the clinical concentration were required to inhibit Ca^{2+} uptake by either the PSS or the normal surface membranes (Ervasti *et al.*, 1989b; Mickelson *et al.*, 1987), making it unlikely that halothane inhibition of surface membrane Ca^{2+} extrusion is responsible for altered PSS muscle Ca^{2+} regulation. Thus, although many aspects of PSS and normal pig surface membrane Ca^{2+} regulation were normal, there was a significant difference between the two groups in the Ca^{2+} transporting activity. While this diminished Ca^{2+} efflux activity could conceivably contribute to altered Ca^{2+} regulation during PSS, it is unlikely to be the primary defect in this syndrome.

Another potentially important difference between purified PSS and normal skeletal muscle surface membranes was the observation that PSS t-tubules had a 30-40% decrease in the B_{max} for dihydropyridine (nitrendipine) binding as well as phenylalkylamine (D888) binding (Ervasti *et al.*, 1989a). The K_d of the PSS and normal dihydropyridine receptor for either type of ligand was similar however (Ervasti *et al.*, 1989a). Related to these observations, is a decrease in other activities associated with the dihydropyridine receptor, as both the voltage-gated slow inward Ca^{2+} current and the intramembrane charge movement were also reduced in PSS skeletal muscle fibers (Lamb *et al.*, 1989). Thus, several independent measurements indicate that the content or activity of the dihydropyridine receptor, so critical to the regulation of excitation-contraction coupling, is decreased in PSS skeletal muscle. Although there is yet no evidence to indicate that the function of this molecule is altered, in view of the alterations in the other key molecule that regulates SR Ca^{2+} release, namely the ryanodine receptor itself, a decreased content of the dihydropyridine receptor may have significant implications for the onset or maintenance of a PSS episode.

SR structure and function in PSS

There have been many studies of the Ca^{2+} transport activity of SR membranes isolated from PSS and normal muscle, as a diminished SR Ca^{2+}-ATPase or Ca^{2+} uptake activity could result in an altered Ca^{2+} regulation in PSS muscle. Important considerations in evaluating the many studies on this topic are, whether the muscles have been subjected to deleterious ante-mortem or post-mortem conditions of elevated temperature or low pH, and that care was taken to ensure that the SR membranes have been carefully purified and characterized by established procedures.

There has been considerable debate regarding the Ca^{2+} pump activity of PSS SR, with studies indicating higher, similar, or lower SR Ca^{2+} pumping activities in PSS muscle. When the considerations of membrane purity and intactness are taken into account, however, most researchers would now agree that there is no defect in the SR Ca^{2+}-ATPase or Ca^{2+} uptake capacity of PSS or MHS muscle (Greaser *et al.*, 1969; Gronert *et al.*, 1979; Mickelson *et al.*, 1986). At clinical concentrations of halothane, both PSS and normal SR Ca^{2+} pumping activities are actually enhanced; halothane concentrations that are more than ten times the clinical concentration are required to inhibit Ca^{2+}-ATPase activity (Gronert *et al.*, 1979; Louis *et al.*, 1992). Thus, an effect of clinical concentrations of halothane on the ability of PSS but not normal SR to transport Ca^{2+} is unlikely to explain the initiation of an episode.

SR membranes isolated from PSS and normal muscle do not differ in total lipid content or protein distribution (including the Ca^{2+}-binding protein calsequestrin) indicating no major difference in membrane structure between the two types of SR (Ervasti *et al.*, 1991; Mickelson *et al.*, 1986). As with surface membranes, it cannot be ruled out that there are small differences in the fatty acid content or the phospholipid distribution of PSS and normal SR membranes. However, most studies now find that the structural properties of PSS and normal SR membranes are the same. For the reasons outlined in the following section, we believe the SR Ca^{2+}-release channel/ryanodine receptor is the site of the primary defect responsible for PSS.

SR Ca^{2+}-release channel abnormalities in PSS

A number of studies have now documented that various aspects of the SR Ca^{2+}-release channel activity are abnormal in PSS and human MH. Depending on the type of preparation or technique utilized, PSS Ca^{2+}-release channel defects are observed as an increased rate of Ca^{2+} release induced by a variety of agents (e.g. Ca^{2+}, caffeine, ATP, halothane) (Kim *et al.*, 1984; Mickelson *et al.*, 1988; Mickelson *et al.*, 1989; Mickelson *et al.*, 1986), a lower luminal Ca^{2+} threshold for induction of Ca^{2+} release (Nelson, 1983; Ohnishi *et al.*, 1983), altered kinetics of single channel activity in planar lipid bilayers (Fill *et al.*, 1990), and altered kinetic properties associated with the binding of ryanodine (Ervasti *et al.*, 1991; Mickelson *et al.*, 1988).

In our own studies of the SR Ca^{2+} release channel, we have primarily utilized heavy SR vesicles purified on sucrose density gradients. These membranes were isolated from the *longissimus dorsi* muscle of homozygous PSS Pietrains, homozygous normal Yorkshires, and heterozygotes derived from several types of mating schemes coupled with the halothane challenge test. To examine Ca^{2+} release, the isolated SR vesicles were passively loaded with $^{45}Ca^{2+}$ and rapidly diluted into media containing activators or inhibitors of the Ca^{2+} channel. Subsequently, Ca^{2+} release is stopped by the addition of the channel inhibitors EGTA, ruthenium red and Mg^{2+}, the remaining $^{45}Ca^{2+}$ content of the SR vesicles measured after filtration, and the rate constant for Ca^{2+} release determined. The Ca^{2+} dependence of the rate constant for SR Ca^{2+} release follows a bell-shaped curve, with activation of Ca^{2+} release occurring in the µM Ca^{2+} range and inhibition of Ca^{2+} release occurring in the mM Ca^{2+} range (Fill *et al.*, 1990). The $Ca^{2+}_{0.5}$ for activation of Ca^{2+} release (0.3µM) and the optimal $[Ca^{2+}]$ for Ca^{2+} release (1-10µM) were similar for PSS and normal SR. However, at any Ca^{2+} concentration, the rate constant for Ca^{2+} release from the PSS SR was approximately twice that of the normal SR (Fill *et al.*, 1990).

Since ryanodine binds to the open state of the SR Ca^{2+}-release channel (Meissner, 1986), determination of the kinetics of [^{3}H]ryanodine binding to PSS and normal SR can provide information about the functional state of the channel. We have found differences in several kinetic parameters of ryanodine binding to PSS and normal SR. The Ca^{2+} dependence of [^{3}H]ryanodine binding was altered, with higher levels of Ca^{2+} being needed to inactivate ryanodine binding (Mickelson *et al.*, 1988). Furthermore, PSS SR had a higher affinity for ryanodine than did normal SR (K_d of 90 *versus* 250nM). The B_{max} for the [^{3}H]ryanodine receptor was similar for the PSS and normal SR (Mickelson *et al.*, 1988). Thus, the increased rate of Ca^{2+} release from the PSS SR was not due to a greater number of channels (Fill *et al.*, 1990).

Evidence for a defect in the PSS Ca^{2+} release channel also comes from studies demonstrating that PSS and MHS muscle have a higher sensitivity to contractures induced by caffeine and halothane. That the site of action of caffeine has been identified as the ryanodine receptor (via an increase in the

Ca^{2+} sensitivity of activation) (Rousseau *et al.*, 1988), and that clinical concentrations of halothane activate SR Ca^{2+} release through the ryanodine receptor (Beeler and Gable, 1985; Louis *et al.*, 1992), all could be explained by the PSS Ca^{2+} release mechanism allowing a greater rate of Ca^{2+} release or having a higher sensitivity to the effects of these reagents. Presumably, the channel-activating effects of these molecules and the very high conductance of this Ca^{2+} release channel are sufficient to raise the PSS muscle sarcoplasmic Ca^{2+} concentration *in vivo* to the contractile activation threshold, in spite of a concomitant activation of the SR Ca^{2+}ATPase, which would attempt to restore resting sarcoplasmic Ca^{2+} concentrations.

Interestingly, heavy SR isolated from a predominantly slow-twitch muscle (*vastus intermedius*) of PSS and normal pigs had the same differences in [^{3}H]ryanodine binding as did the fast-twitch muscle (*longissimus dorsi*) SR (Ervasti *et al.*, 1991). Cardiac muscle SR of PSS and normal pigs, however, displayed identical [^{3}H]ryanodine binding properties, albeit quite distinct from both types of skeletal muscle (Ervasti *et al.*, 1991). Thus, the abnormal PSS muscle Ca^{2+} release channel is expressed in all skeletal muscle fibers but not cardiac muscle, in agreement with the tissue specificity of these two genes derived from mRNA analysis (Otsu *et al.*, 1990). This also indicates that while cardiac muscle is not a site of the primary defect in PSS, all skeletal muscles should presumably be affected. Any apparent differences between muscles in the expression of the PSS defect are presumably due to other factors or other biochemical differences between the muscle fiber types.

An alteration in the PSS SR Ca^{2+} channel activity was also observed in planar lipid bilayers (Fill *et al.*, 1990). Although the open probabilities of the PSS and normal channels were activated by similar Ca^{2+} concentrations, Ca^{2+} concentrations that inactivated the normal channel were unable to inactivate the PSS channel; 10-fold higher Ca^{2+} concentrations were required to inactivate the PSS channels (Fill *et al.*, 1990). These studies thus agreed with the [^{3}H]ryanodine binding studies (Mickelson *et al.*, 1988), in that there was an alteration in the Ca^{2+}-regulated gating of the PSS Ca^{2+}-release channel. It is possible that this Ca^{2+}-dependent inhibition of Ca^{2+} release may serve *in situ* as a means to close the channel after the signal for Ca^{2+} release is completed. Thus, the PSS abnormality could explain not only the increased rates of Ca^{2+} release observed from the isolated PSS SR, but also an increased PSS muscle twitch force, which could be the result of a greater SR Ca^{2+} release (Gallant *et al.*, 1989; Mickelson *et al.*, 1988).

Heavy SR isolated from pigs having only one copy of the PSS gene (i.e., heterozygoytes) exhibits Ca^{2+} release and [^{3}H]ryanodine binding properties that are intermediate between those of the homozygous PSS and homozygous normal animals (Mickelson *et al.*, 1989), further indicating the likely identity of the PSS and the Ca^{2+}-release channel/ryanodine receptor genes. This proposal was subsequently supported by the localization of the ryanodine receptor gene to chromosome 6, within the PSS gene linkage group (Harbitz *et al.*, 1990). Comparison of the cDNA sequence of the ryanodine receptor of a Yorkshire and Pietrain pig by MacLennan and colleagues demonstrated a nucleotide base change (C1843 to T1843), that resulted in the mutation of Arg615 (normal) to Cys615 (PSS) (Fujii *et al.*, 1991). The mutation is highly correlated with PSS in all the major breeds tested to date, with a lod score of greater than 100 (Fujii *et al.*, 1991; Otsu *et al.*, 1991). This aspect of PSS molecular genetics will be covered in more depth by others at this meeting. Thus, a mutation within the skeletal muscle ryanodine receptor gene appears to be responsible for PSS. Although heterogeneity within the human MHS population is likely, linkage of the ryanodine receptor and MH genes on chromosome 19 has been demonstrated (MacLennan *et al.*, 1990), and the Arg615 ryanodine receptor mutation has been found in one family (Gillard *et al.*, 1991); whether there are other mutations in this gene in the MHS human population remains to be discovered.

The Arg615 to Cys615 mutation appears to result in structural as well as functional alterations in the PSS ryanodine receptor. Not only does the PSS ryanodine receptor allele cause the alterations in Ca^{2+} channel activity, it also results in an altered sensitivity of this protein to digestion by trypsin (Mickelson *et al.*, 1992). We have shown recently that the 86 and 99kD ryanodine receptor-derived polypeptides of normal and PSS SR, respectively, are actually derived from the region of the mutation site; the Arg615 to Cys615 mutation removes a trypsin cleavage site in the native molecule resulting in a larger tryptic fragment in the PSS ryanodine receptor (Mickelson *et al.*, 1992). It may be that amino acid residue 615 is in a key regulatory domain of the Ca^{2+} channel as a change of a single amino acid residue results in significant structural and functional alterations of this protein. We, and others, will continue to use this naturally occurring mutation in a channel protein critical to the regulation of skeletal muscle contraction to learn more about the basic physiology and biochemistry of this process.

Mitochondria and PSS

When carefully prepared, mitochondria isolated from PSS and normal pig skeletal muscle do not differ in respiratory capacity, coupling ratios or rate of Ca^{2+} uptake (Cheah and Cheah, 1976; Cheah and Cheah, 1981). However, the PSS mitochondria become swollen and uncoupled at a Ca^{2+} level lower than that of the normal mitochondria (Cheah and Cheah, 1976; Cheah and Cheah, 1981). Also, uncoupler- or anaerobic-induced Ca^{2+} release from the PSS mitochondria was greater than from the normal mitochondria (Cheah and Cheah, 1976; Cheah and Cheah, 1981). In light of the uncertainties concerning mitochondrial Ca^{2+} transport in skeletal muscle Ca^{2+} regulation *in vivo*, the significance of these results using very high Ca^{2+} loads and non-physiological conditions for stimulating Ca^{2+} release is not clear.

In these studies that have reported abnormalities in PSS mitochondria, it has been proposed that fatty acids released from the mitochondria by an endogenous phospholipase or lipase activity are the cause of the Ca^{2+}-induced mitochondrial uncoupling and damage (Cheah and Cheah, 1981). It has also been proposed that the liberated fatty acids, and in particular, polyunsaturated fatty acids, may induce the subsequent release of Ca^{2+} from the SR (Fletcher *et al.*, 1990). Lipid profiles of PSS and normal muscles are similar, however (although another study (Seewald *et al.*, 1991) found some differences), and with the exception of linoleic acid, free fatty acid levels in intact PSS and normal muscle were also not different (Fletcher *et al.*, 1988). Confirmation of these results in other laboratories, and further definition of the possible effects of fatty acids, in particular the polyunsaturated acids, on SR Ca^{2+} release, other membrane activities, and cell functions would clarify this possible defect in PSS muscle.

Some post-mortem implications of skeletal muscle Ca^{2+} regulation

The large Ca^{2+} gradients that exist between the skeletal muscle membrane systems and the sarcoplasm are the result of ATP- or energy-dependent Ca^{2+} transporters. Thus, in post-mortem muscle, it has been hypothesized that as ATP concentrations fall, Ca^{2+} concentrations in the sarcoplasm would rise; this could result from leakage of Ca^{2+} through the Ca^{2+} channels, lowered activity of the Ca^{2+} pumps, or general disruption of membrane structure due to phospholipase and protease activity. Other factors to consider when evaluating whether sarcoplasmic Ca^{2+} might be elevated in post-mortem muscle must include the effects of temperature, phosphate, and pH on the activities of both ATP-dependent Ca^{2+} pumps as well as Ca^{2+} channels. However, as we are not experts in this field, much of the following discussion should be considered as inspired guesses.

To the best of our knowledge, ionized sarcoplasmic Ca^{2+} levels in post-mortem muscle have not been determined by the current techniques of Ca^{2+} indicator dyes or intracellular microelectrodes, and any evidence for changes in sarcoplasmic Ca^{2+} post-mortem would be indirect. There appears to be little doubt, however, that just as in PSS *in vivo* (Lopez *et al.*, 1988), muscles developing into PSE meat post-mortem are subjected to quite high sarcoplasmic Ca^{2+} levels. This would further stimulate contracture, glycogen breakdown, the rate of pH fall and onset of muscle rigor. It also results in damage to the membrane systems responsible for attempting to control the Ca^{2+} concentration (Greaser *et al.*, 1969).

If sarcoplasmic Ca^{2+} levels are increased in normal porcine muscle early post-mortem, it would seem that they are not greatly increased, as this would be detected by muscle contracture and tension development. SR Ca^{2+} release *in vivo is* controlled by the t-tubule membrane potential (Rios and Brum, 1987), and it is not known if there is a significant rate of passive SR Ca^{2+} "leak" into the sarcoplasm independent of t-tubule control. We are unsure for how long the membrane potential of intact muscle post-mortem is maintained at its resting value of approximately -80 mV, or how long it stays above that required to initiate SR Ca^{2+} release. When the action of the (Na^+ + K^+)-ATPase becomes inhibited due to lack of ATP, or if the extracellular K^+ concentration rises, the surface membrane would become depolarized, potentially allowing SR Ca^{2+} release. However, prolonged depolarization is associated with an inactivation of the voltage sensor e-c coupling mechanism, so may not initiate a release of SR Ca^{2+}. In any event, the passive permeability of membranes to Ca^{2+} is very low, and it seems likely that the Ca^{2+}-ATPases of muscles undergoing normal rigor onset would be able to maintain Ca^{2+} gradients for quite some time, until the carcass temperature is quite low, ATP levels are low, and the pH has fallen close to its ultimate. The inability to reaccumulate Ca^{2+} and maintain the Ca^{2+} gradients may explain the shortening that ultimately accompanies rigor development.

For the SR Ca^{2+} pump, ATPase and Ca^{2+} transport activity are optimal at pH 6.8-7.4, however the pump is certainly able to transport Ca^{2+} at lower pH (Fassold and Hasselbach, 1986; Metzler and Berman, 1984). Phosphate, which rises post-mortem, would only serve to stimulate or maintain the storage of Ca^{2+} in the SR (Newbold and Tume, 1981). The rate of SR Ca^{2+} release decreases at lower pH (Louis *et al.*, 1992), and it possible that decreasing sarcoplasmic pH, acting to inhibit SR Ca^{2+} release, may be a factor in muscle fatigue. That PSS SR is not as sensitive as normal SR to this pH inhibition may explain the maintenance of a PSS episode even at low sarcoplasmic pH, when SR Ca^{2+} release should be inhibited (Louis *et al.*, 1992).

The phenomenon of cold-induced contractures of red skeletal (traditionally beef and lamb (Locker and Hagyard, 1963)) and cardiac muscles has gone unexplained for some time. A recent study has suggested that at reduced temperatures the SR Ca^{2+}-release channel actually passes an increased Ca^{2+} current (decreased conductance but increased open probability) (Sitsapesan *et al.*, 1991). Accompanied by a lower SR Ca^{2+} pump content in red muscles, and a decrease in the rate of SR Ca^{2+} pump activity at low temperatures, a cold-induced SR Ca^{2+} efflux could result in an increased sarcoplasmic Ca^{2+} concentration and contracture development. Interestingly, recent practices of very rapid chilling of pig half-carcasses is now observed to cause cold shortening and associated toughening (Feldhusen and Kuhne, 1992). Whether SR Ca^{2+} channel "leakage" is indeed responsible for the cold-shortening effect remains to be proven.

Discussion and future research needs

A variety of questions concerning skeletal muscle Ca^{2+} regulation in PSS and post-mortem muscle remain to be resolved. A few of them are listed below.

1. There is currently overwhelming physiological, biochemical, and molecular genetic evidence that the SR ryanodine receptor mutation is responsible for PSS. Although the initiation of a PSS syndrome by halothane, due to activation of this abnormal Ca^{2+} release mechanism is supported by experimental evidence, it is not yet clear how an episode can be initiated by the other stressors, such as transport, etc. Is it possible that the PSS ryanodine receptor is expressed in very low levels in other tissues, and could cause altered Ca^{2+} regulation in these tissues as well?

2. Is the abnormal ryanodine receptor gene itself responsible for the leanness and muscularity of the stress-susceptible animals, or is/are some other closely linked gene(s) responsible? Although it is possible that muscle hypertrophy results from continued "overstimulation", it is not clear how a gene that causes altered muscle Ca^{2+} regulation could also cause increased muscling, given that elevated sarcoplasmic Ca^{2+} levels are known to initiate muscle proteolysis.

3. Is the resting Ca^{2+} concentration elevated in PSS and MHS muscles? Studies using intracellular microelectrodes and Ca^{2+} indicator dyes do not agree on this issue. If PSS muscle resting Ca^{2+} is indeed elevated what are the expected physiological effects on muscle metabolism, energy utilization, etc.? Are the basal metabolic activities of the PSS and normal pigs different, and is a constant need for energy for muscle Ca^{2+} cycling causing leanness in PSS animals? It is usually considered, though, that the feed conversion of PSS animals is not low.

4. An alternative candidate gene for PSS/MH has been proposed as the hormone-sensitive lipase (Levitt *et al.*, 1990), an enzyme that is primarily responsible for fatty acid liberation from adipose tissue. As yet, however, there are no solid indications that there are alterations of this activity in PSS animals. Would β-adrenergic stimulation and induction of fatty acid release contribute to a PSS episode, or could an altered lipase gene cause leanness in the PSS pigs?

5. The role, if any, that fatty acids, in particular unsaturated fatty acids, may play in effecting either normal or PSS SR Ca^{2+} release is still not well defined. An altered PSS lipase activity has been hypothesized to result in deleterious effects on muscle Ca^{2+} regulation (Cheah and Cheah, 1981; Fletcher *et al.*, 1990). It is not yet clear if the phospholipases or lipases that may release fatty acids are active to any significant extent in muscle *in situ,* or if this process is indeed altered in PSS pigs. Is this the same activity as the hormone sensitive lipase?

6. What role, if any, does sarcoplasmic Ca^{2+} play in the early post-mortem period in non-PSE muscle? In normal pig musculature, it is not known whether there is any significant elevation in sarcoplasmic Ca^{2+} during this time period of ATP concentrations and pH close to their normal values. There is little evidence to suggest that the SR and surface membrane Ca^{2+} pumps are not maintaining the normal Ca^{2+} gradients in pre-rigor muscle. Studies indicating the possibility of membrane damage occurring in post-mortem muscle *in situ* (possibly affecting the Ca^{2+} regulation) are not convincing, in that it can not be ruled out that the membrane damage is occurring during the isolation procedure itself.

7. The contribution that non-PSS stress plays in the post-mortem muscle properties can now be examined in a controlled study. With the development of the molecular test for the ryanodine receptor mutation, it will be easy to remove any possible effects of one or two copies of the PSS gene from all the other stresses occurring at slaughter. What percentage of the carcasses will now be detected as having PSE traits? Although this has been performed previously, misdiagnoses by the halothane test are significant and may have affected the results of previous studies.

8. Much remains to be discovered about the basic mechanisms regulating skeletal muscle excitation-contraction coupling and sarcoplasmic Ca^{2+} concentrations. However, as more information is obtained about the Ca^{2+} pumps and channels that effect skeletal muscle Ca^{2+} regulation, researchers and the pork industry will need to evaluate such information for its potential implications on the performance of the live animal and the qualities of its carcass.

References

Adams, B.A., Tanabe, T., Mikami, A., Numa, S. and Beam, K.G. 1990. Intramembrane charge movement restored in dysgenic skeletal muscle by injection of dihydropyridine receptor cDNAs. Nature 346, 569-72.

Beeler, T. and Gable, K. 1985. Effect of halothane on Ca^{2+}-induced Ca^{2+} release from sarcoplasmic reticulum vesicles isolated from rat skeletal muscle. Biochim. Biophys. Acta 821, 142-52.

Block, B.A., Imagawa, T., Campbell, K.P. and Franzini-Armstrong, C. 1988. Structural evidence for direct interaction between the molecular components of the transverse tubule/sarcoplasmic reticulum junction in skeletal muscle. J. Cell. Biol. 107, 2587-600.

Brandl, C.J., Green, N.M., Korczak, B. and MacLennan, D.H. 1986. Two Ca^{2+} ATPase genes: homologies and mechanistic implications of deduced amino acid sequences. Cell 44, 597-607.

Brandt, N.R. Caswell, A.H. and Brunschwig, J.P. 1980. ATP-energized Ca^{2+} pump in isolated transverse tubules of skeletal muscle. J. Biol. Chem. 255, 6290-8.

Briggs, F.N., Poland, J.L. and Solaro, R.J. 1977. Relative capabilities of sarcoplasmic reticulum in fast and slow mammalian skeletal muscles. J. Physiol. 266, 587-94.

Campbell, K.P., Leung, A.T. and Sharp, A.H. 1988. The biochemistry and molecular biology of the dihydropyridine-sensitive calcium channel. TINS 11, 425-30.

Caswell, A.H., and Brandt, N.R. 1989. Does muscle activation occur by direct mechanical coupling of transverse tubules to sarcoplasmic reticulum? TIBS 14, 161-5.

Catterall, W.A. 1991. Functional subunit structure of voltage-gated calcium channels. Science 253, 1499-500.

Cheah, K.S. and Cheah, A.M. 1976. The trigger for PSE condition in stress susceptible pigs. J. Sci. Fd. Agric. 27, 1137-44.

Cheah, K.S. and Cheah, A.M. 1981. Mitochondrial calcium transport and calcium-activated phospholipase in porcine malignant hyperthermia. Biochim. Biophys. Acta 634, 70-84.

Donoso, P. and Hidalgo, C. 1989. Sodium-calcium exchange in transverse tubules isolated from frog skeletal muscle. Biochim. Biophys. Acta 978, 8-16.

Duncan, C.J. 1978. Role of intracellular calcium in promoting muscle damage: a strategy for controlling the dystrophic condition. Experientia 34, 1531-5.

Ervasti, J.M., Claessens, M.T., Mickelson, J.R. and Louis, C.F. 1989a. Altered transverse tubule dihydropyridine receptor binding in malignant hyperthermia. J. Biol. Chem. 264, 2711-17.

Ervasti, J.M., Mickelson, J.R. and Louis, C.F. 1989b. Transverse tubule calcium regulation in malignant hyperthermia. Arch. Biochem. Biophys. 269, 497-506.

Ervasti, J.M., Strand, M.A., Hanson, T.P., Mickelson, J.R. and Louis, C.F. 1991. Ryanodine receptor in different malignant hyperthermia-susceptible muscles. Am. J. Physiol. 260, C58-C66.

Fabiato, A. 1983. Calcium-induced release of calcium from the cardiac sarcoplasmic reticulum. Am. J. Physiol. 245, C1-C14.

Fassold, E. and Hasselbach, W. 1986. The dependence on internal pH of Ca^{2+} fluxes across sarcoplasmic reticulum vesicular membranes. Eur. J. Biochem. 154, 7-14.

Feldhusen, F. and Kuhne, M. 1992. Effects of ultrarapid chilling and ageing on length of sarcomeres, and tenderness of pork. Meat Sci. 32, 161-71.

Fergusson, D.G. and Franzini-Armstrong, C. 1988. The Ca^{2+} ATPase content of slow- and fast-twitch fibers of guinea pig. Muscle Nerve 11, 561-70.

Fill, M., Coronado, R., Mickelson, J.R., Vilven, J., Ma, J., Jacobson, B.A. and Louis, C.F. 1990. Abnormal ryanodine receptor channels in malignant hyperthermia. Biophys. J. 50, 471-75.

Fletcher, J.E., Rosenberg, H., Michaux, K., Cheah, K.S. and Cheah, A.M. 1988. Lipid analysis of skeletal muscle from pigs susceptible to malignant hyperthermia. Biochem. Cell Biol. 66, 917-21.

Fletcher, J.E., Tripolitis, L., Erwin, K., Hanson, S., Rosenberg, H., Conti, P.A. and Beech, J. 1990. Fatty acids modulate calcium-induced calcium release from skeletal muscle heavy sarcoplasmic reticulum fractions: implications for malignant hyperthermia. Biochem. Cell Biol. 68, 1195-201.

Franzini-Armstrong, C. 1970. Studies of the triad: structure of the junction in frog twitch fibers. J. Cell Biol. 47, 488-99.

Fujii, J., Otsu, K., Zorzato, F., DeLeon, S., Khanna, V.K., Weiler, J.E., O'Brien, P.J. and MacLennan, D.H. 1991. Identification of a mutation in porcine ryanodine receptor associated with malignant hyperthermia. Science 253, 448-51.

Gallant, E.M, Mickelson, J.R., Roggow, B.D., Donaldson, S.K., Louis, C.F. and Rempel, W.E. 1989. Halothane sensitivity gene and muscle contractile properties in malignant hyperthermia. Am. J. Physiol. 257, C781-C786.

Gillard, E.F., Otsu, K., Fujii, J., Khanna, V.K., DeLeon, S., Derdemezi, J., Britt, B.A., Duff, C.L., Worton, R.G. and MacLennan, D.H. 1991. A substitution of cysteine for arginine 614 in the ryanodine receptor is potentially causative of human malignant hyperthermia. Genomics 11, 751-55.

Greaser, M.L., Cassens, R.G., Briskey, E.J. and Hoekstra, W.G. 1969. Postmortem changes in subcellular fractions from normal and pale, soft, exudative porcine muscle. J. Food Sci. 34, 120-24.

Greeb, J. and Shull, G.E. 1989. Molecular cloning of a third isoform of the plasma membrane Ca^{2+} transporting ATPase that is expressed predominantly in brain and muscle. J. Biol. Chem. 264, 18569-76.

Gronert, G.A. 1986. Malignant hyperthermia. In: Engel, A.G. and Banker, B.Q. (eds), Myology, McGraw-Hill, pp. 1763-84.

Gronert, G.A., Heffron, J.J.A. and Taylor, S.R. 1979. Skeletal muscle sarcoplasmic reticulum in porcine malignant hyperthermia. Eur. J. Pharmacol. 58, 179-87.

Gunteski-Hamblin, A.M., Greeb, J. and Shull, G. 1988. A novel Ca^{2+} pump expressed in brain, kidney, and stomach is encoded by an alternative transcript of the slow-twitch muscle sarcoplasmic reticulum Ca-ATPase gene. J. Biol. Chem. 263, 15032-40.

Harbitz, I., Chowdhary, B., Thomsen, P.D., Davies, W., Kaufman, U., Kran, S., Gustavsson, I., Christensen, K. and Hauge, J.G. 1990. Assignment of the porcine calcium-release channel gene, a candidate for the malignant hyperthermia locus, to the 6p11 q21 segment of chromosome 6. Genomics 8, 243-8.

Hidalgo, C., Gonzalez, M.E. and Garcia, A.M. 1986. Calcium transport in transverse tubules isolated from rabbit skeletal muscle. Biochim. Biophys. Acta 854, 279-86.

Iaizzo, P.A., Seewald, M., Oakes, S.G. and Lehmann-Horn, F. 1989. The use of fura-2 to estimate myoplasmic [Ca^{2+}] in human skeletal muscle. Cell Calcium 10, 151-8.

Inesi, G. 1985. Mechanism of calcium transport. Ann. Rev. Physiol. 47, 573-601.

Inui, M., Saito, A. and Fleischer, S. 1987. Purification of the ryanodine receptor and identity with feet structures of junctional terminal cisternae of sarcoplasmic reticulum from fast skeletal muscle. J. Biol. Chem. 262, 1740-7.

Jorgensen, A.O., Shen, A.C.Y., Campbell, K.P. and MacLennan, D.H. 1983. Ultrastructural localization of calsequestrin in rat skeletal muscle by immunoferritin labeling of ultrathin frozen sections. J. Cell Biol. 97, 1573-81.

Kawamoto, R.M., Brunschwig, J.P., Kim, K.C. and Caswell, A.H. 1986. Isolation, characterization, and localization of the spanning protein from skeletal muscle triads. J. Cell. Biol. 103, 1405-14.

Kim, D.H., Sreter, F.A., Ohnishi, S.T., Ryan, J.F, Roberts, J., Allen, P.D., Meszaros, L.G., Antoniu, B. and Ikemoto, N. 1984. Kinetic studies of Ca^{2+} release from sarcoplasmic reticulum of normal and malignant hyperthermia susceptible pig muscles. Biochim. Biophys. Acta 775, 320-7.

Lai, F.A., Erickson, H.P., Rousseau, E., Liu, Q.Y. and Meissner, G. 1988. Purification and reconstitution of the calcium release channel from skeletal muscle. Nature 331, 315-19.

Lamb, G.D., Hopkinson, K.C. and Denborough, M.A. 1989. Calcium currents and asymmetric charge movement in malignant hyperthermia. Muscle Nerve 12, 135-40.

Langer, G.A. 1992. Calcium and the heart: exchange at the tissue, cell, and organelle levels. FASEB J. 6, 893-902.

Lehninger, A.L. 1970. Mitochondria and calcium ion transport. Biochem. J. 119, 129-38.

Levitt, R.C., McKusick, V.A., Fletcher, J.E. and Rosenberg, H. 1990. Gene candidate. Nature 345, 297-98.

Locker, R.H., and Hagyard, C.J. 1963. A cold shortening effect in beef muscle. J. Sci. Fd. Agric. 14, 787-93.

Lopez, J.R., Allen, P.D., Alamo, L., Jones, D. and Sreter, I.A. 1988. Myoplasmic free $[Ca^{2+}]$ during a malignant hyperthermia episode in swine. Muscle Nerve 11, 82-88.

Louis, C.F., Zualkernan, K., Roghair, T. and Mickelson, J.R. 1992. The effects of volatile anesthetics on calcium regulation by malignant hyperthermia susceptible sarcoplasmic reticulum. Anaesthesiol., in press.

MacLennan, D.H., Duff, C., Zorzato, F., Fujii, J., Phillips, M., Korneluk, R.G., Frodis, W., Britt, B.A. and Worton, R.G. 1990. Ryanodine receptor gene is a candidate for predisposition to malignant hyperthermia. Nature 343, 559-61.

Maruyama, K., Clarke, D.M., Fujii, J., Loo, T.W. and MacLennan, D.H. 1989. Expression and mutation of Ca^{2+} ATPases of the sarcoplasmic reticulum. Cell Motility Cytoskeleton 14, 26-34.

McCormack, J.G. and Denton, R.M. 1981. On the role of the calcium transport cycle in heart and other mammalian mitochondria. Biochem. J. 194, 639-43.

Meissner, G. 1986. Ryanodine activation and inhibition of the Ca^{2+} release channel of sarcoplasmic reticulum. J. Biol. Chem. 261, 6300-6.

Meissner, G., Darling, E. and Eveleth, J. 1986. Kinetics of rapid Ca^{2+} release by sarcoplasmic reticulum; Effects of Ca^{2+}, Mg^{2+}, and adenine nucleotides. Biochem. 25, 236-44.

Meissner, G., Rousseau, E., Lai, F.A., Liu, Q.Y. and Anderson, K.A. 1988. Biochemical characterization of the Ca^{2+} release channel of skeletal and cardiac sarcoplasmic reticulum. Mol. Cell. Biochem. 82, 59-65.

Meltzer, S. and Berman, M.C. 1984. Effects of pH, temperature, and calcium concentration on the stoichiometry of the calcium pump of sarcoplasmic reticulum. J. Biol. Chem. 259, 4244-53.

Mickelson, J.R. and Marsh, B.B. 1980. Calcium uptake and release by skeletal muscle mitochondria. Cell Calcium 1, 119-28.

Mickelson, J.R., Ross, J.A., Reed, B.K. and Louis, C.F. 1986. Enhanced Ca^{2+}-induced calcium release by isolated sarcoplasmic reticulum vesicles from malignant hyperthermia susceptible pig muscle. Biochim. Biophys. Acta 862, 318-28.

Mickelson, J.R., Ross, J.A., Hyslop, R.J., Gallant, E.M. and Louis, C.F. 1987. Skeletal muscle sarcolemma in malignant hyperthermia: evidence for a defect in calcium regulation. Biochim. Biophys. Acta 897, 364-76.

Mickelson, J.R., Gallant, E.M., Litterer, L.A., Johnson, K.M., Rempel, W.E. and Louis, C.F. 1988. Abnormal sarcoplasmic reticulum ryanodine receptor in malignant hyperthermia. J. Biol. Chem. 263, 9310-15.

Mickelson, J.R., Gallant, E.M., Rempel, W.E., Johnson, K.M., Litterer, L.A., Jacobson, B.A. and Louis, C.F. 1989. Effects of the halothane sensitivity gene on sarcoplasmic reticulum function. Am. J. Physiol. 257, C787-C794.

Mickelson, J.R., Knudson, C.M., Kennedy, C.F.H., Yang, D.I., Litterer, L.A., Rempel, W.E., Campbell, K.P. and Louis, C.F. 1992. Structural and functional correlates of a mutation in the malignant hyperthermia susceptible pig ryanodine receptor. FEBS Letters 301, 49-52.

Nelson, T.E. 1983. Abnormality in calcium release from skeletal sarcoplasmic reticulum of pigs susceptible to malignant hyperthermia J. Clin. Invest. 72, 862-70.

Newbold, R.P. and Tume, R.K. 1981. Comparison of the effects of added orthophosphate on calcium uptake and release by bovine and rabbit muscle sarcoplasmic reticulum. J. Food Sci. 46, 1327-32.

Nicoll, D.A., Longoni, S. and Philipson, K.D. 1990. Molecular cloning and functional expression of the cardiac sarcolemmal Na^{+}-Ca^{2+} exchanger. Science 250, 562-65.

Ohnishi, S.T., Taylor, S. and Gronert, G.A. 1983. Calcium-induced Ca^{2+} release from sarcoplasmic reticulum of pigs susceptible to malignant hyperthermia. FEBS Letters 161, 103-7.

Otsu, K., Willard, H.F., Khanna, V.K., Zorzato, F., Green, N.M. and MacLennan, D.H. 1990. Molecular cloning of cDNA encoding the Ca^{2+} release channel of rabbit cardiac muscle sarcoplasmic reticulum. J. Biol. Chem. 265, 13472-83.

Otsu, K., Khanna, V.K., Archibald, A.L. and MacLennan, D.H. 1991. Cosegregation of porcine malignant hyperthermia and a probable causal mutation in the skeletal muscle ryanodine receptor gene in backcross families. Genomics 11, 744-50.

Rios, E. and Brum, G. 1987. Involvement of dihydropyridine receptors in excitation-contraction coupling in skeletal muscle. Nature 325, 717-20.

Rousseau, E., LaDine, J., Liu, Q.Y. and Meissner, G. 1988. Activation of the Ca^{2+} release channel of skeletal muscle sarcoplasmic reticulum by caffeine and related compounds. Arch. Biochem. Biophys. 267, 75-86.

Sanchez, J. and Stefani, E. 1978. Inward calcium current in twitch muscle fibers of the frog. J. Physiol. 283, 197-209.

Seewald, M.J., Eichinger, H.M. and Iaizzo, P.A. 1991. Malignant hyperthermia: an altered phospholipid and fatty acid composition in muscle membranes. Acta Anaesthesiol. Scand. 35, 380-86.

Sitsapesan, R., Montgomery, R.A.P., MacLeod, K.T. and Williams, A.J. 1991. Sheep cardiac sarcoplasmic reticulum calcium-release channels: modification of conductance and gating by temperature. J. Physiol. 434, 469-88.

Smith, J.S., Coronado, R. and Meissner, G. 1985. Sarcoplasmic reticulum contains adenine nucleotide-activated calcium channels. Nature 316, 446-9.

Somlyo, A.P. and Somlyo, A.V. 1986. Electron probe analysis of calcium content and movements in sarcoplasmic reticulum, endoplasmic reticulum, mitochondria, and cytoplasm. J. Cardiovasc. Pharmacol. 8, S42-S47.

Takeshima, H., Nishimura, S., Matsumoto, T., Ishida, H., Kangawa, K., Minamino, N., Matsuo, H., Ueda, M., Hanaoka, M., Hirose, T. and Numa, S. 1989. Primary structure and expression from complementary DNA of skeletal muscle ryanodine receptor. Nature 339, 439-45.

Tanabe, T., Takeshima, H., Mikami, A., Flockerzi, V., Takahashi, H., Kangawa, K., Kojima, M., Matsuo, H., Hirose, T. and Numa, S. 1987. Primary structure of the receptor for calcium-channel blockers from skeletal muscle. Nature 328, 313-18.

Tanabe, T., Beam, K.G., Powell, J.A. and Numa, S. 1988. Restoration of excitation-contraction coupling and slow calcium current in dysgenic muscle by dihydropyridine receptor cDNA. Nature 336, 134-9.

Valdivia, C., Valdivia, H.H., Potter, B.V.L. and Coronado, R. 1990. Ca^{2+} release by inositol-trisphosphorothioate in isolated triads of rabbit skeletal muscle. Biophys. J. 57, 1233-43.

Wang, K.K.W., Villalobo, A. and Roufogalis, B.D. 1992. The plasma membrane calcium pump: a multiregulated transporter. Trends Cell Biol. 2, 46-52.

Discussion

*Lundström - Do the heterozygotes have partially normal calcium-release channels, and partially abnormal channels?

*Mickelson - That is the question that we would like to answer. Since this calcium-release channel is a tetramer, and it appears that in heterozygotes both of the alleles of the ryanodine receptor are being translated, so, you could end up with a tetramer consisting of all possible combinations of both types of ryanodine receptor. All four subunits being the normal type, all four subunits being the abnormal type, but also having mixtures, one and three, two and two, three and one. That could possibly be resolved at the single-channel level, but we haven't been able to do it yet, by ryanodine binding. All we know is that the grand average of these heterozygotes is intermediate between the two types of homozygotes.

*Puolanne - Would you expect differences within a breed, or between breeds in normal pigs?

*Mickelson - We've looked at several types of breeds of normal pigs now, the Yorkshires are the ones we've used at Minnesota, and also a very fatty line that's peculiar to Minnesota, called the Minnesota Number 1. Both of those types had the same ryanodine binding activities. Now we've been able to generate a line of Pietrain pigs that is homozygous for the normal allele of the ryanodine receptor, and that has the same ryanodine receptor activity as the other types of normal pigs. So, all normal genotypes we've examined have the same property, which is distinct from the PSS gene properties.

*Vögeli - I have a question about the calcium channels in humans and swine. What are the differences? Are there different types MH calcium-release channels in humans that are not detected in swine? I don't understand that, from the literature.

*Mickelson - The biochemical and electrophysiological defects observed in the MH human calcium-release channels are different from the defect we see in the porcine calcium-release channels. MH human channels don't have that abnormal calcium regulation that we see with the PSS channel. But, by the same token, the MHS human channels are more sensitive to caffeine. So, there are minor differences in the nature of the defect between human and porcine ryanodine receptors. Now, at the molecular level, the arginine 615 mutation has only been seen in about 4% of the MHS human families tested. That doesn't rule out that there aren't other mutations present in the ryanodine receptor in MH humans.

*Lister - First, can I say what a pleasure it was to hear such a complicated story expressed so well. I am delighted that you have also come to the same conclusions and accepted the puzzles that have plagued the rest of us for so long. It seems that until something happens in PSS muscle, it is perfectly normal in every respect. You cannot detect any differences in metabolism, or biochemistry, or whatever. And I am also delighted to discover that we still need to know what the switch mechanism is.

*Mickelson - I think there's a lot of data out there that was generated years ago that has not been explained. And until we explain that data, we won't have a complete picture of what's going on with PSS, or MH, or whatever. This data was gathered by competent researchers, and just because we couldn't explain it 10 years ago doesn't mean we should forget about it.

Cell Membrane Fluidity and Pig Meat Quality

E. Rock

Institut National de Recherches Agronomiques; Station de Recherches sur la Viande, 63122 Ceyrat, France

Introduction

Cell membrane fluidity is a parameter reflecting many simultaneous interactions between different membrane components. The fact that membrane components can move, more or less freely, because of their weak electrostatic associations, is assumed to be part of this definition (Shinitzky, 1984; McMurchie, 1988). These characteristics are the basis of the fluid mosaic model proposed by Singer and Nicolson (1972).

Cell membrane fluidity reflects a general physical state of cell membranes, influenced or influencing, through the membrane functions, the physiological state of the cell. Cell membrane fluidity is largely studied in living cells, and a sum of knowledge has been acquired on some of the many possible molecular mechanisms underlying this process. Some of these understandings result from studies of cells in particular physiological states (high or low temperature, ageing), or from studying diseased cells (atherosclerosis). However, little is known about the post-mortem changes in cell membrane fluidity. The results developed later in this paper constitute preliminary approaches to that subject. This report will therefore deal with post-mortem changes in lipid fluidity of sarcoplasmic reticulum membranes isolated from normal, and malignant hyperthermia, or PSE-prone, pigs. Before developing this particular point, general features of cell membranes, and the effect of some well-known modulators of cell membrane fluidity will be reviewed.

Cell membranes

Biological membranes are structural components that individualize the cell on one hand, and, on the other hand, compartmentalize its internal space into organelles such as the nucleus, the mitochondria and the endoplasmic reticulum. One of the particular features of biological membranes is their organization, which is common to almost all living cells. They are built as sheet-like structures, 60 to 100 Å thick, consisting mainly of lipids and proteins. The architecture of bilayer structure in cell membranes results from hydrophobic and hydrophilic interactions between the diverse amphipathic molecules. The membrane lipid components can be divided into phospholipids (PL) and cholesterol. Other lipids that must be cited, although they are not structural components, are the free fatty acids, the triacylglycerols and the cholesterol esters. These molecules contribute to change the bulk lipid fluidity of the membranes.

The phospholipids

The phospholipids are a group of molecules characterized schematically by a polar head group linked to two fatty acid chains. The polar heads consist of different phosphorylated alcohols which define the different phospholipids, namely, phosphatidylcholine (PC), phosphatidylethanolamine (PE), phosphatidylserine (PS), and phosphatidylinositol (PI). Two other structurally different phospholipids are: 1) diphosphatidylglycerol (DPG), also called cardiolipin (CL), which is mainly found in mitochondrial membranes; and is characterized by the presence of four fatty acid chains instead of two; and 2) sphingomyeline (SpH) whose backbone is a ceramide (sphingosine) instead of a glycerol.

Some of the phospholipids are primarily distributed in the inner membrane leaflet (e.g. PC), while others are mainly found in the outer one (e.g. PE), reflecting one aspect of the asymmetry of biological membranes. The role of the phospholipids in the fluidity level of membranes is relatively important, as they have different phase transition temperatures, which can be, for instance, from 20 to 30°C lower in PC as compared with PE molecules, both containing the same fatty acid chains. However, because of the great variety of phospholipids, and in their fatty acyl composition, natural membranes are always in a liquid crystalline phase. One can also notice that some phospholipids are rigidifiers (e.g. PC) whereas some others are fluidizers (e.g. PE, SpH) and the ratios of PE:PC or SpH:PC can often be used as indicators of membrane fluidity state. Their contributions to influence, or change, membrane fluidity are limited because of their synthesis rate. In some situations, it was shown that membrane fluidity can be modulated by changes in the composition of phospholipid head groups. That is the case of successive methylation of PE, which is thus converted to PC. This methylation process, catalysed by PL-methyl transferases, increases the overall fluidity of cell membranes. Besides chemical-based classification, these different head groups also determine many properties of the membrane through their size, and their charged elements.

The fatty acids

Two positions on the glycerol backbone of a PL are occupied by the fatty acids. The sn-1 position is most often fitted with saturated fatty acids whereas unsaturated fatty acids are often detected in the sn-2 position. These fatty acid chains, linked to polar heads of the phospholipids form the hydrophobic core of the bilayer. Most of these fatty acids are synthesized from palmitic acid (C16:0), which can be elongated and/or unsaturated before the incorporation into the structural PL. The chain length varies from 14 to 22 carbons and it may have 0 to 6 double bonds. The introduction of double bonds tends to considerably decrease the viscosity, or to increase the fluidity. Thus, the degree of unsaturation is considered to be a potential modulator of cell membrane fluidity. Changes in this parameter may occur either during *de novo* synthesis, or via the dietary fatty acids. It has also been shown that the acyl chains of intact PL can be desaturated *in situ*. Another pathway for changing the composition of fatty acid chains implies the action of PLases A1 and A2, which hydrolyse the fatty acid chain in sn-1 and sn-2 positions respectively; the lyso-phospholipids can then be re-acylated. The reacylation process requires fatty acid-Coenzyme A, which is formed with fatty acids derived from *de novo* synthesis or from dietary sources.

Cholesterol

This particular molecule is found in all mammalian cells and is mainly localized in plasma membranes. The cholesterol content differs not only between the different types of membranes in a given cell, but also in cells from different tissues. Its interaction with phospholipid components may modulate membrane functions, such as permeability, or even membrane enzyme activity. However, these interactions are rather complex; it was thus suggested that, within sarcoplasmic reticulum membranes, cholesterol may be excluded from regions surrounding the Ca-ATPase proteins. Even so, it is possible that cholesterol may still affect protein function by interacting with their boundary phospholipids.

With regard to fluidity, it was shown that cholesterol either fluidizes or rigidifies pure phospholipid liposomes depending on whether the temperature is above or below the phase transition temperature of the phospholipid. Under physiological condition, cholesterol decreases the fluidity state of the membranes. The cholesterol to phospholipid ratio (Chol/PL) can thus be used as indicator of cell membrane fluidity. The very high variation of this ratio within different type of cells can be related to the fact that all membranes do not need the same level of fluidity for their optimal function, and in this field, cholesterol molecules can be replaced by other factors such as the length or the degree of unsaturation of fatty acyl chains.

The cholesterol is provided to cell membranes either by *de novo* synthesis or by lipoproteins. It is symmetrically distributed in the two halves of the bilayer. The metabolism of this molecule, through the esterification and de-esterification processes, provides an efficient way to maintain a constant Chol/PL ratio, which in turn maintains a relatively constant level of membrane fluidity.

The proteins

Protein content in cell membranes may vary from 40 to 50% of weight. They are classified as peripheral or integral ones with respect to their position in the membrane; some of them are loosely bound and others are completely embedded into the bilayer. Except for proteins anchored to the cytoskeleton, these molecules diffuse in the membrane. These motions are sometimes necessary for their functions. In that context, it is often described that, in a membrane, the lipids provide a fluid matrix and impermeable barrier for protein functions which are responsible for such specific properties as signal transduction, or membrane transport. Aside from this current concept, it was also shown that interactions occurred between the proteins and the lipids. In terms of fluidity modulation, proteins behave somewhat like cholesterol; in natural membrane, they decrease the bulk fluidity of the membrane. This effect is very strong on their boundary layer of lipids, also called the "lipid annulus", which is highly immobile. This particular interaction with the lipid annulus is essential for the functional properties of membrane proteins. On the other hand, experimental data also suggest that the overall physical state of membrane may have consequences on protein function, however, no general rules have been defined. It is likely that a combination of changes in the conformation state, translational diffusion, rotational motion and the position of the protein within the bilayer are involved. The position of membrane-bound proteins in the bilayer, especially the integral proteins, results from an equilibrium of hydrophilic and hydrophobic forces which tend to respectively extrude from, or include these proteins in the membrane. This steady-state position can be modified by changes in the physical properties of the surrounding lipids as Borochov and Shinitzky suggest in the "vertical displacement hypothesis" (Borochov and Shinitzky, 1976). This hypothesis agrees with some experimental observations. For instance, it was shown that upon an artificial increase or decrease of membrane fluidity, the membrane-bound phospholipase A2 uses either exogenous, or endogenous lipids, as substrates, respectively. This suggests that, depending on the fluidity state, the enzyme may be more or less exposed to the outer face of the membrane.

Measurements and significance of the fluidity parameter

The fluidity or microviscosity of a cell membrane can be determined using spectroscopic techniques such as nuclear magnetic resonance (NMR), electron spin resonance (ESR) and fluorescence spectroscopy. In contrast to the NMR technique, ESR and fluorescence spectroscopy require the use of probes to determine the motional characteristics of the membrane lipids. The most common probes are nitroxyde radicals and diphenyl hexatriene (DPH) for ESR and fluorescence spectroscopy respectively. Inherent problems in using spin or fluorescent probes are that: 1) they may perturb the system; 2) their precise location is not well defined; and 3) they only report an average data. However, important information can be obtained, at least for relative studies. For instance, the existence of increased fluidity gradient from the polar heads to the hydrophobic core of the bilayer, first demonstrated with the ESR technique, was then confirmed by non-perturbing NMR spectroscopy.

The significance and, more precisely, the interpretation of experimental data is rather complex in relation to the complexity of the system, i.e. a cell membrane. Schematically, one could say that these techniques describe the ease of movements of the reporter molecule in the membrane. This takes into account the structural or thermodynamic parameter (orientation of the lipid molecules relative to the membrane plane) and the dynamic parameter. The latter parameter is actually a complex combination of the lateral diffusion of the phospholipids, their rapid wobbling motions and the segmental motion of the fatty acyl chains. In commonly used steady state fluorescence anisotropy, it is admitted that for fluorescence anisotropy values below 0.200 the dynamics parameter contributes predominantly, whereas structural parameters are taken into account above this value (Pottel *et al.*, 1983).

Sarcoplasmic reticulum membrane fluidity of pig skeletal muscle

Introduction

Variations of membrane fluidity are often related to changes in membrane functions that occurred physiologically in living cells (Shinitzky, 1984). Cell death also induces profound modifications in the cell metabolism. In muscle tissue, the post-mortem metabolism has an important incidence on parameters defining the meat quality (Bendall, 1960). In pigs, prominent problems of meat quality are associated with the development of pale, soft, and exudative (PSE) meat. The PSE condition actually results from a combination of a rapid reduction of intramuscular pH, and an elevation of muscle temperature, which in turn solubilizes the sarcoplasmic proteins and decreases the water-holding capacity (Bendall and Wismer-Pederson, 1962). Another characteristic is that animals producing PSE meat show an abnormal sensitivity towards different stresses. Moreover, a close relationship was shown between the animals developing malignant hyperthermia (MH) syndrome and their susceptibility to develop PSE meat. The MH syndrome, which has also been described in humans, is mainly characterized by an increase in body temperature associated with skeletal muscle contracture when the individuals are anesthetized with halothane, a fluorinated anesthetic (Eikelenboom and Minkema, 1974). This characteristic has been used to detect susceptible pigs and to help to provide evidence for the inheritance of this myopathy. Basic studies undertaken on the MH syndrome have revealed defects in the fluxes of calcium ions and associated abnormalities at the level of membranes that regulate myoplasmic calcium homeostasis (Gronert *et al.*,1988). Research has mostly been focalized on sarcoplasmic reticulum (SR) membranes because of the qualitative and quantitative aspects of these membranes. This research led to the demonstration of a functional modification in the calcium-release process (Ohnishi *et al.*, 1983). This abnormality has been associated with a modification at the structural level of the calcium channel (the ryanodine receptor) (Mickelson *et al.*, 1992), probably resulting from the mutation discovered in the associated RYR gene (Fujii *et al.*, 1991). These aspects on functional properties of SR membrane proteins will be developed elsewhere by other lecturers.

The development of our work on the lipidi portions of SR membranes resulted from observations and experimental data dealing with the "general membrane defect" hypothesis underlying the MH syndrome. It has been shown that functions and/or structural defects may occur not only in the SR membranes of MH muscle (Rock *et al.*, 1990; Ohnishi *et al.*, 1986), but also in sarcolemmal (Rock and Kozak-Reiss, 1987; Mickelson *et al.*, 1987), T-tubules (Ervasti *et al.*, 1989), and mitochondrial membranes (Cheah and Cheah, 1981). Other studies also suggest that cell membranes of blood (Thatte *et al.*, 1987; Fink *et al.*, 1992), and liver (Iaizzo *et al.*, 1991) tissues can be affected in MH individuals. Lastly, recent works have provided evidence of defects in muscular lipid metabolism (Fletcher *et al.*, 1989). These authors proposed that a mutation in the gene of hormone dependant lipase (LIPE) could be at the origin of these defects. This gene is localized in the same chromosome and near the RYR gene cited above (Levitt *et al.*, 1990).

Together, these recent developments on the eventual molecular mechanism(s) of MH syndrome cannot necessarily be associated with the ones implied in the transformation of muscle into PSE meat. Within this framework of knowledge, more particularly on SR membranes, we have projected to investigate possible relationships between MH and PSE syndrome. The preliminary results developed below show that: i) a change of lipid composition occurs post-mortem in SR membranes; ii) cholesterol seems to play an important role in these changes; and iii) these changes can affect important functions of SR membranes, particularly those membranes isolated from individuals showing both the capacity to react positively to the halothane test and to produce PSE meat.

Results

Post-mortem evolution of SR membrane fluidity

Lipid fluidity was determined with the fluorescence spectroscopy technique, using diphenyl hexatriene as a probe, according to procedures described by Shinitzky and Inbar (1976). Many observations in our laboratory indicated that lipid fluidity decreased post-mortem in SR membranes isolated from *M. longissimus dorsi* muscle. Together, and independently of breed or nutritional factors, a significant ($P<0.05$) decrease of 0.003 to 0.016 DPH fluorescence anisotropy units (r) occurred from early (7 to 10 min) to 45 min post-mortem. These measurements were performed on muscle excized out of the carcass and kept in a cold room. Kinetic measurements showed a mean decrease of 0.012 ± 0.002 r units/h in normal muscle. This phenomenon can be slowed down by freezing and keeping the excised muscle in liquid nitrogen for at least 4h. However, storage of muscle at -80°C for 15 days induces a very high increase in the rigidity state of isolated SR membranes (0.160 ± 0.006 *versus* 0.135 ± 0.003; $n=3$). In that case, no variation of SR membrane fluidity was detected between those samples prepared from muscle frozen at 7 or 30 min post-mortem.

Another interesting point is the level of fluidity reached by SR membranes isolated from halothane-positive pigs. Table 1 shows that MH-susceptible SR membranes have a lower fluidity state than normal ones after 4h post-mortem time. In that experiment, all MHS pigs from the Pietrain breed developed PSE meat, as the pH values were below 5.8.

Table 1. Post-mortem evolution of SR membrane fluidity state. DPH fluorescence anisotropy values were determined in membranes isolated from 3 control (MHN) and 3 Malignant Hyperthermia susceptible (MHS) pigs. *Longissimus Dorsi* muscle was excised from the carcass and kept at 10°C before using it for membrane isolation

PIGS	MHN		MHS	
	15 min	4 h	15 min	4 h
#1	0.158	0.159	0.157	0.170
#2	0.157	0.161	0.159	0.169
#3	0.157	0.167	0.155	0.171
Mean ± S.D.	0.157±0.000	0.162±0.003	0.157± 0.001	0.170±0.001

Lipid analysis of SR membranes

Similar results were observed in another set of experiments using five positive and five negative cross-bred pigs (Pen Ar Lan breed). Analysis was performed at the lipid level to better define the low level of fluidity detected in MHS membranes. The results shown in Table 2 indicate that the main differences reside in the cholesterol to phospholipid (Chol/PL) and in the phosphatidylethanolamine to phosphatidylcholine (PE/PC) ratios. No significant differences were observed either in total, or in the other class of phospholipid content, or in the degree of unsaturation of their fatty acid chains. However, a significantly higher level of palmitate and arachidonate were detected in MHS membranes (data not shown). The higher value in PE/PC ratio in MHS membranes actually resulted from a significant decrease and increase in their PC and PE contents respectively. This observation led us to analyse the PL-N-methylation process, which is one of the two metabolic pathways of PC synthesis (Pelech and Vance, 1984). Using radioactively labelled S-adenosyl methionine (SAM), we measured its incorporation in the different PL according to procedures described by Kuncl *et al.* (1985). The results depicted in Figure 1 show a significant decrease of incorporation into monomethylphosphatidylethanol-amine (MMPE) and an associated decrease into PC.

Table 2. Biochemical analysis of SR membranes isolated from *L. dorsi* muscle of five Normal (MHN) and five MH-susceptible (MHS) pigs, 30 min post-mortem. DPH fluorescence anisotropy and cholesterol content were determined using native vesicles. The different classes of phospholipids and their fatty acid chain content were measured after extraction of total lipids. From these data, we calculated phosphatidylethanolamine: phosphatidylcholine ratio, and the saturated to unsaturated ratio of the fatty acids

Types	Fluorescence anisotropy	Chol/PL	Degree of unsaturation	PE/PC
MHN	0.152 ± 0.005	0.148 ± 0.030	1.98 ± 0.09	0.40 ± 0.04
	**	**	n.s.	**
MHS	0.166 ± 0.008	0.231 ± 0.020	1.79 ± 0.10	0.53 ± 0.04

No significant difference was observed in the level of SAM incorporation at either dimethylphosphatidylethanolamine (DMPE), or the neutral lipids level. These results strongly suggest that this enzymatic process is affected in MHS membranes. Moreover, the decreased level of MMPE products, whose substrate is PE, could also account for the accumulation of the latter phospholipid in MHS membranes.

PE and PC molecules from normal and MHS membranes were purified using the HPLC technique and their fatty acid composition was determined after transmethylation and separation by gas chromatography. Table 3 shows that, in both phospholipid molecules isolated from MHS membranes, there is generally an increase in unsaturated fatty acids, with a concomitant decrease in saturated ones. This leads to a lower value of the PE/PC ratio. In MHS membranes, contrarily to PC molecules, PE molecules show a significantly lower amount of palmitic acid and a higher content of arachidonic acid, which contributes to the larger decrease of the unsaturation degree observed in these phospholipids (Table 3).

Table 3. Degree of unsaturation of fatty acyl chains of phoshatidylcholine (PC) and phoshatidyletanolamine (PE) of normal and MH-susceptible SR membranes. The same membranes described in Table 2 were used for this study

	Types	Saturated	Unsaturated	Ratio
PC	MHN	34.1	63.4	0.54
	MHS	32.8	65.4	0.50
PE	MHN	24.9	62.2	0.40
	MHS	20.7	64.1	0.32

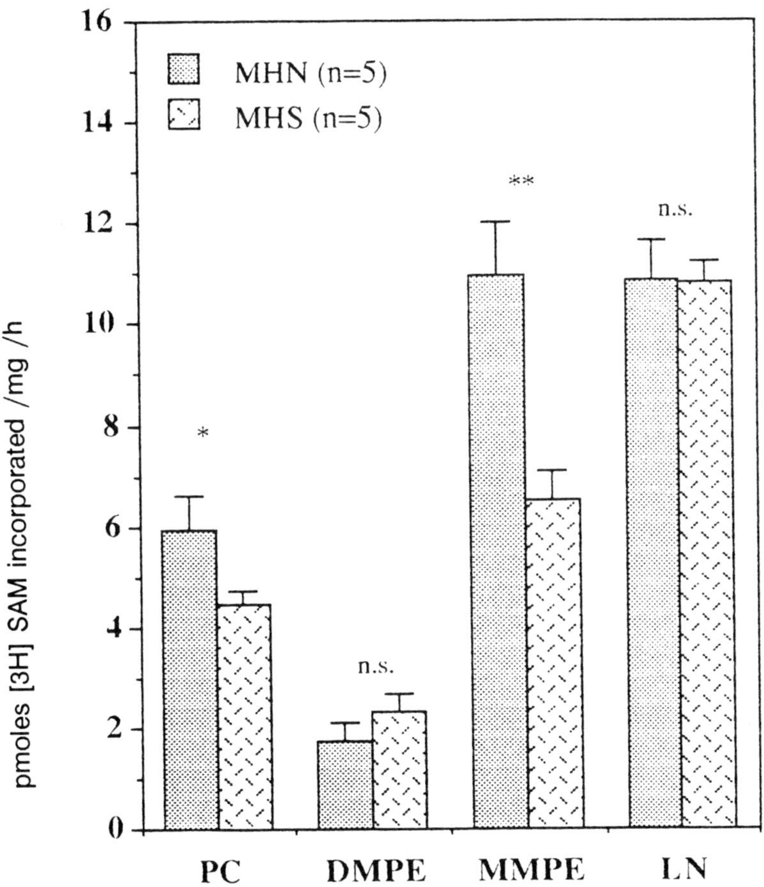

Figure 1. Phospholipid N-methlytransferase activity of SR membranes from control (MHN) and malignant hyperthermia susceptible (MHS) pigs.

Sarcoplasmic reticulum properties, in relation to the MH and PSE syndromes

In the following experiment, we used 39 halothane tested pigs. The *Longissimus dorsi* muscle was removed from the carcass 45 min post-mortem, then a carcass pH value was determined. From the excized muscle, one piece was kept at 4°C for water-holding capacity (WHC) determination and another piece was frozen in liquid nitrogen. The latter muscles were transported to the laboratory on dry ice and kept at -30°C until the preparation of SR membranes. After purification, the membranes were frozen in liquid nitrogen and kept until use at -80°C. Information from the halothane test, the pH and WHC values enabled us to classify the pig muscles into four groups, namely, H^+P^+, H^+P^-, H^-P^+ and H^-P^-, where H^+/H^- indicate that the pig reacted positively or negatively to the halothane test and P^+/P^- indicate that the resulting meat was, or was not, PSE. For this latter discrimination, we chose a limiting pH value of 5.8. We noticed that pH values correlated very closely with WHC data. Finally, four different SR membrane preparations from each category were used to produce the following results.

Determinations of fluidity state and cholesterol to PL ratio

Figure 2 shows that all the membranes used in this experiment had essentially the same fluidity level. The DPH fluorescence anisotropy value of 0.160 observed in this study indicates that a maximum level was reached by all the membranes. The determination of cholesterol/PL ratio in these membranes indicates that MH-susceptible pigs producing PSE meat had the highest value, whereas no significant differences were observed in either of the three other types (Fig. 3). If we take into account only PSE characteristic, then significantly higher levels of this ratio are found in P^+ pigs, compared with P^- ones (Fig. 3).

Calcium permeability and ryanodine binding measurements:

Calcium permeability was measured by determining the calcium level inside the vesicles after incubation of the membranes with 5mM ^{45}Ca for 1h. The results shown in Figure 4 clearly demonstrate that vesicles from H^+P^+ muscles retain less calcium than those obtained from the three other types.

The amount of ryanodine receptors were determined in experimental conditions using low concentration of calcium (1μM), low ionic strength (100mM KCl) and 100nM [^{3}H] ryanodine. Specific binding was calculated from the differences measured in the absence or in the presence of 10 μM unlabelled ryanodine. Once more, the membranes obtained from H^+P^+ muscles show the lowest amount of binding. The amount of ryanodine bound to the SR membranes from H^-P^+ animals is also low, compared with H^+P^- and H^-P^- ones (Fig. 5). As with the cholesterol to phospholipid ratio, the membranes from PSE developing muscles (P^+) show a significant decrease in the binding of ryanodine when compared with normal meat producing (P^-) muscle membranes (Fig. 5).

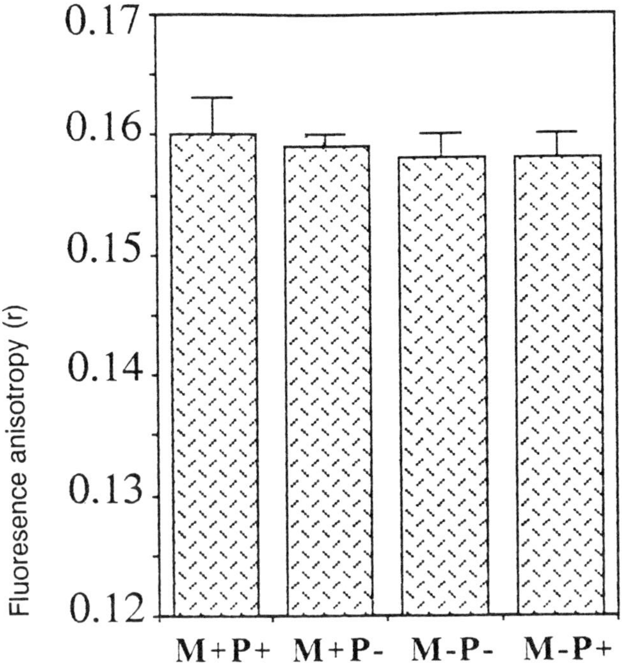

Figure 2. DPH flourescence anisotropy determined in SR membranes of pigs characterized on the basis of halothane sensitivity (M^+/M^-), and meat quality (P^+ = PSE; P^- = not PSE).

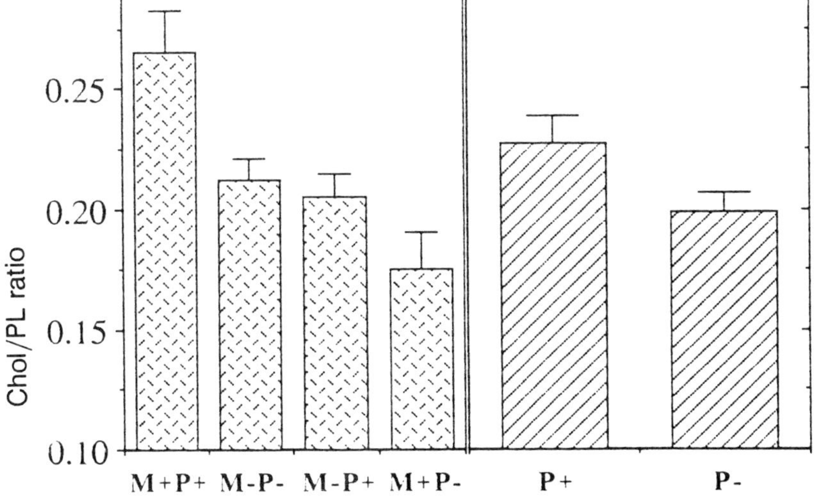

Figure 3. Cholesterol to phospholipid ratio in SR membranes isolated from pigs characterized on the basis of halothane sensitivity (M^+/M^-), and meat quality (P^+ = PSE; P^- = not PSE)

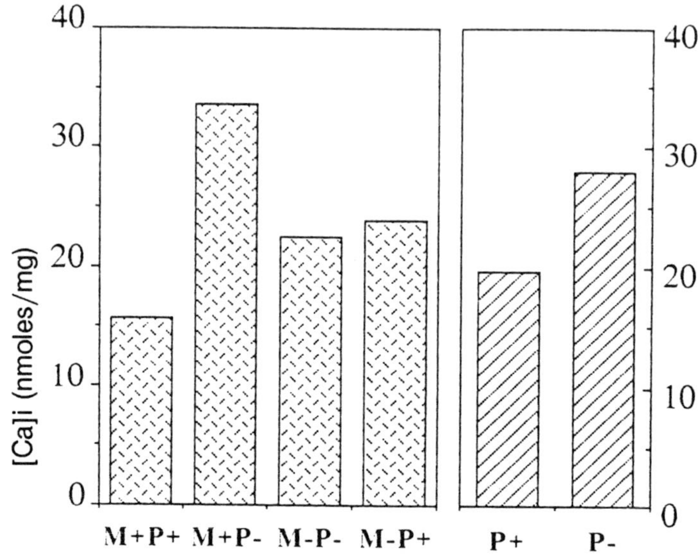

Figure 4. Intravesicular calcium determination in SR membranes isolated from pigs characterized on the basis of halothane sensitivity (M^+/M^-), and meat quality(P^+ = PSE; P^- = not PSE).

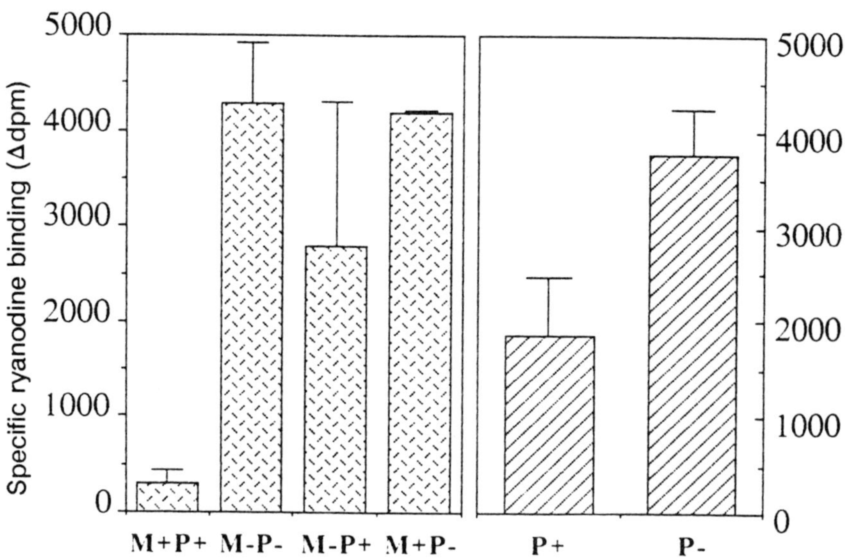

Figure 5. Ryanodine binding measurements in SR membranes isolated from pigs characterized on the basis of halothane sensitivity (M^+/M^-), and meat quality (P^+ = PSE; P^- = not PSE).

Discussion and perspectives

Sarcoplasmic reticulum membranes isolated from post-mortem muscle are in a more rigid state. It remains unclear whether this process occurred during muscle homogenization, or already existed *in situ*. This post-mortem rigidification is enhanced in membranes from halothane positive or MH susceptible muscle. Based on that observation, comparative studies indicate that the decrease of the fluidity state results mainly from an increase in the cholesterol content, a well-known membrane rigidifier. In MHS membranes, the abnormal rigidity level is enhanced by a higher PE/PC ratio. These PL molecules also show a higher level of the degree of unsaturation, which could represent a compensatory phenomenon to counteract the rigidification process. However, from the increase in the cholesterol content, such a high level of unsaturated fatty acids should decrease the microviscosity of these membranes. Such a result was already obtained with SR membranes isolated from freshly post-mortem excized muscle (Rock *et al.*, 1990). This higher fluidity in MHS pigs was also observed in sarcolemmal membranes by means of microscopic (Vignon *et al.*, 1989) and electron spin resonance spectroscopic (Thomas *et al.*, 1991) techniques. Moreover, PL composition of SR membranes isolated from biopsied MHS pig muscle was also shown to have a lower PE/PC ratio than the control samples. On the other hand, *in situ* measurements in plasma membranes of lymphocytes from MHS pigs also show a higher fluidity state when compared to membranes isolated from control animals (Adam *et al.*, 1989). Together, these results strongly suggest that MHS membranes, at least in living cells, could be in a more fluid state than the membranes from normal pigs. Within this hypothesis, the higher level of Chol/PL ratio observed in post-mortem MHS membranes indicates that some undefined abnormality occurs in the post-mortem interaction of cholesterol with membrane components. We also noticed, as many other authors have, that (Ca^{2+}- Mg^{2+})-ATPase activity is not affected in MHS membranes. As this integral protein has been shown to be sensitive to cholesterol (Madden *et al.*, 1981), we can assume that: i) cholesterol molecules interacting, post-mortem, with SR membranes are probably different from those acting as classical structural components; and ii) this new pool of molecules interacts with bulk phospholipids rather than with proteins or their lipid annulus. The relationship between cell membrane fluidity and PSE syndrome is as complex as the one between MH susceptibility and PSE development. However, from the preliminary results presented here, two groups were differentiated; one formed by MH- susceptible pigs developing PSE meat (H^+P^+), and the other group formed by the three other kinds of muscle. The results obtained from H^+P^+ as compared with H^-P^- membranes showed that cholesterol partition is higher in the former group although the fluidity state had reached a same plateau value in both kinds of membranes. This result is similar to the one observed in membranes from MHS pigs, as discussed above. On the other hand, the same fluidity values described in all four groups of pigs suggest that the kinetic, rather than extreme values of the rigidification process should be analysed. This work hypothesis is all the more interesting since the Chol/PL ratio changes evolve differently in H^+P^+ muscle membranes. Moreover, the undifferentiated fluidity state between the different groups of pigs confirm that cholesterol is not the only factor influencing the fluidity state of the membranes.

The highest level of Chol/PI ratio in H^+P^+ membranes as compared with H^+P^- ones indicate that: i) the MH process alone does not induce the modification in cholesterol partition in MHS membranes; and ii) PSE inducing phenomenon acts synergistically at that level; in other words, two different mechanisms are likely implied in these two syndromes. The results from Klont (1991), and the relatively high numbers of H^-P^+ confirm that PSE syndrome is closely, but not solely, related to the stress state of the animals before slaughter. Compared with the other groups, H^+P^+ membranes also show a low value in the ryanodine binding parameter, associated with enhanced calcium permeability. This data could result from a greater proteolytic activity, resulting from an increase of the intracellular calcium ions as proposed by Cheah (Cheah *et al.*, 1984).

In conclusion, cell membrane fluidity has been shown to play an important role in the different stages of cell life. As expected, our results suggest that cell death modifies the homeostasis of cell membrane fluidity. The resulting effect of this modification is a rigidification process where the role of cholesterol seems to be preponderant. The post-mortem consequences in the metabolism and in the membrane partitioning of cholesterol are influenced by both MH susceptibility of the muscles, and PSE meat development. Further analysis is now necessary to better define, at the molecular level, this post-mortem cholesterol-membrane interaction. Studies of the composition and comparative kinetics of this post-mortem cholesterol-membrane interactinon in membranes isolated from the four groups of pigs should provide an interesting model.

Acknowledgements

I would like to thank Drs X.Vignon, G. Gandemer, G. Monin, E. Gueux, Y. Rayssiguier, M.-A.Thomas, J. Viret, F. Schwaegele and S. Maak for their helpful discussions and for their scientific collaboration. Parts of the research works were supported by grants from the "Institut National de Recherches Agronomiques" (INRA), the "Association Francaise contre les Myopathies" (AFM) and the Organisation for Economic Co-operation and Development (fellowship under OECD Project on Biological Resource Management).

References

Adam, H. *et al*. 1989. Differences in the phospholipid composition of a sarcosomal preparation between MHN and MHS pigs. In: Hofmann, J.G. and Schmidt, A. (eds), Malignant Hyperthermia - An Update. Berlin, Verlag Volk und Gesundheit, pp. 240-46.

Bendall, J.R. 1960. Post-mortem changes in muscle, in: Bourne, G.H. (ed.), Structure and Function of Muscle, Vol III, Academic Press, New York, pp. 227-74.

Bendall, J.R. and Wismer-Pedersen, J. 1962. Some properties of the myofibrillar proteins of normal and watery pork muscle. J. Food Sci. 27, 144-49.

Borochov, H. and Shinitzky, M. 1976. Vertical displacement of membrane proteins mediated by changes in microviscosity. Proc. Natl. Acad. Sci. USA 73, 4526-30.

Cheah, K.S. and Cheah, A.M. 1981. Mitochondrial calcium transport and calcium-activated phospholipase in porcine malignant hyperthermia. Biochim. Biophys. Acta 634, 70-84.

Cheah, K.S. *et al*. 1984. Relationship between Ca^{2+} release, sarcoplasmic Ca^{2+}, glycolysis and meat quality in halothane-sensitive and halothane-insensitive pigs. Meat Sci. 10, 117-30.

Eikelenboom, G. and Minkema, D. 1974. Prediction of pale, soft, exudative muscle with a non-lethal test for the halothane-induced porcine malignant hyperthermia syndrome. Neth. J. Vet. Sci. 99, 421.

Ervasti, J.M., Mickelson, J.R. and Louis, C.F. 1989. Transverse tubule calcium regulation in malignant hyperthermia. Arch. Biochem. Biophys. 269, 497-506.

Fink, H.S., Maak, S., Von Lengerken, G. and Till, V. 1992. Abnormalities in the regulation of blood platelet free cytosolic calcium in malignant hyperthermia. II. Pig platelets. Cell Calcium 13, 157-62.

Fletcher, J.E. *et al*. 1989. Triglycerides, not phospholipids, are the source of elevated free fatty acids in muscle from patients susceptible to malignant hyperthermia. Eur. J. Anesthesiol. 6, 355-62.

Fujii, J., Otsu, F., de Leon, V.K., Weiler, J., O'Brien, P.J. and MacLennan, D.H. 1991. Identification of a mutation in porcine ryanodine receptor associated with malignant hyperthermia. Science 253, 448-51.

Gronert, G.A., Mott, J. and Lee, J. 1988. Aetiology of malignant hyperthermia. Brit. J. Anaesthesiol. 60, 253-67.

Iaizzo *et al*., 1991. Enhanced mobilization of intracellular Ca^{2+} induced by halothane in hepatocytes isolated from swine susceptible to malignant hyperthermia. Anesthesiol. 74, 531-38.

Klont, R.E. 1991. Effects of preslaughter anesthesia on meat quality characteristics from pigs of different halothane genotypes. Proceeding of the 37th ICoMST, Kulmbach-Germany, pp. 402-04.

Kuncl, R.W., Drachman, D.B. and Kishimoto, Y. 1985. Phospholipid methylation in skeletal muscle membranes. Muscle Nerve 8, 426-34.

Levitt, R.C., McKusik,V.A., Fletcher, J.E. and Rosenberg, H. 1990. Gene candidate (Lett). Nature 345, 297-98.

Madden, T.D., Klng, M.D. and Quinn, P.J. 1981. The modulation of Ca^{2+}-ATPase activity of sarcoplasmic reticulum by membrane cholesterol - the effect of enzyme coupling. Biochim. Biophys. Acta 641, 265-69.

McMurchie, E. 1988. Dietary lipids and the regulation of membrane fluidity and function. In: Aloia, R.C., Curtain, C.C. and Gordon, L.M., (eds), Advances in Membrane Fluidity- Physiological regulation of membrane fluidity. Alan R. Liss Inc. New York.

Mickelson, J.R. *et al.* 1987. Skeletal muscle sarcolemma in malignant hyperthermia: evidence for a defect in calcium regulation. Biochim. Biophys. Acta 897, 364-76.

Mickelson J.R. *et al.* 1992. Structural and functional correlates of a mutation in the malignant hyperthermia-susceptible pig ryanodine receptor. FEBS Lett. 301, 49-52.

Ohnishi, S.T., Taylor, S. and Gronert, G.A. 1983. Calcium-induced calcium release from sarcoplasmic reticulum of pigs susceptible to malignant hyperthermia. The effects of halothane and dantrolene. FEBS Lett. 161, 103-7.

Ohnishi, S.T. *et al.* 1986. Abnormal membrane properties of sarcoplasmic reticulum of pigs susceptible to malignant hyperthermia: mode of action of halothane caffeine, dantrolene and two other drugs. Arch. Biochem. Biophys. 247, 294-301.

Pelech, S.L. and Vance, D.E. 1984. Regulation of phosphatidylcholine biosynthesis. Biochim. Biophys. Acta 779, 217-51.

Pottel, H., Van der Meer, W. and Herreman, W. 1983. Correlation between the order parameter and the steady-state fluorescence anisotropy and an evaluation of membrane fluidity. Biochim. Biophys. Acta 730, 181-5.

Rock, E. and Kozak-Reiss, G. 1987. Effect of halothane on the Ca^{2+}-transport system of surface membranes isolated from normal and malignant hyperthermia pig skeletal muscle. Arch. Biochem. Biophys. 256, 703-7.

Rock, E. *et al.* 1990. Abnormal fluidity state in membrane of malignant hyperthermia pig skeletal muscle. Arch.Biochem. Biophys. 281, 36-40.

Rock, E., Motta, C., Vignon, X. and Kozak-Ribbens, G. 1992. Fluidity state of lymphocytes plasma membrane in MHS pigs and humans, (submitted).

Shinitzky, M. (ed.) 1984. Physiology of Membrane Fluidity Vol. 1. CRC Press Inc. Boca Raton. Florida.

Shinitzky, M. and Inbar, M. 1976. Microviscosity parameters and protein mobility in biological membranes. Biochim. Biophys. Acta 433, 133-49.

Singer, S.J. and Nicolson, G.L. 1972. The fluid mosaic model of the structure of cell membranes. Science 175, 720-31.

Thatte, H.S. *et al.* 1987. Temperature-dependent abnormalities of the erythrocyte membrane in porcine malignant hyperthermia. Biochem. Med. Metab. Biol. 38, 366-77.

Thomas, M-A., Rock, E. and Viret, J. 1991. Membrane properties of the sarcolemma and sarcoplasmic reticulum of pigs susceptible to malignant hyperthermia. Action of halothane. Clinica Chim. Acta 200, 201-10.

Vignon, X., Kozak-Reiss, G. and Rock, E. 1989. Interaction of halothane with normal and MH skeletal muscle cells and membranes: ultrastructural studies. In: Hofmann, J.G. and Schmidt, A. (eds), Malignant Hyperthermia - An Update. Berlin, Verlag Volk und Gesundheit, pp. 205-208.

Discussion

*Mickelson - Could you clarify whether the membranes in which you see the altered cholesterol to phospholipid ratio are on muscle that was extracted immediately post-mortem, or whether you started to see the changing cholesterol:phospholipid ratio at increasing times post-mortem?

*Rock - The altered cholesterol to phospholipid ratio exists both in muscle extracted immediately post-mortem and hours later. It means that a change in that ratio occurred in post-mortem muscles. What we propose in that study is that the kinetic of this post-mortem change is accelerated in MH muscles. Thus measurements made in the membranes extracted 15 min post-mortem show almost no differences between normal and MH types whereas the MH membranes extracted from 1h post-mortem muscle were significantly less fluid than the normal ones. This observation led us to hypothesize that MH membranes could even be more fluid before the death of the tissue. The finding that plasma membranes of living lymphocytes of MH pigs are effectively in a more fluid state than the control ones improve this hypothesis.

*Demeyer - Just a very technical detail. How did you determine the phospholipids? Was determination based on fatty acid analysis, or other designs?

*Rock - HPLC.

*Demeyer - So, there's no possibility that fatty acids may be destroyed during the ageing process, and that would affect your results, because you isolated the phospholipid itself.

*Rock - Right. First we isolated the phospholipid, after the storage.

*Demeyer - The point is that it is based on fatty acids. So, it could be that some fatty acids were destroyed during the storage.

*Rock - Probably, or oxidized.

*Demeyer - Well, that would not be an enzyme activity in fact, but an oxidation of fatty acids.

*Puolanne - How readily do protons, or lactic acid, diffuse through the membranes? What is the time factor post mortem, fiber to fiber?

*Essén-Gustavsson - Good question, but we can't answer it. If these stress-susceptible pigs have larger fibers, and you take a piece of muscle, you will of course have less membranes per gram of muscle, but that won't influence your results in any way will it? The way you express your results is important, per unit of DNA, or per gram or whatever.

Quality of Fresh Pork - Review

K.O. Honikel

Federal Centre for Meat Research, E.-C. Baumann-Strasse 20,
D-8650 Kulmbach, Germany

Introduction

Quality has different meanings to different people. These multiple views are some of the reasons why PSE pork could develop so extensively, and over so many years. Salesmen and consumers alike favoured large hams in carcasses, little backfat and very lean fresh pork, accepting the pale and exudative meat. Furthermore the subjective appreciation of pale and lean PSE pork was mixed up with the suitability of pork for a certain purpose and both were called quality. Carcass evaluation (grading) was also thought to be an expression of pork quality. However, the percentage of lean meat is primarily a quantitative, not a qualitative measure. Due to this Babylonian confusion with the term "quality," I want to first define the meaning of meat quality.

Definition of quality

In a scientific approach, quality must be defined as an objective description of the properties of a product. In order to do this, one needs methods to detect characteristics objectively, that means in a reliable, definitive, and reproducible manner, by applying proper analytical procedures at the right time and at the right part of the good. The sum of the characteristics measured describes the quality of the food.

Quality characteristics, however, may have different meanings for different people, and for different purposes. Thus, each characteristic is modified by a suitability factor of an appropriate magnitude. These factors may have a negative or a positive sign. Examples should clarify this:

1. A muscle with an ultimate pH-value of 6.2 is not very well suited for fresh pork in retail packages. In this instance a pH of 6.2 would be qualified as a negative characteristic. For a meat processor who produces finely comminuted frankfurter-type sausages, a meat cut with a pH of 6.2 is rather suitable, as a high pH improves water and fat binding. His suitability factor would have a positive sign.

2. Producers of a "light," ready-to-eat meal will probably be rather indifferent to the pale colour of PSE-pork when used as a product ingredient. Therefore, the suitability factor will be 1. However, the importer of PSE pork from Bavaria to mediterranean countries highly esteems the pale colour, and its suitability factor is well above 1.

In conclusion: **pork quality (PQ)** is a sum of objectively measured characteristics. It is a neutral description of a pork cut.

$$PQ = \sum_{i=1}^{n} C_i$$

The **suitability of pork (PS)** is the product of all characteristics modified by suitability factors (f).

$$PS = \sum_{i=1}^{n} f_i C_i$$

The **appreciation (A)** of consumers may contain some objective elements, but, in summary, it is neither PQ nor PS.

$$\mathbf{PQ \neq A \neq PS}$$

The modern consumer includes in his appreciation a number of socio-economical and environmental issues which may influence the pork quality but which are not quality characteristics themselves. These production criteria may be regarded as **production** (management or manufacturing) **quality (MQ).** Some of the variables that could be included in MQ are; a) the number of animals per farm (manure and ammonia production); b) animal housing (in stable, or in the open); c) treatment with growth promoters or supplement feed (from the third world); d) transport and slaughter of pigs (animal welfare); and e) chilling of carcasses and meat transport.

The factors of MQ are important issues, as the consumers in Western societies are concerned about these aspects. Consequently, MQ aspects are major parts of brand-name pork programs. The realization that the production conditions influence the quality of the end product was one of the most progressives step forward in pig production, with regard to pork quality, in the last decade.

In summary: production quality influences pork quality. The European consumer is more and more interested in production quality than he is, or ever was, informed in product quality. With his growing interest in these things, his "appreciation" has an influence on production quality (e.g. Scharrel pigs, brand-name pork) and therefore

A→MQ→PQ and PS.

Factors of quality

In Table 1, groups of quality characteristics are presented. The lists are not comprehensive, but they represent the most important quality criteria.

Table 1. Quality characteristics

Nutritional factors		Technological factors
Proteins and their composition		Water-holding capacity
Fats and their composition		Protein content and its status
Vitamins		Fat content and its status
Minerals		Connective tissue content
Digestibility		Tenderness
		pH-value
		Color
Hygienic factors		Sensory factors
Microorganisms		Color, hue
Factors of shelf life	pH value	Color brightness
	a_w-value	Marbling
	Degree of heating	Odor
	Temperature of storage	
		Flavor
Residues	Nitrosamines	Juiciness
	Antibiotics	Consistency
	Hormones	Tenderness
	Other pharmaceuticals	
Contaminants	Pesticides	
	Mycotoxins	
	Heavy metal	
	Nuclides	

According to Hofmann, K. (1986).

Nutritional factors seem at a first glance to be easily and comprehensively determined in pork. The negative image of pork in the public, and with nutritionists, proves the opposite. Pork is regarded as fatty, and full of residues and toxins.

The hygienic factors comprise microbiological aspects and problems of health. These factors include hazardous substances like pharmaceutical residues and their metabolites, environmental compounds, or man-made residues resulting from processing (e.g. smoking, nitrosamines).

Even if many consumers do not admit it, sensory factors are the decisive ones for consuming meat. These factors are difficult to measure objectively, but legions of scientists try to develop reliable and reproducible sensory methods.

Technological factors are the most important for all levels of processing. Most of the measured characteristics on carcasses and cuts serve this purpose.

Influence of environmental factors (production quality)

I will refer to the criteria of pork production and processing, as the environmental factors of influence on quality characteristics. Many aspects of the influence of breeds, genetics and handling have been reported in the preceeding papers. This paper will focus mainly on the treatment beyond abbatoir gates. The treatment of the animals before stunning, during stunning and debleeding will be discussed. Handling of carcasses during scalding and dehairing may influence meat quality. Also, post-mortem chilling and cutting procedures are important in this respect.

Many environmental factors of influence on pork quality are described in the guidelines for "certified pork in Germany" which the Central Marketing Agency for Agricultural Products (CMA), together with leading scientists and managers of the meat industry, has compiled (CMA, 1991).

I. Criteria on breeds

The following requirements exist as guidelines for certified pork:

1. Defined breeds are allowed only; at least one parent must be homozygous, with regard to the halothane-negative gene. This is controlled also after slaughter by 2.
2. Due to the inherited stress resistance, 80% of all pigs from one farm must have a pH_1 above 6.0.
3. The grading must show 50 - 60% of meat in the carcass.

II. Feeding requirements

1. No growth promotors above 40kg live weight are permitted.
2. Animals treated with medicines (besides necessary vaccinations) are removed from the pigs for the brand-name program.
3. External controllers have 24h a day access, without prior notice, to stable and farms.
4. In some co-operatives, farmers have to draw a bill of exchange of DM 10000 (or higher), which is lost if they are found to be in violation of co-operative rules.

III. Criteria from farm to abbatoir gate

1. No feeding of animals for 12h before transport is accepted.
2. On the lorry, a space of $0.5m^2$/animal is required; no electric prods are allowed.
3. Transport must be carried out in social groups.

IV. Criteria in the slaughterhouse

1. Each animal must have access to water.
2. Holding and moving in social groups is required.
3. The rectal temperature must be below 39.2°C.

V. Criteria of stunning and sticking

Proper stunning, fulfilling regulations for animal welfare, is required.

1. Electrical stunning on the head must reach 1.25A within the first second.
2. Sticking must be done within about 8s after the end of electric stunning.
3. $C0_2$ stunning requires 80% $C0_2$ as soon as possible, and sticking must occur as soon as possible.
4. Debleeding should be done in a lying position.

VI. Criteria of scalding and dehairing

1. Between sticking and scalding, there should be a minimum time of 5 min.
2. The temperature of the water in the scalding tank must be above 60°C.
3. Scalding in a hanging position is recommended.

VII. Criteria of chilling

The time/temperature conditions of all muscles in a carcass are shown in Table 2. Additionally, after cutting, a temperature of below 4°C is required.

All these criteria are laid down in the requirements (CMA, 1991). Beyond this list, I would like to evaluate in more detail some aspects that were investigated in very recent publications about stunning, scalding and chilling.

Table 2. Quality characteristics of certified pork

Characteristic		45 min	1.5h	4h	24h
pH 12.-14. rip	LD[a]	> 6.0	> 5.8	> 5.5	5.4-5.85
ham	SM[a]	> 6.0	> 5.9	> 5.5	5.4-5.85
T (°C)	LD	< 40	< 35	10-20	< 7
	SM	< 40	< 36	< 22	< 7
Electr. conduct.	LD	-	< 5	-	< 5
(mS/cm)	SM	-	< 5	-	< 7
Color brightness	LD	-	< 50	-	< 53
(L*)	SM	-	-	-	< 50
Drip loss (%) 24h-48h					< 4

[a]LD = *M. longissimus dorsi*; SM = *M. semimembranosus*.

Woltersdorf (1992) has studied the necessity for proper electric stunning, and some common mistakes. In Germany, manual electric stunning of pigs is the most widespread method of stunning. In order to guarantee an efficient stunning, which leads to immediate unconsciousness of the pig as required by animal welfare regulations, several criteria are defined:

1. The electrodes must be applied in a way that the current passes the brain at the shortest distance.
2. The animal must be unconscious within 1s. This requires a current of at least 1.25A and a voltage of 240V.
3. Within 30s after end of flow of current the animal must be debleeding.
4. Pork quality will be improved by shorter time intervals between the end of stunning and sticking.

A faulty stunning procedure conflicts with animal welfare and it may deteriorate meat quality by an accelerated glycolysis. Various instruments for stunning are available for use. In some slaughter houses animals are held by a kind of trap. In most slaughter houses pigs move freely in a cabin and are caught by the stunner with the clamp. The author investigated the various positions of the electrodes on the head on the degree of current flow.

"Sharp" electrodes lead to a better flow of current. All electrode positions near the eye and ear were very effective. The electrodes positioned on the neck and cheeks were poor, with regard to the flow of electric current. One, or even both, electrode(s) in these positions exhibited a reduced current flow. The current must reach 1.25A at 1s and a sum of 12-15A×s (Coulomb) were necessary for a extensive period of unconsciousness beyond 30s. Preferably, this amount (12-15 Coulomb) was obtained within about 6s.

After checking all possible combinations the author concludes that in order to fulfill the requirements:

1. The skin of the animal must be wet.
2. The animal must be fixed in a kind of trap.
3. Electrodes with "sharp" tips are positioned at the eye or ear area on both sides of the head.
4. A voltage of 240V is necessary, and the electricity amount must be 12-15A×s (Coulomb) within 4-6s.
5. As soon as possible after stunning, the animal must be debleeding.
6. Too much electricity (above 15A×s) speeds up the post mortem changes (electrical stimulation!), and insufficient electricity, or a "slow" increase (12-15A×s in more than 10s) are contradictory to animal welfare requirements.

Troeger (1992) investigated the scalding and dehairing steps in the slaughter process of pigs. The shelf-life of pork, from the microbiological point of view, is regarded as shorter than that of meat from beef or sheep. Satisfactory explanations are missing, as meat in the interior of a live animal should be nearly sterile. Troeger provided an convincing explanation in his paper "Influence of scalding and dehairing techniques on the contamination of pork carcasses." Pigs are usually scalded and dehaired, contrary to cattle and sheep. In most slaughterhouses scalding is done in water tanks of about 60°C, into which the carcasses are immersed. Scalding on hanging carcasses is still rare. During immersion, water penetrates into the sticking hole. If, either during the scalding or dehairing period, the carcasses are revolved, the penetrated non-sterile water is massaged through the carcasses via blood vessels and the lymphoid system.

The author used a auxotrophic *Bacillus subtilis* strain 164 as a marker which he put as spores into the scalding tank. He investigated various scalding and dehairing systems. Scalding while hanging caused little penetration of *Bac. subtilis* spores. Scalding in tanks with parallel dehairing by revolving the carcass allowed him to find *Bac. subtilis* counts in loin, shoulder and ham muscles. The spores must have been massaged into the muscles, far away from the sticking hole at the throat. These results have added a further argument against the optically unpleasant scalding tank. Scalding in the hanging position will be one future technique that will improve the shelf-life of pork.

Our own recent results were concerned with the chilling regimes of PSE-prone pork muscles. Pork muscles with a pH_1 of 5.45 were chilled under extremely fast, or slow conditions. Figure 1 shows the results, with regard to drip loss. Chilling to 34°C within 51 min post mortem prevented the increase of drip loss typical for PSE pork. If 34°C are reached after 100 min or later (at pH_1 5.45 in the muscle) then exudation is extreme. Fast chilling causes drip losses similar to the amounts found in normal pork (pH_1 = 6.2). This confirms earlier findings when we found that the velocity of temperature decrease must follow the pH changes (Honikel, 1989).

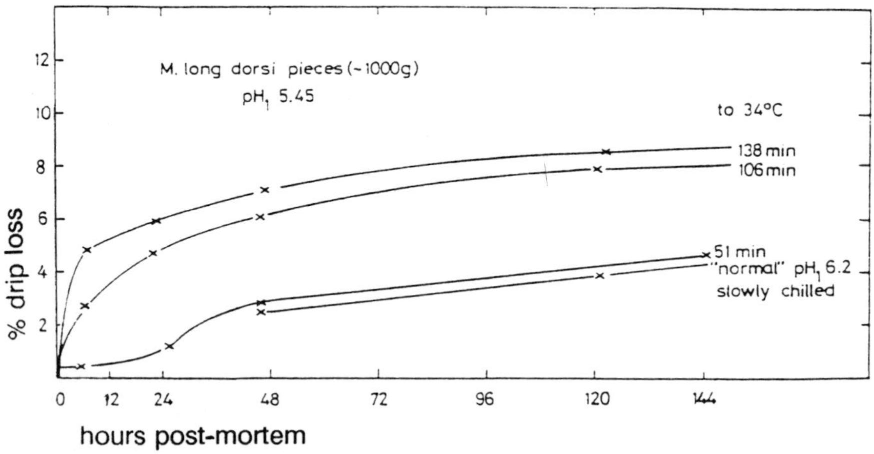

Figure 1. Influence of early post-mortem chilling to 34°C on drip loss of a PSE-prone *M. longissimus dorsi* with a pH_1 of 5.45. In comparison, a slowly (138 min to 34°C) chilled *M. long. dorsi* of pH_1 6.2 is shown.

Characteristics of high quality fresh pork

With all the requirements for production of animals and handling of meat the certified pork must possess the following characteristics at the balance in the abbatoir. The composition of carcass and meat must fall within the following guidelines:

1. Carcass weight must be > 80kg, but < 100kg.
2. Meat percentage by grading must be between 50 and 60%.
3. The fat content (chemically determined) of *M. longissimus dorsi* must be above 1.5%.

With these requirements, lean carcasses are still produced, but a minimum of intramuscular fat is demanded.

Other quality characteristics are collected in Table 2. Due to the various post-mortem changes within a carcass, the requirements vary between muscles. But the characteristics must be easily measurable; that means a destructive cutting of a carcass must not occurr. Therefore the easily accessible *M. longissimus dorsi* and *M. semimembranosus* are recommended as representative muscles.

At 24h post-mortem the pH should not be lower than 5.4. Excessively low pH leads to high drip losses (New Hampshire effect). A pH_u of about 5.85 shortens the shelf-life. The measurement of all characteristics mentioned in Table 2 is not required at all times post-mortem. We recommend that the underlined measurements be performed on every carcass. The other parameters should be measured at intervals of a few hundred carcasses. Looking at Table 2, one realizes that at 45 min post-mortem, no measurement of conductivity, color brightness or reflection values with the grading probes are listed. This is due to the fact that these characteristics have not fully developed at 45 min post-mortem, as shown in Figure 2.

Hygienic conditions of fresh pork

Thanks to the strict requirements of the slaughter process, the requirements for microbiological counts can be also rather demanding. Table 3 shows the maximum counts during the process of slaughter and handling. At 2h post-mortem, when the carcasses have usually reached the final position in the chiller and will not be touched further by man, the total counts may not exceed 10^4 counts/cm^2. On packaging of a retail cut, the total counts must not exceed 10^6 counts/cm^2.

Conclusions

This paper was written to provide a comprehensive view of pork quality, listing the factors of environmental influence and the required characteristics for certified brand-name pork as provided by guidelines compiled by CMA (CMA, 1991).

In order to improve the present quality of pork, a comprehensive approach is required that recognizes the justified demands of consumers, and includes breeding, as well as slaughter technology, and hygiene.

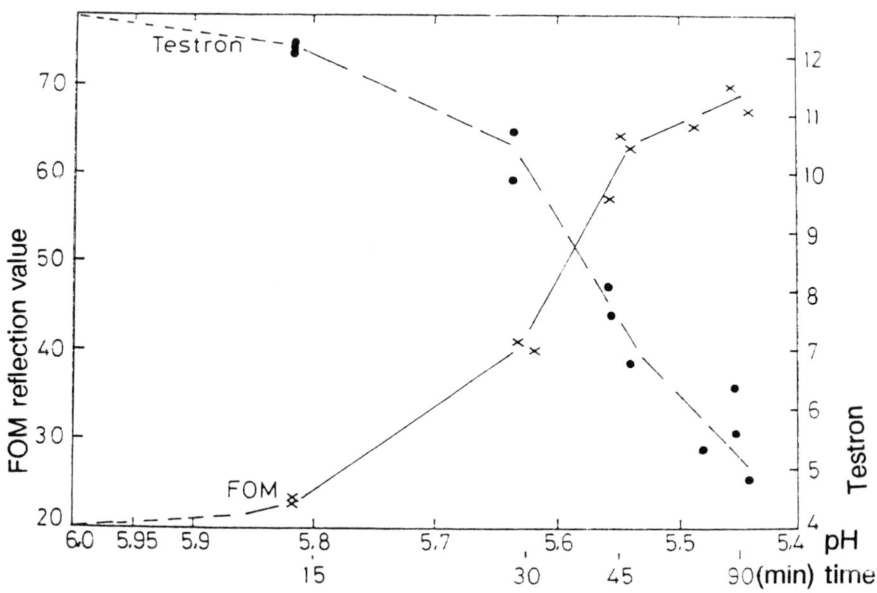

Figure 2. Change of reflection value (Fat-O-Meter readings) and impedance (conductivity) measurements (Testron readings) with post mortem pH fall (according to Sack *et al.* 1985).

Table 3. Required microbial counts in certified pork

Time of step of processing	Total counts/cm^2 [a)]	Enterobacteria/cm^2 [a)]
2h post-mortem	$<10^4$	-
24h post-mortem	$<5\times10^4$	-
On a carcass before cutting	$<5\times10^5$	-
On a meat cut	$<5\times10^4$	-
On a retail cut before packaging	$<10^6$	$<10^4$

[a)]Median of at least 20 carcasses with four defined positions. The measurements must be taken at least once a month, at different times of the day and every time on a different day of the week.

References

CMA 1991. Qualitäts- und Prüfbestimmungen für das Prüfsiegel "Deutsches Qualitätsfleisch aus kontrollierter Aufzucht", Centrale Marketing Gesellschaft der deutschen Aprarwirtschaft mbH, Bonn.

Hofmann, K. 1986. Ist Fleischqualität meβbar. In: (ed. Institut für Chemie und Physik) Chemisch-physikalische Merkmale der Fleischqualität, Kulmbacher Reihe, Band 6, Bundesanstalt für Fleischforschung, Kulmbach, S. 1.

Honikel, K.O. 1989. The Meat aspects of water and food quality. In: Harmann, T.M. (ed.), Water and Food Quality. Elsevier Applied Science, London.

Sack, E., von Canstein, B., Fischer, K. und Scheper, J. 1985. Einfluβ des Meβzeitpunkts auf den FOM-Reflexionswert bei Schweinehälften. Mitteilungsblatt der Bundesanstalt für Fleischforschung, Heft 87, p. 6266.

Troeger, K. 1992. Einfluβ der Brüh-/Enthaarungstechnik auf den Keimgehalt von Schweineschlachtkörpern, Mitteilungsblatt der Bundesanstalt für Fleischforschung 31, in press.

Woltersdorf, W. 1992. Mängel der manuellen Elektrobetäubung beim Schwein. Mitteilungsblatt der Bundesanstalt für Fleischforschung 31, in press.

Discussion

*Lundström - I would like to ask you how you define "holding and moving in social groups." Is that only in the pen, or in the herd, how do you define it?

*Honikel - It's defined as how they are held on the farms, in groups. They should be transported, and also held in the slaughterhouses in such groups. I must say it is sometimes rather difficult, but, we try to enforce it. And, all groups and all co-operatives that observed the rules, realized in, let's say within a few weeks, that this is a huge advantage, and really tried to keep the pigs in the same groups as they were in on the farm.

*Lundström - Do you really have the fat percent above 1.5%?

*Honikel - No, these are the rules that we, in the beginning, thought we could enforce, but we can't do it anymore, so we just recommend it now. I would say, however, that if we go on selecting for lean pigs, then eventually we will have only two kinds of pigs: lean pigs, and pigs that are too lean.

*van der Wal - Dr. Honikel, you said in your lecture, that debleeding should be done in a lying position. Do you mean the sticking handling? Then, I agree with you, because shackling a carcass is very time-consuming, so sticking in the lying position can only bring a shorter time between stunning and debleeding. But, do you also mean that the entire debleeding process should be done in a lying position?

*Honikel - Not in all cases. I know instances where the animals can be stuck within 3s after the end of stunning. They fall out, and immediately they are stuck. And then a conveyor belt carries them along, in some instances until there is no blood coming out, which takes about 4 min. Then they are shackled, and go into the scalding tank. Some places keep the animals on the conveyor belt for only one and a half minutes, but, in all cases, the animals are dead. They don't have any heart beat when they are shackled. We don't know where we should put the limit, but, it is demanded that as soon as possible after the stunning is finished, the animal should be debleeded, because all of the stress hormones caused by the stunning would otherwise go all around the carcass. When you debleed them as soon as possible, there will not be such a high concentration of the stress hormones going around the body, and that's why we have this demand.

*van der Wal - Can you shackle them immediately after sticking? Does it make any difference in the amount of blood that is lost from the carcass?

*Honikel - No, I don't think so. We are now investigating if there really is a difference if you debleed in the lying position or not. Our experience, and the results we have so far, show that there is no difference if you let them debleed in the lying position, or in the shackled position.

*Smulders - My question is about how realistic the scheme is, and how feasible it is for the industry to achieve quality control. I was wondering how realistic it is to demand of an industry that pH at 45 min post-mortem, in all muscles, is greater than 6.0, and at 4h post-mortem, larger than 5.5, unless you provide a scheme in which you can actually measure the pH in all these muscles? And, more importantly, unless you provide procedures, in general, for integrated quality control, the experience in Holland is that it is hardly possible to expect these things from processors. Even more so, unless you have integrations, meaning the primary production almost linked to the supermarket chain, it is very difficult, if not impossible, to demand many of the things you have recommended. I know that the ideal situation is something other than what may be achieved within the near future, but, I would like your opinion on these things, because, in Holland we have tried, and it seems like the integrations work very well indeed. But, there are more pork processors than the integrated ones, and these people have major difficulties. So, I'd like to have your opinion on how realistic it is to have a fixed pH in all muscles at two fixed times. Do you provide any procedures for people to follow? And, how difficult is it for non-integrated pork processors to comply with these demands?

*Honikel - To start with the last part, we tried to enforce it in integrated systems. This is actually the ideal that was behind it from the beginning. Scientists are usually idealistic persons, and so we tried to force some companies into it. And I agree with you, we don't meet all the time with success. But, we have met with success in several co-operatives, which, on realistic terms, we never would have believed in the beginning. For instance, you could say measuring pH is not feasible on the slaughter line. If you run a slaughter line with 300 pigs an hour, it's impossible, I agree. But, these co-operatives realized that speeds of more than 150 pigs per hour are not feasible with a quality scheme, and they are now down to speeds of 120 to 130 pigs, and these groups indeed measure the pH in each loin.

*Smulders - But you required, it's distinct, "in all muscles".

*Honikel - OK, in all muscles, they can't do it, there are dozens of muscles. But they measure it in the loin, and some other muscles too, once in a while. OK, it may be expressed incorrectly, but, a number of groups measure once in the loin, and another measurement for instance in the ham, and other muscles every few minutes or so. Our demands are certainly quite harsh, but we meet with success. So far, it's certainly only a few percent, only 2 or 3% of the pigs we slaughter in Germany, and I'm quite sure we never will reach 20%, but these groups are guiding the whole market to a higher level of quality.

*Frøysten - You mentioned meat percentage in classification. How's that done? It doesn't quite agree with the meat percentage we operate with in the Nordic countries. We use the Hennessey probe.

*Honikel - It's the FOM.

*Frøysten - The FOM? So, that's a little bit different from the Hennessey rating. Because a lot of our pigs are above 60%.

*Honikel - Yes, the average in Germany is also about 58, or 59%. But with this special quality scheme, they are all much lower.

*Frøysten - What kinds of breeds do they use then, because you didn't mention anything on the breeds, nor the criteria of breeds in your paper.

*Honikel - There are a number of breeds. We did not want to enforce the breeds on the different groups. The most widely used are from the German hybrid program. The are all commercial crosses, or hybrid crosses, most of them.

*Puolanne - It was interesting to see the pH limit set at 5.85, because, I think the rigor mortis can be different in pork when the pH_u is over 5.85. I have noticed this on many occasions. I don't know the reason, but it seems to me that it is. I have also seen it when studying hot-boned pork.

*Honikel - That was also the reason why we picked that value.

Processed Pork - Technological Parameters Related to Type of Raw Material - Review

A.J. Møller[1], G. Bertelsen[2] and A. Olsen[1]

Royal Veterinary and Agricultural University, Department of Dairy and Food Science, 2000 F, Copenhagen, Denmark.
[1]Howitzvej 11; [2]Thorvaldsensvej 40

Introduction

The diversity of pork has long allowed for competition in two segments of the meat industry, the fresh meat, and the further processed meat areas. In Europe, pig meat has by far the biggest share in meat production and in many countries, roughly half the meat available is processed into meat products such as bacon, ham and sausages. Accelerated processing (pre-rigor or hot-processing) has achieved major interest in recent research as a processing method for increasing the potential of meat protein functionality, and thereby improved technological characteristics of the raw material.

Product faults due to a change in the raw meat material from pigs are mainly due to pH-related properties (PSE, DFD and low ultimate pH). For many years, pig breeding has concentrated on properties of direct economical importance for producers such as growth rate and feed efficiency. Through genetic selection, the lean meat content has been increased, due to a general increased market for leaner meat with lower calorie density and less fat. Increased leaness has also been demonstrated by growth promoters and by using intact male pigs for meat production. Recently, several studies have been aimed at producing healthier pork through animal diet modifications for enhanced content of unsaturated fat in pork products.

The published literature on processed pork is vast and this review is restricted to recent research within the following areas: accelerated processing; pH related properties (DFD omitted); leaner pork; and dietary fat.

Abbreviations are used for the following muscles: AD = *M. adductor*; BF = *M. biceps femoris*; LD = *M. longissimus dorsi*; SM = *M. semimembranosus*; ST = *M. semitendinosus*; SS = *M. supraspinatus*.

Accelerated processing

The economic advantage of pre-rigor processing prior to carcass chilling is mainly due to reduced energy and labor costs (Cuthberson, 1980). Pre-rigor, hot-boned meat also has superior functional properties, e.g. water binding and fat-emulsifying properties, as compared with post-rigor meat (Hamm, 1986). To benefit from the functional properties of pre-rigor meat, processing must be conducted immediately after slaughter, particularly when post-mortem glycolysis is fast, as in porcine muscles. The development of rigor mortis can be largely prevented by the addition of salt in the pre-rigor state (Hamm, 1986) but the added salt needs to penetrate the muscle tissue before the development of rigor. For whole meat, this may be achieved by stich-pumping followed by tumbling or massaging at low temperature.

Cured products

Abu-Bakar *et al.* (1983), reported reduced overall yield for hot-processed, cured, pork bellies compared with conventional processing. However, Neel *et al.* (1988) observed no significant change in precentage smokehouse yields, or in slicing yields among post-mortem treatments, including electrical stimulation, rapid chilling and fabrication within 1h or 24h post-mortem. Sensory attributes appear unaffected by time of processing, except for crispiness ratings, which are lower for hot-processed bacon (Abu-Baker *et al.*, 1983; Neel *et al.*, 1988). Cut distortions may be associated with hot-processed pork bellies unless a bacon press is used (Mandigo and Henrickson, 1967).

Taylor *et al.* (1982), reported similar yields for pre- or post-rigor cured Wiltshire bacon and acceptable bacon produced within 5 days from slaughter.

In terms of yields or major quality traits of bacon, pre-rigor processing does not seem to have a great advantage or disadvantage, as only minor changes were observed when compared with conventional processing.

Cured cooked products

Equal or increased cooking yield (CY) has been shown when ham is produced from pre-rigor meat (Mandigo and Henrickson, 1966; Motycka and Betchel, 1983; Troeger and Woltersdorf, 1987; Laack, van *et al.*, 1989). At pH values below 5.9, the benefit of using pre-rigor meat for hams are greatly reduced (Troeger and Woltersdorf, 1987). Addition of phosphate to the brine increased CY for hams independently of time at processing (Laack, van *et al.*, 1989).

Advantages of hot-processing are suggested to be most marked in products with inferior water-binding potential (Smulders and van Laack,

1988). Fast glycolysing muscles generally show lower CY values, however, the yield enhancing effect of pre-rigor processing was reported to be unaffected by pH_1 values of 6.60 or 6.29 used for characterizing slow and fast glycolysing muscles respectively (van Laack, 1989). Sensory evaluation of color and palatability traits showed no appreciable effect derived from the raw material, whether used for hot or cold processing (van Laack *et al.*, 1989). Other reports have shown improved juiciness, tenderness, flavor and color from pre-rigor processing of hams (Troeger and Woltersdorf, 1987; Chow *et al.*, 1986). Increased nitroso pigment content for hot processed ham is suggested to be caused by an accelerated rate of cure diffusion in pre-rigor meat (Arganosa and Henrickson, 1969).

Tumbling or massaging is well known to facilitate the extraction of salt soluble proteins which, after heat coagulation, form a bind matrix to increase yield, cohesiveness and juiciness. Chow *et al.* (1986), studied the main effects (e.g. electrical stimulation, rigor condition and tumbling) and their interactions on various quality characteristics of canned pork shoulder. No effect of rigor condition appeared on CY but the amount of increased CY due to tumbling was greater in pre-rigor than in post-rigor muscles.

Cooked sausages

The myofibrillar system plays a dominant role in water binding and related quality parameters such as tenderness, juiciness and color in comminuted meat products (for review, see Hamm, 1986; Offer and Trinick, 1983). Addition of salt to a muscle homogenate causes an increase in water-holding capacity (WHC) that is much more pronounced in pre-rigor muscle than in homogenates made from post-rigor muscle. Neither cold shortening, nor rigor contraction at elevated temperatures, which may occur in hot boned meat, has a significant influence on the amount of water released upon heating of a sausage batter prepared from such a material (Honikel *et al.*, 1981; Kim *et al.*, 1985). Unlike WHC, the tenderness of meat is strongly influenced by the degree of overlapping of myofilaments, but cold-induced toughening (Offer and Trinick, 1983) was largely reduced by curing before cooking (Møller, 1991).

Troeger and Woltersdorf (1987), used the pH value before processing as an indicator of rigor development in hot boned meat used for a frankfurter-type sausage production. At pH values above 5.9 they observed only slight variations in cooking loss but a further drop in the pH of the raw material caused the percentage of jelly and fat deposit to increase significantly above its former values.

The NaCl concentration necessary to maintain the pre-rigor salting effect has been studied. Hamm (1986), reported that in order to attain a higher water-binding capacity in pre-rigor beef homogenates, the NaCl concentration must be at least 1.8%. This result is in general agreement with Puolanne and Terrell (1983), who found a 2% NaCl concentration necessary in pre-rigor salted pork. Bernthal *et al.* (1989), observed the largest pre-rigor salting effect with at least 2.0% NaCl but 0.5 and 1.0% NaCl in pre-rigor homogenates had advantages over similar and higher NaCl concentrations in post-rigor homogenates.

Fresh sausages

Flavor stability is a major problem that limits storage life of fresh sausages (Drerup *et al.*, 1981). Judge and Aberle (1980), reported that pre-rigor grinding of either light or dark pork semitendinosus muscle reduced the rate of lipid oxidation during storage at 2°C. Salt (2% NaCl) was shown to promote oxidation in both pre-rigor, and post-rigor ground muscle, but pre-rigor ground and salted muscle oxidized more slowly than muscle that was ground and salted post-rigor. From the latter study, pH was negatively correlated with thiobarbituric acid reactive substances (TBARS). The higher pH values associated with grinding and salting pre-rigor may be due to the effect of NaCl and increased chilling rate on inhibition of post-mortem glycolysis, combined with a prolonged aerobic metabolism supported by the oxygen introduced during pre-rigor grinding (Judge and Aberle, 1980).

pH-related properties

The technological quality of pork is generally known to be quite variable and mainly due to differences in rate and extent of the post-mortem decline in pH values in the muscle fiber. A fast rate of pH fall leads to pale, soft and exudative (PSE) meat because of extensive protein denaturation, whereas limited pH changes result in dark, firm and dry (DFD) meat (Bendall and Wismer-Pederson, 1962). Muscle glycogen level at slaughter also causes variation in pork quality, since it may result in the technological defects of acid meat or the so-called, Hampshire-type meat (Monin and Sellier, 1985). Several reviews have dealt with this topic, either in general (Hofman, 1988) or with the focus on specific quality parameters such as WHC (Hamm, 1986; Offer and Trinick, 1983; Bendall and Swatland, 1988), tenderness (Newbold and Harris, 1972), or color (Seideman *et al.*, 1984).

Cured products

Bacon is one of the most variable products in terms of overall composition. Lean:fat ratio of bacon is generally considered to be the most important quality attribute of the product, lean pork being preferred. Observations on processing and cooking shrink have indicated that fat bellies show higher yields after processing but also result in greater cooking losses when fried (Smith *et al.*, 1975).

When cuts of meat with reduced WHC, as in PSE meat, are cured, the net gain during curing is less than normal. When whole sides are cured and the muscle membranes left intact, as in the Wiltshire process, the loss of fluid from the meat is reduced (Wismer-Pederson, 1968). More recent research showed PSE carcasses yielded 1.6% less mature bacon and lost 1% more drip than normal carcasses (Smith and Lesser, 1982).

The effect of the PSE condition on the appearance of cured bacon is not so clear. During curing, nitrite is added to meat and the pigments present are converted to nitrosomyoglobin. According to Wismer-Pedersen (1968), meat with a reduced WHC absorbs more pickle than normal meat when cured, and frequently Wiltshire bacon of normal quality is produced. Taylor *et al.* (1973), found that PSE bacon was paler than normal bacon when measured by reflectometer but visible color differences were only seen in the two-toned ST muscle. In the study of Warriss and Akers (1980), an inverse relationship between pH_1 and the appearance of Wiltshire bacon was found, an improved score for bacon quality being associated with lower pH_1 values in the carcass.

Cooked cured products

The PSE condition is less desirable for ham manufacturing when large pieces of muscle are used and hence the condition of the raw material is of particular importance. The reduced WHC leads to an increase in percentage gelatinous cook-out (% jelly) in the cans due to a higher than normal degree of aggregation of the meat proteins, especially for pasteurized, canned ham (Wismer-Pedersen, 1968). Polyphosphates are commonly used to improve WHC and technological yield (TY) and PSE ham processed with polyphosphates showed significant improvement (Davis *et al.*, 1975). In a study by Honkavaara (1988), PSE pork ($pH_1 \le 5.8$) and non-PSE pork ($5.8 < pH_1 \le 6.4$) was examined for cooked cured ham production. PSE pork resulted in decreased, although acceptable sensory scores and in decreased TY at 94.0 compared with 105.9% for non-PSE ham. More recently, Müller (1991), reported similar relationship between pH and quality of ham. At increased pH values, the yield of cooked ham became higher, while the number of hams with pores and holes as well as the amount of juice exudation declined. Corresponding to the higher yield, hams with increased pH also resulted in a more tender product. Processing of ham from raw material with pH between 5.8 to 6.4 was considered as the optimum range.

Some recent reports have concentrated on relating variation in pH_u to changes in technological quality of Paris ham processing (Sellier *et al.*, 1988; Fernandez *et al.*, 1991). Paris ham is a type of cured ham, generally cooked in molds. The ultimate pH of meat depends mainly on muscle glycogen content at the time of slaughter and various factors affecting glycogen content were reviewed recently (Fernandez and Tornberg, 1991). In earlier studies, the glycogen content in muscle was significantly increased by feeding pigs a high sugar diet before slaughter with the aim of reducing DFD pork for bacon processing (Fernandez and Tornberg, 1991). The sides from the sugar-fed pigs had a higher gain during tank curing and a lower weight loss during maturation, due to the accumulation of reducing sugars in the meat. When the meat was heat processed (canned ham), the effect of the lower pH in the sugar-fed group showed an increased shrinkage. The effect of sugar on weight loss was suggested to be associated with the raw meat (Wismer-Pedersen, 1959).

Breed is one of the factors influencing muscle glycogen content. Comparisons involving Hampshire pigs indicate that these have an extended duration of glycolysis because of high initial levels of muscle glycogen, especially in fast-white muscles (Essén-Gustavsson and Fjelkner-Modig, 1985; Monin *et al.*, 1987). This unusual high glycogen content can lead to a lower pH_u in the ham resulting in pale color and low WHC. It has been suggested that muscle glycogen content also influences the technological quality of pig meat independently of its direct effect on pH_u (Fernandez *et al.*, 1991). Glycogen in muscle is capable of binding 2-4g of water per gram. This water is likely to be freed during processing, which could lead to an excess of free water during cooking. A close relationship was indicated between TY of Paris ham and muscle residual glycogen ($r = 0.60$, $P < 0.01$), while lower correlations appeared between TY and both protein content and pH ($r = 0.4$, $P < 0.01$) (Fernandez *et al.*, 1991).

Cooked sausages

In addition to decreased WHC, the ability of meat to emulsify fat, as in frankfurter and bologna processing, is impaired by the increased acid condition of PSE muscle (Hamm, 1986).

Camou and Sebranek (1991), recently reported on thermally induced gelation characteristics of salt-soluble proteins from PSE compared with normal meat. Gel strength of PSE extracts was 45% of that obtained from normal meat at equivalent protein concentration, indicating that native protein from PSE muscle was considerably less functional even after solubilization. In the study of Camou and Sebranek (1991), the effect of heating rates on gel strength were studied. Slow heating rate (17°C/h), compared with a fast heating rate (93°C/h), resulted in higher gel strength in both PSE and normal samples.

Honkavaara (1989) evaluated the economic consequences of using PSE meat in the preparation of an all-pork, cooked, emulsion-type sausage. Fifteen percent more trimmings from PSE carcasses should be added than to those from normal carcasses to bind the same amount of water. The chemical composition and sensory quality of the PSE-formulated sausages did not differ from those of normal sausages (Honkavaara, 1989). Thus, the PSE muscles may be utilized if the condition is recognized and the differences in behavior are taken into account during formulation and processing.

Fermented sausages

In raw cured products, such as fermented dry sausages, the tendency of PSE meat to lose water has been suggested as a potential advantage. While a low pH is desirable in the products, the use of only PSE meat in dry sausages can cause drying faults because too much water is released (Wirth, 1986). Townsend *et al.* (1980), reported dry sausage made from PSE pork showed higher weight loss and required only 50-60% of the time required for drying products from normal pork, but sausages made from PSE meat only resulted in more pale color and poor consistency. In a study by Honkavaara (1988), PSE pork ($pH_1 \leq 5.8$), and non-PSE pork ($5.8 < pH_1 \leq 6.4$) was used with beef in a 48/52 mixture for fermented dry sausage manufacture. Percentage weight loss was higher in PSE than in non-PSE sausages, but PSE sausages were preferred in texture, flavor and color. Wirth (1986), reported that PSE meat can be used up to a level of 30% of the total meat without any recognizable changes during the dry sausage ripening.

Leaner pork production

Fat reduction in pork has been successfully achieved by genetic selection. Increased muscularity, especially in stress-susceptible pigs, is associated with larger muscle fibers with lower capillary density and a transformation of αR fibers to αW fibers (Ashmore *et al.*, 1972; Essén-Gustavsson *et al.*, 1992). A major question is therefore whether or not technological problems, mainly related to the PSE condition, are an inevitable consequence when pork becomes leaner. As an alternative to genetic improvement in growth and carcass quality, recent research has focused on the use and administration of growth promoters, such as porcine growth hormone, porcine somatotropin (pST), and β-adrenergic agonists (bAA). The advantage of these compounds in improving carcass quality are well documented (Etherton *et al.*, 1987; Warriss *et al.*, 1990) However, limited work has been published about the effect of those compounds on meat technological characteristics or quality of processed meat. The use of intact males has also gained increased attention for producing leaner pork.

Technological characteristics

Apart from PSE meat, the effects of breed improvement on other meat quality parameters are less known, being confined mostly to chemical composition. The phenotypic changes in meat quality in breeds used in Denmark (Landrace, Yorkshire, Duroc and Hampshire) during 1983-1987 has been evaluated (Barton-Gade, 1990). During this period, meat content increased (1-2%, depending on breed) along with a decrease in intramuscular fat%, which on average were related to a slightly poorer tenderness in LD. This relationship may partly be influenced by leaner carcasses being more prone to cold shortening (Barton-Gade *et al.*, 1987; Møller and Vestergaard, 1987). During the period 1983-1987, protein content in the SM muscle increased, resulting in a higher potential for yield in e.g. bacon and canned ham manufacturing. The total heme pigment concentration decreased throughout the period for BF and LD, possibly indicating a change in fiber composition towards glycolytic fast-twitch fibers and an increased risk of PSE meat in long-term selection for leaner pork (Barton-Gade, 1990).

Porcine somatotropin (pST)
In fresh meat, pST has shown to increase protein and moisture content, whereas lipid concentration is decreased (Beerman *et al.*, 1988). Mourot *et al.* (1992) reported that the percentage polyunsaturated fatty acids was significantly increased by pST in backfat and lipids within LD, whereas no clear trend was observed concerning fatty acid composition of AD muscle. Work on characterizing muscle fiber types in pigs receiving pST has shown an increase in muscle fiber size for all three muscle fiber types but inconsistent changes in muscle fiber distribution (Solomon *et al.*, 1990; Solomon *et al.*, 1991). From the latter studies, pST resulted in hypertrophied (giant) fibers.

At present, limited information is available about the processing characteristics of pST-treated pork. Some researchers have noticed an increase in the incidence of PSE meat with pST treatment (Solomon *et al.*, 1990; Solomon *et al.*, 1991), while other studies show that the rate of initial pH fall after slaughter is not affected by pST administration (Mourot *et al.*, 1992; Demeyer *et al.*, 1988). No consistent effect of pST on ultimate pH values appear from earlier studies (Mourot *et al.*, 1992; Demeyer *et al.*, 1988; Ender *et al.*, 1989).

Treatment of animals with pST has shown only minor influences on cooking yields and proximate composition. Prusa *et al.* (1990), and Boles *et al.* (1991), reported no difference in yield from ham due to pST treatment, whereas Halloran *et al.* (1991) found lower final yields for hams from pST-treated animals. In cured, cooked SM muscles, Boles *et al.* (1991) observed no significant effect on protein and lipid content but a significant increase in moisture content was noted. Halloran *et al.* (1991) found higher moisture and protein but lower fat content in hams produced from the pST group of animals.

Color and sensory attributes are not significantly influenced by pST (Prusa *et al.*, 1990; Boles *et al.*, 1991; Halloran *et al.*, 1991), although Halloran *et al.* (1991) reported Warner-Bratzler shear force values were higher in cured, cooked loins obtained from pigs receiving pST. Prusa *et al.* (1990) observed no effect on tenderness in hams from pST treatment, whereas Canadian-style bacon received a lower tenderness score.

β-adrenergic agonists (bAA)

A change in fiber type distribution may also be involved by using bAA to obtain increased lean in pork. Thus, a decreased proportion of type IIA fibers and an increased proportion of type IIB fibers are reported (Aalhus *et al.*, 1991; Oxbjerg, 1992). The effect of reducing fat content by bAA (salbutamol) on lean meat quality has recently been evaluated by Warriss *et al.* (1990), in order to identify possible interactions between genotype, muscle type (LD, SM, AD, SS) and response to the agonist. These workers reported reduced total heme pigment concentration in all muscles, but only LD showed decreased level of intramuscular fat in treated animals. bAA did not promote PSE development but tended to deplete glycogen in muscles resulting in slightly elevated pH_u values. Warriss *et al.* (1990) reported that, due to the reduced fatness from bAA treatment, there was an increased propensity for the fat layer to separate from the underlying lean, however, there was no effect on fat firmness. An increased content of linoleic acid of subcutaneous fat is reported to be caused by bAA (Dazzi *et al.*, 1991). The effect of bAA on meat tenderness has only been reported on fresh meat, showing generally an increased toughness probably caused by a decrease in endogenous proteinases responsible for meat tenderization during aging (Tarrant, 1987).

Boar meat

It is well established that using boars instead of castrates in pig meat production results in lower production costs and increased carcass leanness. The main reason for not commercializing boar pork is the problem of tainted carcasses, although boar taint occurs in only a limited proportion of the boar carcasses. The main contributing compounds to boar taint are considered to be androstenone (fat-soluble) and skatole (fat- and water-soluble). A closer relationship is reported between skatole and subjective impression of boar taint than between androstenone and boar taint (Lundström *et al.*, 1988). An objective screening test for skatole has been developed in Denmark (Mortensen and Sorensen, 1984), and is now in use as a fully automated system operating on the slaughter line. For the producers, eventually receiving penalties for delivering boars above the acceptable level of skatol in the fat, the major problem still is that the mechanism for skatol production and deposition in pig tissues is not fully understood.

Compared with breed, the effect of sex has revealed smaller effects on eating and processing quality of pork. Barton-Gade (1987) reported no difference between the sexes, with respect to pigment content and PSE/DFD status. Due to a generally reduced backfat thickness, boars show increased fat/lean separation and reduced fat firmness, as a result of a higher content of linoleic acid (Barton-Gade, 1987; Wood *et al.*, 1988). The processing quality of the lean meat from boars is slightly poorer, mainly caused by a less protein and higher water content (Barton-Gade, 1987). In canned hams, where the protein content of the raw material determines the yield of product, the reduced protein in boar meat result in 2-3% less yield compared with gilts and 0.8% less yield than for castrates (Barton-Gade, 1987).

Consumer acceptability of boar pork appears to be very different among studies. The inconsistency may be due to differences among consumers in sensitivity to tainted meat and/or to higher frequency and intensity of taint in the raw material. Differences between meat from boars and controls may be reduced by processing (for review see Malmfors and Lundström, 1983). Results reported on boar bacon generally have been positive. In brine-cured pork bellies, however, Lundström *et al.* (1988), reported a larger proportion of both untainted and tainted boar meat less pleasant in odor during preparation and eating, while the greater leanness of boar meat was appreciated by the consumers. Desmoulin and Bonneau (1982) reported that overall acceptability of processed products (eaten cold) from boars was only slightly reduced compared with controls. These differences were mainly due to a few boars having high androstenone levels. When boar carcasses were selected accoring to different levels of androstenone, boar products of cooked ham and dry sausage in which residual androstenone content exceeded 1μ/g were unacceptable. Flavor ratings of both cooked ham and dry sausage did not differ significantly for controls and for boars below 0.5μ/g androstenone in samples. For frankfurter sausage, Plimpton *et al.* (1976) stated that sausage mix containing up to 50-70% tainted boar meat was acceptable and that boar pork with a moderate taint level could possibly be used as the sole meat component.

The improvement in consumer acceptance of boar meat after processing is suggested as being due to odor disappearance during processing, and cold consumption of meat products minimizing odor release (Desmoulin and Bonneau, 1982). Added spices may mask unpleasant flavors and the boar odor problem can be reduced through dilution with untainted meat. The economic benefits of using boars for meat production are dependent on the proportion of tainted boar carcasses and whether or not these are utilized efficiently.

Dietary fat

Saturated fat is considered a primary cause of hypercholesterolemia (Mattson and Grundy, 1985). Human diets high in mono-unsaturated and polyunsaturated fat have been associated with decreases in the incidence of coronary heart disease. A practical method for increasing mono-unsaturates and polyunsaturates in the human diet is to modify the dietary regimen of meat-producing monogastric animals. Recently, there have been several studies aimed at increasing the ratio between mono-unsaturated and saturated fatty acids (M/S ratio) and between polyunsaturated and saturated fatty acids (P/S ratio) in pork through animal diet modifications, and to further investigate the effect on quality of pork and pork products (Acuff *et al.*, 1990; Miller *et al.*, 1990; Rhee *et al.*, 1990a; 1990b; Shackelford *et al.*, 1990a; 1990b; 1990c; Zirpin *et al.*, 1990; Kuhn *et al.*, 1991).

Lipid oxidation is one of the major causes of deterioration in the quality of meat and meat products, and oxidative deterioration of meat products can directly affect many quality parameters such as color, flavor, texture, nutritive value and safety (Pearson *et al.*, 1983). There have been several attempts to increase the stability of lipids in pork and pork products by adding vitamin E to the diet (Buckley and Connolly, 1980; Asghar *et al.*, 1989; Buckley *et al.*, 1989; Monahan *et al.*, 1990, 1992).

Cured products

An elevated level of polyunsaturates in the diet seems to affect the quality of bacon negatively. In an experiment by Shackelford *et al.* (1990a), a 10% addition of canola oil in the diet, which increased the amount of polyunsaturated fatty acids in the neutral lipids in the bacon from 8.1% in the control to 21.1% for canola oil, decreased slicing yields, and off-flavors were detected most frequently (65% of the panelists compared with 5% for the control bacon). On the contrary, Mazhar *et al.* (1990) did not find any effect of the diet on the processing yield of bacon produced from pigs fed diets containing 20% intact canola, 20% ground canola or 0% canola, but, after 8 weeks of cold (2°C) storage, vacuum-packed bacon from pigs fed ground canola had significant higher thiobarbituric acid reacting substances (TBARS) values than bacon from pigs fed either intact canola or without any canola in the diet.

Bacon produced from pigs fed diets containing elevated levels of mono-unsaturated fatty acids, however, does not appear to be inferior in oxidative stability (Shackleford *et al.*, 1990a; Ziprin *et al.*, 1990). In the study of Shackelford *et al.* (1990a), a 10% addition of sunflower or safflower oil in the diet resulted in 20-25% more oleic acid in the neutral lipids from the bacon. The dietary supplementation of sunflower and safflower did not result in off-flavors or reduced processing yields, but the high oleic acid contents were found to decrease crispiness. Ziprin *et al.* (1990), evaluated the oxidative stability during 4 months freezer (-20°C) storage of bacon from pigs fed either a control sorghum-soya bean diet or a similar diet containing 12% high-oleic sunflower oil. The extent of the accumulation of TBARS during storage was not affected by the dietary treatment, regardless of the packaging method (waxed freezer paper or vacuum-packed).

Feeding supplemental vitamin E (800mg/day/animal) to pigs 1 week before slaughter does not appear to affect the oxidative stability of bacon (vacuum-packed or packed in a oxygen-permeable packaging material) during freezer storage probably due to the presence of nitrite, which can also act as an antioxidant (Buckley and Connolly, 1980).

Cured cooked products

Feeding pigs with an elevated level of monounsaturates does not affect the palatability of pasteurized hams produced from the pigs (Shackelford *et al.*, 1990c), whereas an elevated level of polyunsaturates presumably results in problems with off-flavors. Shackelford *et al.* (1990c) fed pigs a control diet of corn and soya bean meal and four similar test diets that contained a 10% addition of either animal fat, safflower oil, sunflower oil or canola oil. The dietary treatment resulted in an increase in oleic acid in hams from the sunflower- and the safflower-oil treatment and in an increased proportion of polyunsaturated fatty acids in the muscle tissues from pigs fed canola oil. Dietary treatment did not affect any yield parameters, but off-flavors were most frequently detected in samples from the canola oil. In this study no storage experiment was included, which could reveal differences in the oxidative stability due to dietary treatment. Olsen *et al.* (1973) however, included chill storage in an experiment with pasteurized hams processed from pigs fed either a corn diet or a corn-roasted full-fat soyabean diet. Feeding the soyabean diet, which results in a higher content of polyunsaturated fatty acids, did not reduce the palatability of the chill-stored product, but the hams were stored in cans, presumably with a very low residual oxygen content in the head-space, minimizing oxidation in the product.

Cooked sausages

Increasing the M/S or P/S ratio through animal diet modification results in inferior sensory quality (more dry, less springy, less flavorful etc.) of high fat (> 20%) frankfurters (Shackelford *et al.*, 1990a; 1991) whereas low-fat (10%) frankfurters produced from the test diets (added 10% safflower oil, sunflower oil or canola oil) were comparable in quality with the control (Shackelford *et al.*, 1991). The oxidative stability measured by determination of thiobarbituric acid reacting substances (TBARS) of vacuum-packed frankfurters stored at 2°C for 24 weeks were not affected by dietary treatment, and the products were acceptable during the storage period (Shackelford *et al.*, 1991). St.John *et al.* (1986), however, found an increased tendency to off-flavors during 6 weeks chill storage of frankfurters produced from pig meat fed a test diet containing 20% canola oil compared with a control diet without any canola oil.

Patties, restructured products and pre-cooked pork products

Cooked meats are very susceptible to lipid oxidation, and the rapid onset of rancidity in those products was first recognized by Tims and Watts (1958), who used the term Warmed Over Flavor (WOF) to describe the phenomenon. WOF can also be encountered in grounded or restructured raw meat (Sato and Hegarty, 1971), where the muscle membranes have been disrupted making the phospholipids more susceptible to oxidation.

A high content of polyunsaturates in the diet have been shown to increase lipid oxidation in raw (ground) and in cooked muscles. Thus Monahan *et al.* (1992) showed that cooked pork patties from pigs fed diets containing 3% soya oil were significantly more susceptible to lipid oxidation during chill (4°C) storage than the same product from pigs fed a 3% tallow diet. In muscles from pigs fed the soya diet, the C18:2 content in the polar lipid fraction, containing the membranal phospholipids responsible for the initial stages of rancidity in cooked meat (Igene and Pearson, 1979; Gray and Pearson, 1987), was 25% higher than in the tallow diet, which may be responsible for the increased susceptibility of the cooked muscle to oxidation.

Dietary supplementation of α-tocopherol both short-term (2 weeks before slaughter) and long-term (10 weeks before slaughter) have been shown to improve the oxidative stability of raw pork patties and of cooked pork products during refrigerated or frozen storage (Buckley *et al.*, 1989; Monahan *et al.*, 1990, 1992). The extension of shelf-life through the incorporation of α-tocopherol acetate via the diet is probably due to stabilization of the membranal lipids (Asghar *et al.*, 1989; Buckley *et al.*, 1989; Monahan *et al.*, 1990).

With the intention of increasing the M/S ratio in raw products (e.g. patties and restructured chops) and cooked products Rhee *et al.* (1990a, 1990b), incorporated 12% high-oleic sunflower oil (HOSO) into a swine diet. Sensory properties and oxidative stability measured by TBARS values during chill storage were not adversely affected by changes in fatty acid composition. In fact, cooked roasts from the HOSO groups had lower TBARS values after 4 days chill storage. The reason for this unexpected result might be a substantial amount of tocopherols in the crude HOSO, reducing the susceptibility of the muscle tissue toward lipid oxidation.

Future developments

A major question in future work is to clarify whether further leanness - as obtained by genetic selection or eventually by growth promoters - is inevitably linked to decreased functional properties of muscle. Although not consistently approved, some indications of reduced technological characteristics due to increased muscularity has appeared such as increased potential for PSE/DFD, decreased heme protein and intramuscular fat content. Further development in on-line measurements for quality control - including boar odour - would allow more optimal utilization of raw material and also of muscles not of standard quality. Reduced fat, an elevated level of unsaturates and less use of additives are among the main issues for improving the consumer's image of processed pork. In our concern for quality improvements of processed pork the importance of culinarity, however, should hopefully not be overshadowed by the prevailing trend in recent years toward an increased awareness of dietary and health. In future work, juicy, tasty and tender pork may still be major goals for the processing industry.

Acknowledgements

This research is part of the FØTEK program, sponsored by the Danish Ministry of Education and Research. The authors would also like to thank Claus Jensen, for his help with the literature search, and Maud Hammershøy, for her help with the preparation of the manuscript.

References

Aalhus, J.L., Gariepy, C., Murray, A.C., Jones, S.D.M. and Tong, A.K.W. 1991. Stunning and shackling influences on quality of porcine *longissimus dorsi* and *semimembranosus* muscles. Meat Sci. 29, 323.

Abu-Bakar, A., Reagan, J.O., Wynne, R.L., Davis, C.E., Townsend, W.E. and Carpenter, J.A. 1983. Chemical, physical and sensory attributes of pre- and post-chill processed hams. J.Food Sci. 48, 1886.

Acuff, A.R., Lucia, L.M., LePage, L.E., Vanderzant, C. and Rhee, K.S. 1990. Microbiological characteristics of pork chops and ground pork from swine fed a high mono-unsaturated fat diet. Meat Sci. 28, 237.

Arganosa, F.C. and Hendrickson, R.L. 1969. Cure diffusion through pre- and post- chilled porcine muscles. Food Technol. 23, 75.

Asghar, A., Gray, J.I., Booren, A.M., Miller, E.R., Pearson, A.M., Ku, P.K. and Buckley, D.J. 1989. Influence of dietary vitamin E on swine growth and meat quality. Proc. 35th Int. Congr. Meat Sci. and Technol. Copenhagen, Denmark, p. 1065.

Ashmore, L.R., Thompkins, G. and Doerr, L. 1972. Postnatal development of muscle fiber types in domestic animals. J. Anim. Sci. 34, 37.

Barton-Gade, P.A. 1987. Meat and fat quality in boars, castrates and gilts. Livestock Prod. Sci. 16, 187.

Barton-Gade, P.A. 1990. Danish experience in meat quality improvement. IV World Congr. on Genetic Applied to Livestock Production. Edinburgh, Ireland, p. 511.

Barton-Gade, P.A., Beyerholm, C. and Borup, W. 1987. Influence of different chilling procedures on the eating quality of pork chops. Proc. 33rd Int. Congr. Meat Sci. and Technol., Helsinki, Finland, p. 181.

Beerman, D.H., Armbruster, G., Boyd, R.D., Reneke, K. and Fagin, K.D. 1988. Comparison of the effects of two recombinant forms of porcine somatotropin (pST) on pork composition and palatability. J. Anim. Sci. 66 (suppl. 1), 281.

Bendall, J.R. and Wismer-Pedersen, J. 1962. Some properties of the fibrillar proteins of normal and watery pork muscles. J.Food Sci. 27, 144.

Bendall, J.R. and Swatland, H.J. 1988. A review of the relationship of pH with physical aspects of pork quality. Meat Sci. 24, 85.

Bernthal, P.H., Booren, A.M. and Gray, J.I. 1989. Effect of sodium chloride concentration on pH, water-holding capacity and extractable protein of pre-rigor and post-rigor ground beef. Meat Sci. 25, 143.

Boles, J.A., Parrish, F.C., Skaggs, C.L. and Christian, L.L. 1991. Sensory and processing properties of cured *semimembranosus* muscle from stress-susceptible pigs treated with porcine somatotropin. J. Anim. Sci. 69, 4049.

Buckley, D.J. and Connolly, J.F. 1980. Influence of alpha-tocopherol (vitamin E) on storage stability of raw pork and bacon. J. Fd. Prot. 43, 265.

Buckley, D.J., Gray, J.I., Asghar, A., Price, J.F., Crackel, R.L., Booren, A.M., Pearson,A. and Miller, E.R. 1989. Effects of dietry antioxidants and oxidized oil on membranal lipid stability and pork product quality. J. Food Sci. 54, 1193.

Camou, J.P. and Sebranek, J.G. 1991. Gelation characteristics of muscle proteins from pale, soft, exudative (PSE) pork. Meat Sci. 30, 207.

Chow, H-S., Ockerman, H.W., Cahill, V.R. and Parrett, N.A. 1986. Evaluation of cured, canned pork shoulder tissue produced by electrical stimulation, hot processing and tumbling. J. Food Sci. 51, 288.

Cuthbertson, A. 1980. Hot processing of meat: A review of the rationale and economic implications. In: Lawrie, R. (ed.), Developments in Meat Science. Appl. Science Publishers London. p. 61.

Davis, G.W., Smith, G.C., Carpenter, Z.L. and Cross, H.R. 1975. Relationship of quality indicators to palatability attributes of pork loins. J. Anim. Sci. 41, 1305.

Dazzi, G., Madarena, G., Campesato, E., Campanini, G. and Chizzolini, R. 1991. β-agonist treatment of heavy pigs. Effects on growth, carcass and meat quality parameters. Fleischwirtsch. 71, 179.

Demeyer, D., Verbeke, R. and Van de Worde, G., Fabry, J., Deroover, E. and Dalrymple, R.H. 1988. Effect of recombinant porcine somatotropin on carcass and meat quality of Belgian and Dutch pigs. In: van der Wal, P., Nieuwhof, G.J. and Politiek, R.D. (eds), Biotechnology for Control of Growth and Product Quality in Swine. Implications and Acceptability, Pudoc, Wageningen, p. 191-9.

Desmoulin, B. and Bonneau, M. 1982. Consumer testing of pork and processed meat from boars: The influence of fat androstenone level. Livest. Prod. Sci. 9, 707.

Drerup, D.L., Judge, M.D. and Aberle, E.D. 1981. Sensory properties and lipid oxidation in pre-rigor processed fresh pork sausage. J. Food. Sci. 46, 1659.

Ender, K., Lieberenz, M., Poppe, S., Hackl, W., Pflughaupt, G. and Meisinger, D. 1989. Effect of porcine somatotropin (pST) treatment on growing-finishing pigs: Carcass characteristics. J. Anim. 67 (suppl. 1), 523.

Essén-Gustavsson, B. and Fjelkner-Modig, S. 1985. Skeletal muscle characteristics in different breeds of pigs in relation to sensory properties of meat. Meat Sci. 13, 33.

Essén-Gustavsson, B., Karlström, K. and Lundström, K. 1992. Muscle fibre characteristics and metabolic response at slaughter in pigs of different halothane genotypes and their relation to meat quality. Meat Sci. 31, 1.

Etherton, T.D., Wiggins, J.P., EVock, C.M., Chung, C.S., Rebhun, H., Walton, P.E. and Steele, N.C. 1987. Stimulation of pig growth performance by porcine growth hormone; determination of the dose-response relationship. J. Anim. Sci. 64, 433.

Fernandez, X. and Tornberg, E. 1991. A review of the causes of variation in muscle glycogen content and ultimate pH in pigs. J. Muscle Foods 2, 209.

Fernandez, X., Lefaucheur, L., Gueblez, R. and Monin, G. 1991. Paris ham processing: Technological yield as affected by residual glycogen content of muscle. Meat Sci. 29, 121.

Gray, J.I. and Pearson, A.M. 1987. Rancidity and warmed over flavor. Adv. Meat Res. 3, 221.

Halloran, J.D., Rogers, R.W., Mikel, W.B. and Althen, T.G. 1991. Processing characteristics of pork as influenced by porcine somatotropin (pST) administration to growing finishing swine. J. Food. Sci. 56, 859.

Hamm, R. 1986. Functional properties of the myofibrillar system and their measurements. In: Bechtel P.J. (ed.), Muscle as a Food. Academic Press, London. p.135.

Hofman, K. 1988. pH. A quality criterion for meat. Fleischwirtsch. 68, 67.

Honkavaara, M. 1988. Influence of PSE pork on the quality and economics of cooked, cured ham and fermented, dry sausage manufacture. Meat Sci. 24, 201.

Honkavaara, M. 1989. Effect of PSE pork on the processing properties of cooked meat products. Fleischwirtsch. 69, 1573.

Honikel, K.O., Hamid, A., Fischer, E. and Hamm, R. 1981. Influence of post-mortem changes in bovine muscle on the water-binding capacity of beef. Post-mortem storage of muscle at various temperatures between 0° and 30°C. J. Food Sci. 46, 23.

Igene, J. and Pearson, A.M. 1979. Role of phospholipids and triglycerides in warmed-over flavor development in meat model systems. J. Food Sci. 44, 1285.

Judge, M.D. and Aberle, E.D. 1980. Effect of pre-rigor processing on the oxidative rancidity of ground light and dark porcine muscles. J. Food Sci. 45, 1736.

Kim, C.J., Honikel, K.O. and Hamm, R. 1985. Veränderungen im Schweinemuskel nach dem Schlachten und ihr Einfluss auf das Wasserbindungsvermögen von Fleisch und Muskel Homogenaten. Fleischwirtsch. 65, 645.

Kuhn, B., Haeuser, A., Frigg, M., Voegeli, P. and Prabucki, A. 1991. The influence of vitamin E and lipoprotein allotypes on oxidation stability of pork. J. Anim. Breed and Gen. 108, 227.

Lundström, K., Malmfors, B., Malmfors, G., Stern, S., Petersson, H., Mortensen, A.B. and Sorensen, S.E. 1988. Skatole, androstenone and taint in boars fed two different diets. Livestock Prod. Sci. 18, 55.

Malmfors, B. and Lundström, K. 1983. Consumer reactions to boar meat - A review. Livest. Prod. Sci. 10, 187.

Mandigo, R.W. and Henrickson, R.L. 1966. The influence of pre-chill processing techniques on bacon. Food Technol. 20, 186.

Mandigo, R.W. and Henrickson, R.L. 1967. Influence of hot processed pork carcasses on cured ham. Food Technol. 21, 1262.

Mattson, F.H. and Grundy, S.M. 1985. Comparison of dietary saturated, mono-unsaturated and polyunsaturated fatty acids on plasma lipids and lipoproteins in man. J. Lipid Res. 26, 194.

Mazhar, A., Busboom, J.R., Field, R.A., Rule, D.C., Heald, T., Russell, W.C. and McCormick, R.J. 1990. Functional characteristics, fatty acid composition, and palatability of bacon from pigs fed canola. J. Food Sci. 55, 575.

Miller, M.F., Shackelford, S.D., Hayden, K.D. and Reagan, J.O. 1990. Determination of the alteration in fatty acid profiles, sensory characteristics and carcass traits of swine fed elevated levels of monounsaturated fats in the diet. J. Anim. Sci. 68, 1624.

Møller, A.J. 1991. The tenderizing effect of NaCl on cold-shortened pork. Proc. 37th Int. Congr. Meat sci. and Technol., Kulmbach, Germany, p. 758.

Møller, A.J. and Vestergaard, T. 1987. Effect of delay time before chilling on toughness in pork with high or low initial pH. Meat Sci. 19, 27.

Monahan, F.J., Buckley, D.J., Gray, J.I., Morrissey, P.A., Asghar, A., Hanrahan, T.J. and Lynch, P.B. 1990. Effect of dietary vitamin E on the stability of raw and cooked pork. Meat Sci. 27, 99.

Monahan, F.J., Buckley, D.J., Morrisey, P.A., Lynch, P.B. and Gray, J.I. 1992. Influence of dietary fat and alfa-tocopherol supplementation on lipid oxidation in pork. Meat Sci. 31, 229.

Monin, G. and Sellier, P. 1985. Pork of low technological quality with a normal rate of muscle pH fall in the immediate post-mortem period: The case of the Hampshire breed. Meat Sci. 13, 49.

Monin, G., Mejenes-Quijano, A. and Talmant, A. 1987. Influence of breed and muscle metabolic type on muscle glycolytic potential and meat pH in pigs. Meat Sci. 20, 149.

Mortensen, A.B. and Sorensen, S.E. 1984. Relationship between boar taint and skatole determined with a new analysis method. Proc. 30th European Meeting of Meat Res. Workers, Bristol, England, p. 394.

Motycka, R.R. and Bechtel, P.J. 1983. Influence of pre-rigor processing, mechanical tenderization, tumbling method and processing time on the quality and yield of ham. J. Food Sci. 48, 1532.

Mourot, J., Bonneau, M., Charlotin, P. and Lefaucheur, L. 1992. Effects of exogenous porcine somatotropin (pST) administration on pork meat quality. Meat Sci. 31, 219.

Müller, W.-D. 1991. Cooked cured products. Fleischwirtsch. 71, 544.

Neel, S.W., Reagan, J.O., Mabry, J.W. and Miller, M.F. 1988. Effect of rapid chilling and accelerated processing on the physical and sensory characteristics of cured pork bellies. J. Food Sci. 53, 1905.

Newbold, R.P. and Harris, P.V. 1972. The effect of pre-rigor changes on meat tenderness. A review. J. Food Sci. 37, 337.

Offer, G. and Trinick, J. 1983. On the mechanism of water binding in meat: The swelling and shrinking of myofibrils. Meat Sci. 8, 245.

Olsen, D.G., Parrish, J., Rust, R.E. and Miner, B.E. 1973. Effect of feeding roasted soybeans on cured pork palatability. J. Anim. Sci. 37, 49.

Oxbjerg, N. 1992. Effects of body weight, β-adrenergic agonists, and porcine growth hormone on muscle growth and muscle histochemical and metabolic properties in growing pigs. PhD thesis. Royal Veterinary- and Agricultural University, Copenhagen, Denmark.

Pearson, A.M., Gray, J.I., Wolzak, A.M. and Horenstein, N.A. 1983. Safety implications of oxidized lipids in muscle foods. Food Technol. 37 (7), 121.

Plimton, R.F., Ockerman, H.W., Cahill, V.R. and Hilt, E. 1976. The effect of boar odour level, meat block dilution and the addition of fennel spice on the acceptability of cooked frankfurters processed from boar pork. 22nd European Meeting of Meat Res. Workers, Malmö, Sweden, p. A7:1.

Prusa, K.J., Sebranek, J.K., Love, J.A. and Miller, L.F. 1990. Quality attributes of various processed meats from pigs treated with porcine somatotropin. J. Food Sci. 55, 929.

Puolanne, E.J. and Terrell, R.N. 1983. Effects of rigor state, levels of salt and sodium tripolyphosphate on physical, chemical and sensory properties of frankurter-type sausages. J. Food Sci. 48, 1036.

Rhee, K.S., Davidson, T.L., Cross, H.R. and Ziprin, Y.A. 1990a. Characteristics of pork products from swine fed a high mono-unsaturated fat diet: Part 1 - Whole muscle products. Meat Sci. 27, 329.

Rhee, K.S., Ziprin, Y.A. and Davidson, T.L. 1990b. Characteristics of pork products from swine fed a high mono-unsaturated fat diet: Part 2 - Uncured processed products. Meat Sci. 27, 343.

Sato, K. and Hegarty, G.R. 1971. Warmed-over flavor in cooked meats. J. Food Sci. 36, 1098.

Seideman, S.C., Cross, H.R., Smith, G.C. and Durland, P.R. 1984. Factors associated with fresh meat color: A review. J. Food Quality 6, 211.

Sellier, P., Mejenes-Quijano, A., Marinova, P., Talmant, A., Jacquet, B. and Monin, G. 1988. Meat quality as influenced by halothane sensitivity and ultimate pH in three porcine breeds. Livestock Prod. Sci. 18, 1710.

Shackelford, S.D., Miller, E.R., Haydon, K.D., Lovegren, N.V., Lyon, C.E. and Reagan, J.O. 1990a. Acceptability of bacon as influenced by the feeding of elevated levels of monounsaturated fats to growing-finishing swine. J. Food Sci. 55, 621.

Shackelford, S.D., Miller, M.F., Haydon, K.D. and Reagan, J.O. 1990b. Effects of feeding elevated levels of monounsaturated fats to growing-finishing swine on acceptability of low-fat sausage. J. Food Sci. 55, 1497.

Shackelford, S.D., Reagan, J.O., Haydon, K.D. and Miller, M.F. 1990c. Effects of feeding elevated levels of monounsaturated fats to growing-finishing swine on acceptability of boneless hams. J. Food Sci. 55, 1485.

Shackelford, S.D., Reagan, J.O., Haydon, K.D., Lyon, C.E. and Miller, M.F. 1991. Acceptability of low-fat frankfurters as influenced by the feeding of elevated levels of monounsaturated fats to growing-finishing swine. Meat Sci. 30, 59.

Smith, W.C. and Lesser, D. 1982. An economic assessment of pale, soft exudative musculature in the fresh and cured pig carcass. Anim. Prod. 34, 291.

Smith, G.C., West, R.L. and Carpenter, Z.L. 1975. Factors affecting desirability of bacon and commercially processed pork bellies. J. Anim. Sci. 41, 54.

Smulders, F.J.M. and van Laack, H.L.J.M. 1988. Carcass yield, meat quality and functional properties of pork as affected by hot processing procedures. Proc. 34th Int. Congr. Meat Sci.and Technol., Brisbane, Australia, p. 613.

Solomon, M.B., Campbell, R.G., Steele, N.C. 1990. Effect of sex and exogenous porcine somatotropin on *longissimus* muscle fiber characteristics of growing pigs. J. Anim. Sci. 68, 1176.

Solomon, M.B., Campbell, R.G., Steele, N.C., Caperna, T.J. 1991. Effects of exogenous porcine somatotropin administration between 30 and 60 kilograms on *longissimus* muscle fiber morphology and meat tenderness of pigs grown to 90 kilograms. J. Anim. Sci. 69, 641.

St.John, L.C., Buyck, M.J., Keeton, J.T., Leu, R. and Smith, S.B. 1986. Fatty acid profiles and sensory and carcass traits of tissues from steers and swine fed an elevated mono-unsaturated fat diet. J. Food Sci. 51, 1144.

Tarrant, P.V. 1987. Muscle biology and biochemistry. 33rd Int. Congr. Meat Sci. and Technol., Helsinki, Finland.

Taylor, A.A., Dant, S.J. and French, J.W.L. 1973. Processing Wiltshire bacon from PSE-prone pigs. J. Food Technol. 8, 167.

Taylor, A.A., Shaw, B.G., Jolley, P.D. and Nute, G.R. 1982. Hot curing Wiltshire bacon. J. Food Technol. 17, 339.

Tims, M.J. and Watts, B.M. 1958. Protection of cooked meat with phosphates. Food Technol. 12, 240.

Townsend, W.E., Davis, C.E., Lyon, C.E. and Mescher, S.E. 1980. Effect of pork quality on some chemical, physical and processing properties of fermented, dry sausages. J. Food Sci. 45, 622.

Troeger, K. and Woltersdorf, W. 1987. Hot boning and hot meat production from pig carcasses. Fleischwirtschaft 67, 707.

van Laack, H.L.J.M. 1989. The quality of accelerated processed meats - An integral approach. PhD Thesis, University of. Utrecht, The Netherlands.

van Laack, H.L.J.M., Smulders, F.J.M. and Roon, P.S. van. 1989. Functional properties of pre-rigor pork: Hot vs. cold processing of cooked shoulders and hams. J. Food Sci. 54, 307.

Warriss, P.D. and Akers, J.M. 1980. The effect of sex, breed and initial carcass pH on the quality of cure in bacon. J. Food Technol. 15, 629.

Warriss, P.D., Brown, S.N., Rolph, T.P. and Kestin, S.C. 1990. Interactions between the β-adrenergic agonist salbutamol and genotype on meat quality in pigs. J. Anim. Sci. 68, 3669.

Wirth, F. 1986. The technology of processing meat not of standard quality. Fleischwirtsch. 66, 1256.

Wismer-Pedersen, J. 1959. Some observations on the quality of cured bacon in relation to ante-mortem treatment. Acta Agr. Scand. 9(1), 69.

Wismer-Pedersen, J. 1968. The pH dependent water-binding capacity of pork muscle in relation to processing. In: Sybesma, van der Wal and Walstra (eds). Recent points of view on the condition and meat quality of pigs for slaughter, Schoonoord, Zeist, The Netherlands. p. 53.

Wood, J.D., Enser, M. and Moncrieff, C.B. 1988. Effects of carcass fatness and sex on the composition and quality of pigment. 34th Int. Congr. Meat Sci. and Technol. Brisbane, Australia, p. 562.

Ziprin, Y.A., Rhee, K.S. and Davidson, T.L. 1990. Characteristics of pork products from swine fed a high mono-unsaturated fat diet: Part 3 - A high-fat cured product. Meat Sci. 28, 171.

Discussion

*Smulders - I've got a question on the use of β-agonists as it relates to meat quality. I haven't done any work in pork with β-agonists, but I have worked with veal calves, and it seems to indicate the same things, that the meat gets tougher, and that it also gets lighter, which is a very valuable thing for the veal industry. Fortunately, mark my words, fortunately, we are not using the stuff since we've got a control program, much to the dismay of some people. But, I can never really understand why the meat becomes lighter, unless it's simply a dilution effect. Would you care to expand on other possible mechanisms, I had thought about the transport of iron in the primary production phase being interfered with by β-agonists. But I'm not sure what mechanism explains why meat gets lighter through the use of β-agonists. Have you got any ideas on that?

*Møller - Yes, I have mentioned the work of Wallace. He suggested that the reason probably was that the intermediate fiber types were changed to the white fiber type. So, the white ones contain less pigment, and, that could be the reason. This is the only explanation I have. So a change in fiber type, if produced, could probably decrease the heme content, and the amount of pigment. But, I don't know, it is just speculation.

*Smulders - Then it could be a dilution effect.

*Møller - I think it has been observed in Danish pork as well, but Paul Henckel can respond to this.

*Henckel - We can't deny the dilution effect, because there's a problem of time. If you increase the area of fibers very rapidly, myoglobin will probably not be produced quite as fast. So, it might be an intermediate effect. It might become redder later on. And, there could be a growth effect too. There might be a compensatory effect in the larger fibers to improve the diffusion of oxygen. We've only studied the short-term effects of the β-agonists, so we don't know yet. But, there is a change in the fiber type distribution with β-agonists. In pigs, that interferes both with the myosin molecule, and the metabolism. And, the same effect is not seen with pST therapy. With pST, the myosin isn't changed, only the metabolism.

*Puolanne - Salmon, for instance, use white fibers as protein reservoirs. They use these proteins when coming up the rivers. At that time, they don't eat at all, so, the diameter of the white fibers decreases tremendously, during their travels. I wonder if this has anything to do with meat? On a different topic, I would like to ask about your experiences with hot-boning, and curing hot-boned pork. What are your experiences at the industrial level? At the laboratory, it is very easy to get results, even here in Europe, but also in countries where the post-mortem reactions are slower.

*Møller - Well, in Denmark, we have not been doing hot boning very much. Some years ago, we had a plant hot-boning beef, but it was only for fresh meat production, not really for processed meat. But, this plant does not exist anymore, so, at the moment, we have absolutely no hot-boning, or accelerated processing in Denmark. That's it. On the experimental level, there has been some interest. I think that there is interest in the industry, but they don't have enough courage to go on. The industry is interested in the process, there's no doubt about that. They could have absolute control of the situation, from the beginning, at slaughter, right to the final product. That must be the main reason, for it is so difficult to start this process up. I think we could have some interest in the future, when we have to reduce the additives. It is a way to increase functionality, so that is one of the favorable things we can mention, but there is a lot of problems, and they know that.

*Puolanne - Yes, there's a lot of problems, and maybe everything we do with pork must be done faster than we do now.

Water and Fat Distribution in Pork in Relation to Sensory Properties

E. Tornberg, A. Andersson, Å. Göransson and G. von Seth

Swedish Meat Research Institute, P.O.B. 504, S-244 24 Kävlinge, Sweden

Introduction

Lean muscle tissue of pork contains about 75% water and 1-2% lipids. The water content, and its distribution within the meat, affects not only the sensory properties of the cooked meat, but is also of great importance economically, when losses or uptake of water occur. The influence of intramuscular fat content (IMF) on the sensory properties of pork muscle is a matter subject to much debate in the literature. In this paper, the water and fat distribution in porcine *M. longissimus dorsi* (LD) of different quality, raw and cooked, will be dealt with, in relation to the sensory properties of the cooked meat.

It must be emphasized that most of the water in the raw muscle is free and only a minor amount (4-5% of the total) can be considered as bound. This water is often called hydration water and is restricted in motion, due to the proximity of protein molecules (Wismer-Pedersen, 1971; Hamm, 1975). The water is largely held by the meat structure itself, having pores so small (< 100μm), that capillary forces come into play. An excellent review on water-holding in meat has recently been published by Offer and Knight (1988). For a deeper understanding of this field, it is recommended.

A powerful tool for studying the distribution of water in the muscle is the use of the non-invasive ^{1}H-pulse-NMR (nuclear magnetic resonance) technique. Water protons in meat have shorter transverse relaxation times (T_2) than in bulk water. This is explained by a fast exchange between free-water and the hydration water adjacent to the proteins. A long relaxation time suggests a long diffusion distance of the free water protons to the exchange site. This means that larger pores of water within the structure have a greater chance of obtaining relaxation times in the proximity of free water values than water in smaller pores. Therefore, the water distribution in pores of different sizes can be studied.

The multi-exponential decay of the T_2 of water protons in pork muscle has been reported in the literature (Renou *et al.*, 1985; Fjelkner-Modig and Tornberg, 1986b; Larsson and Tornberg, 1988; Renou *et al.*, 1989; Borisova and Oreshkin, 1992). Two dominating, discernible relaxation processes are usually observed. The major fraction (≈80%) of muscle water has a T_2 between 35-50ms, whereas the rest of the water relaxes in the range of 100-150ms. The percentage of water relaxing with the shortest relaxation time can be considered as mainly intracellular water, since a very high fraction of the muscle volume is occupied by the myofibrils (≈80%). Lillford *et al.* (1980) pointed out, however, that these discrete water domains do not necessarily have to arise from the structural domains seen in histological pictures of meat. The multi-exponential decay of T_2 is generally explained in terms of heterogeneous mass distribution down to ≈10μm. However, Tornberg and Larsson (1986) have shown by comparing the percentage of water having a T_2 ≈100-150ms with the percentage of water around the fiber bundles (evaluated by microscopy) that the latter could be predicted using the pulse-NMR method with an 80% probability.

Aberrations in the quality of pig meat, such as PSE and DFD, have also been registered using ^{1}H-pulse-NMR-measurements (Renou *et al.*, 1985 and 1989; Larsson and Tornberg, 1988). If the meat is pale and the drip loss is high, it is usually called PSE meat, whereas dark meat of enhanced ultimate pH (pH > 6.0 in LD) is called DFD meat. Larsson and Tornberg (1988) found with ^{1}H-pulse-NMR measurements that PSE meat had the shortest relaxation time for the water inside the fiber bundles, compared to normal and DFD meat. The DFD meat, on the other hand, had the shortest relaxation time for the water around the fiber bundles. These findings are not in agreement with the results of Renou *et al.* (1985), who claimed that the longest relaxation time (that which is supposed to reflect the water outside the fiber bundles) is primarily related to protein denaturation (PSE-criteria), whereas the shortest relaxation time correlated best with ultimate pH (DFDcriteria).

Fresh pork is always cooked prior to consumption. Cooking induces structural changes that decrease the water-holding capacity of the meat. Studies of the water distribution of cooked meat, using ^{1}H-pulse-NMR, have been done by Fjelkner-Modig and Tornberg (1986b); Tornberg and Larsson (1986); and Borisova and Oreshkin (1992). It was noted both by Fjelkner-Modig and Tornberg (1986b), and by Tornberg and Larsson (1986), that the T_2 for the major fraction of the water in both beef and pork was lowered after cooking from around 40ms to around 30ms. However, the relative proton populations representing the two major relaxation times did not change drastically, when comparing raw with cooked, drained meat.

The only paper so far to relate data on the water distribution of meat, using pulse-NMR, with sensory properties is the one by Fjelkner-Modig and Tornberg (1986b). They found that the juiciness and tenderness of fried LD could be predicted fairly well by the water distribution in the raw meat. However, due to the influence of breed and cooking temperature, different variables are best for the prediction. In that study, meat of normal quality and from pure breeds was studied. In the present investigation, the water distribution in raw and cooked meat from pure breeds and cross-breeds, when normal, PSE and DFD, will be compared and related to the sensory properties of the cooked meat.

The content of intramuscular fat (IMF) in LD and especially its influence on the sensory properties has been reported in the literature, whereas the distribution of the intramuscular fat in the meat has been studied to a lesser extent. Fjelkner-Modig and Tornberg (1986a), found that the fat depots were mainly situated between fiber bundles and, to some extent, in between fibers. The diameter of the fat cells varied between 15 and 180μm, but a diameter of 30-60μm was most frequently noted.

In the literature, there is a controversy as to whether the IMF significantly influences the sensory properties of the meat or not. Some researchers state that about 2% IMF is necessary for good taste characteristics, whereas IMF above 3% has little further improvement on tenderness (Bejerholm and Barton-Gade, 1986; DeVol *et al.*, 1988). Fjelkner-Modig and Tornberg (1986a), when studying meat from pure breeds, found that only the breed having the lowest level of IMF, i.e. the Swedish Landrace (1.4%), had any significant influence on the tenderness of LD. The relationship, though, had a low degree of explanation. Several other researchers have shown that the IMF content has no significant effect on eating quality (Rhodes, 1970; Skelley *et al.*, 1973). Thus, another investigation was undertaken, as presented in this paper, in which the influence of IMF on the eating quality of LD was studied. The samples of LD were chosen in such a way that the variation in IMF (0.5 to 3%) was as large as possible under the circumstances prevailing in Sweden, while other factors, which might affect the eating quality, such as pH_1, carcass weight, percentage lean, pH_{24} and internal light reflectance (FOP_{24}) were kept as constant and normal as possible. By this experimental design the influence of IMF should be maximized, to ensure the eventual impact of IMF on the sensory properties of LD.

Experiment

Materials

The water-distribution experiment

M. longissimus dorsi (LD), in the region of the last rib, was taken from 40 pigs of differing meat quality, i.e. normal, PSE and DFD. Pure breeds (Swedish Landrace, Swedish Yorkshire and Hampshire) 3 days post-mortem, as well as cross-breeds (50% Swedish Landrace + Swedish Yorkshire and 50% Hampshire), 1 day post-mortem were used.

The IMF experiment

29 porcine LD muscles with an IMF level ranging from 0.5 to 3.0% were selected. Only commercial cross-breeds (Hampshire (Swedish Yorkshire x Swedish Landrace)), gilts and castrates with a slaughter weight and percentage of lean varying from 73 to 87kg and 53 to 64%, respectively, were used. The selected pigs were cooled in a normal commercial manner. To ensure avoidance of warm- and cold-shortened meat, the lowering of the carcass temperature was monitored by thermocouples. After cutting, the meat was vacuum-packed and stored at 4°C for 3 days.

Methods

Chemical analysis

For the IMF-experiment, the chemical composition was determined in a 4cm section of the LD muscle removed at the last rib. The content of IMF (SBR method, NMKL, 1974), protein (Kjeldahl, as modified by Nilsson, 1968), connective tissue (Stegeman, 1958 as modified by Weber, 1973), and water (Nilsson, 1969) were analysed.

For the water-distribution experiment, only the water and fat contents of the muscle samples were determined, using the methods given above.

Meat-quality traits

For the water distribution experiment the following meat-quality parameters were measured: ultimate pH (Portamess pH-meter 651 with a glass electrode, Ingold 404), internal light reflectance (fiber optic probe (FOP), MK1, MRI, TBL, Leeds, England) and drip loss (Lundström *et al.*, 1984).

In the IMF experiment, pH measurements were made 45min (pH_1) and 24h (pH_{24}) post-mortem using, a portable pH-meter. To ensure normal meat, only meat with $pH_1 > 6.0$ and pH_{24} between 5.35 and 5.55 was chosen. Internal light scattering was measured with a fiber optic probe (FOP, MKIII, MRI, TBL, Leeds England) instrument 24h (FOP_{24}) post-mortem. FOP_{24} was kept between 27 and 46 to ensure normal meat.

Meat-structure traits in the water-distribution experiment

Microscopy. Microphotographs of cross-sectional cuts from LD were taken with a Nikon Optiphot light microscope. The meat samples were first fixed overnight with 4% glutaraldehyde and then washed in phosphate buffered saline for 1h. Embedding was carried out with 5% agar according to Kotter (1955) and the embedded samples were then frozen and mounted with the muscle fibers perpendicular to the knife in a cryostat microtome (type TE, SLEE, London).

The distance, x, between fiber bundles was determined using an image analysing system LAB EYE 3.07 (Innovativ Vision AB, Sweden) measuring 20 to 150 places on each micrograph. The distance between each measuring point on the micrograph was, on average, 20µm. Two micrographs per sample were taken at a magnification of 32x. The average of the squared distance, x^2, was calculated.

Sarcomere length. For the sarcomere length measurements, a piece 15 x 5 mm was taken from the centre of a slice of the LD. The piece of muscle was further cut into four smaller parts and fixed according to the procedure used by Cross *et al.* (1981). The lengths of the sarcomere diffraction bands were recorded using a helium-neon laser (Voyle, 1971).

Proton-pulse-NMR. Two small rods of fresh meat, approximately 7mm long and $35mm^2$ in cross-sectional area, were put into NMR tubes with the fibers perpendicular to the tube wall. The samples were thermostatically held at 25°C for 15 min prior to the NMR-measurements.

After the NMR measurements on the fresh meat in the NMR-tube, it was cooked in a water bath for 15min at 68°C. The samples were allowed to cool and thermostatically held at 25°C for 15 min before the next NMR measurement. The percentage of free water, having a relaxation time $T_2 \geq 1s$, was regarded to be the cooking loss determined by NMR-measurement (named NMR-cooking loss).

The water distribution was recorded using a proton-pulse NMR instrument (Bruker, Minispec, PC/20). The transverse relaxation time (T_2) of the water protons within the meat was recorded at a frequency of 20MHz at 25°C using the Carr-Purcell-Meibom-Gill method (Meibom and Gill, 1958). The T_2-recordings were made for τ-spacing 4000µS and at each measurement, 49 scans were accumulated. Data was collected in the time range of 8 up to 1350ms. The time lapse between two consecutive measurements was 10s, to allow the nuclear magnetization to return to an equilibrium value.

The relaxation data was analysed in multiexponential decay by curve decomposition using a microcomputer (Luxor ABC 806) as described by Tornberg and Nerbrink (1984) and Tornberg and Larsson (1986).

Sensory analysis and cooking loss

At 4 days post-mortem, the pork loins were assessed by a nine-member taste panel. Loin slices 1.5cm thick were fried 2 x 3 min (180°C) to an internal temperature of 68°C and served immediately to the assessors. Tenderness, flavour and juiciness were judged on a 9-point scale, where 1 = very tough, weak flavour and dry and 9 = very tender, strong flavour and very juicy. In the IMF-experiment only tenderness is given in the results.

In the water-distribution experiment, frying loss, given as the percentage loss through weighing the meat before and after frying, was registered. Cooking loss was determined on samples treated as follows: A stick of fresh meat, 1 x 1 x 2cm, was cooked in a test tube, dipped into a water bath of 68°C for 15 min. Weighing the meat before and after cooking gave the percentage cooking loss.

In the water-distribution experiment, the sample used for sensory analysis was cut from the LD muscle 4cm from the last rib and ahead. In the IMF experiment, the LD muscle was divided into two different samples, one 8cm section taken immediately ahead of the last rib and one 8cm section taken 16cm ahead of the last rib. These two samples will henceforth be referred to as the posterior and anterior pieces, respectively, and they were assessed as two different samples. The IMF content was estimated in each piece in accordance with a quadratic equation describing the IMF content at different positions along the LD (von Seth, unpublished results).

Statistical analysis

Data was analysed statistically using the SYSTAT programme (SYSTAT, 1987) using t-test, linear regression analysis and analysis of variance.

Results and discussion

The water-distribution experiment

Raw meat. In Table 1, all the measured meat quality and structural traits of the raw meat are gathered for the three different quality groups: normal (N), PSE and DFD. The classification of the meat with regard to the three meat quality groups (N, PSE and DFD) was based on the FOP and pH values. Pig meat with FOP values ≥ 55 was considered PSE and pigs with ultimate pH ≥ 6.0 as DFD. Of the samples collected from the 40 pigs, eight were found to be PSE and six DFD. As can be seen from Table 1, both FOP and drip loss differed significantly between the groups, whereas normal and PSE meat had a similar ultimate pH. There was no significant difference in sarcomere length between the different quality groups, but the variation in sarcomere length was greatest in the PSE group. This observation is in agreement with some of the contradictory results found in the literature. Honikel and Kim (1986), for example, found no contractions occurring for severe PSE muscle, whereas Cloke *et al.* (1981) found PSE muscle far more sensitive to severe contraction than normal muscles.

A three-component T_2-relaxation behaviour was mostly observed in the ^{1}H pulse-NMR-measurements, which indicates that there are three water regions with populations P21, P22 and P23, that can be quantified. The longest relaxation time was classified as the "expelled water" and was excluded from the results (not exceeding 1% of the water). The major fraction (70-94%) of the muscle water had a relaxation time (T23) between 35-50ms. The rest of the water relaxed in the range of 70-180ms (T22). These results are in accordance with those found in the literature.

Table 1. Raw meat quality and structural traits for the three meat quality groups normal, PSE and DFD. Mean values (x) and standard deviations (SD) are given

Number of samples	Normal		PSE		DFD		Significance level [1)]		
	x	SD	x	SD	x	SD	N-PSE	N-DFD	PSE-DFD
	(26)		(8)		(6)				
Ultimate pH	5.45	0.15	5.41	0.12	6.18	0.18		***	***
FOP value	35.1	10.3	68.4	12.3	13.8	5.0	***	***	***
Drip loss (%)	2.7	1.0	4.4	1.4	1.0	0.3	***	***	***
Sarcomere length (μm)	1.78	0.08	1.78	0.18	1.76	0.04			
Pulse-NMR									
T22 (ms)	131.9	17.9	145.0	18.0	98.4	14.5	P=0.08	***	***
T23 (ms)	43.9	2.8	40.8	2.9	43.2	3.5	**		
P22 (%)	14.8	5.2	12.2	4.4	16.3	5.6			
P23 (%)	84.7	5.2	87.4	4.4	83.5	5.7			

[1)]Significant differences between meat quality groups: **= P≤0.01, ***= P≤0.001.

The PSE group had, on average, the fastest T23 and differed significantly in comparison with the other two meat qualities. The DFD group had the fastest T22 and the PSE group had a tendency to have the slowest T22. In some PSE samples, i.e. three out of eight duplicates, four relaxation processes were found. This indicates more structural inhomogenities for this type of meat quality and is probably due to the fact that part of the proteins in PSE meat are aggregated.

The meat quality parameters characterizing the raw meat, i.e. the drip loss, the FOP-values and the ultimate pH, were best correlated with the T22-relaxation time of all the data obtained from the pulse-NMR-measurements (drip: r = 0.60***, FOP: r = 0.61 ***, pH: r = -0.67 ***). Evidently, the T22 relaxation time reflects a structural property of importance to meat quality.

As it was reasoned in the introduction, the T_2-relaxation time reflects the diffusion distance of a proton to the exchange site of the hydration water. We have chosen three samples from each quality group (N, PSE and DFD). For these samples the average distance between fiber bundles, x, was evaluated from micrographs of cross-sectional cuts from LD with an image analysing system. Assuming the proton migration to the exchange site is diffusion-controlled, the average time, t, for the molecule to travel that distance, x, when the diffusion coefficient is D, is given by Einstein (1956), in accordance with the formula:

$$t=\frac{x^2}{2D}$$

This average time, t, should in fact be related to the T22 relaxation time registered in the pulse-NMR-measurements. Thus, we have plotted the T22 relaxation time for the nine samples in Figure 1, as a function of the squared distance, x^2. According to the formula, this should give a straight line, which is obtained in Figure 1, with a correlation coefficient of r = 0.85***. This is a very promising result, showing the importance of the water compartment outside the fiber bundles for meat quality and the possibility to more quantitatively measure it by proton-pulse NMR. However, further studies are needed before these relationships can be fully understood.

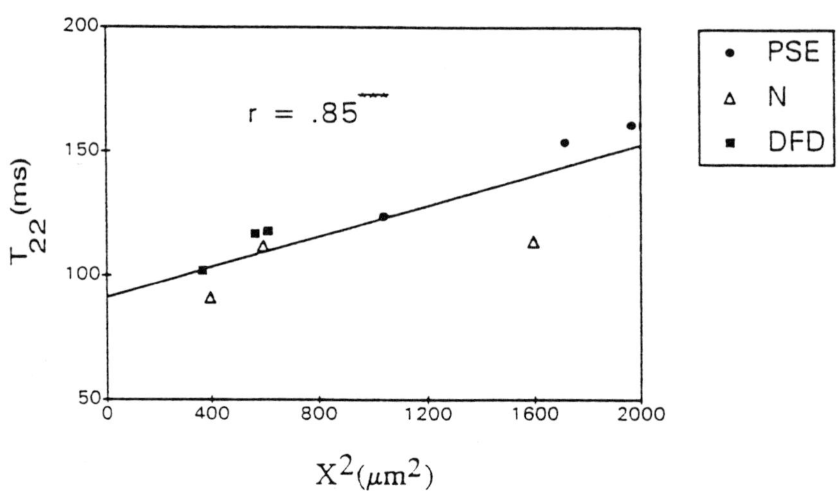

Figure 1. The T22 relaxation time (ms) measured, using ^{1}H-pulse NMR, as a function of the squared distance, x^2 (μm^2), between fiber bundles in raw porcine LD muscle of different qualities (N, PSE, DFD).

Cooked meat

In Table 2, the sensory properties and the two differently measured cooking losses are collected for the three meat quality groups. Unfortunately, only two samples of DFD were available for sensory analysis, so significant differences are difficult to obtain. It is, in any case, the DFD meat that gives rise to separate meat flavor and tenderness compared with N and PSE meat. The two latter qualities do not differ significantly in flavor, tenderness or juiciness. Dransfield *et al.* (1985) reported an increased pork tenderness for meat above pH = 5.9, which is in accordance with the results presented here. The less-developed meat flavor of DFD meat has also been observed by Fjelkner-Modig and Ruderus (1983). There is a controversy in the literature concerning the tenderness of PSE meat, as compared with normal meat. Some authors report a higher tenderness in PSE pork (Kemp *et al.*, 1976), whereas others find a better tenderness in normal meat (Buchter and Zeuthen, 1971; Barton-Gade and Bejerholm, 1985). We have found no difference at all in any of the sensory attributes tested between PSE and normal.

The two types of cooking losses are substantially less for DFD meat than for normal and PSE meat. As in the case of the sensory properties, the PSE and the normal meat do not exhibit significantly different behavior on heating. Suess and Honikel (1987), also noted that PSE and normal meat did not differ in losses, when cooked to different cooking temperatures, whereas DFD meat had a much better water-holding capacity.

Table 2. Sensory properties and cooking losses of meat cooked to 68°C for the three meat-quality groups: normal, PSE and DFD. Mean values (x) and standard deviations (SD) are given

Number of samples	Normal		PSE		DFD		Significance level[1]	
	x	SD	x	SD	x	SD	N-DFD	PSE-DFD
	(n = 22)		(n = 5)		(n = 2)			
Meat flavor	5.4	0.4	5.4	0.5	4.5	0.6	*	*
Tenderness	5.8	0.8	5.8	1.4	6.7	0.7		
Juiciness	5.5	0.6	5.1	0.7	5.1	1.7		
Frying loss (%)	24.0	4.3	24.0	5.4	19.5	5.0		
Cooking loss (%)	25.5	4.5 (n = 26)	25.5	3.5 (n = 8)	15.0	5.4 (n = 6)	***	***

[1] Significant differences between meat quality groups: * = P≤0.05, ** = P≤0.01, *** = P≤0.001.

Table 3. Meat structural traits of meat cooked to 68°C for the three meat-quality groups: normal, PSE and DFD. Mean values (x) and standard deviations (SD) are given

Number of samples	Normal x	Normal SD	PSE x	PSE SD	DFD x	DFD SD	Significance level[1] N-PSE	N-DFD	PSE-DFD
	(26)		(8)		(6)				
Sarcomere length (μm)	1.69	0.08	1.69	0.14	1.53	0.14		**	*
Pulse-NMR T22 (ms)	123.6	13.6	122.5	15.5	107.9	16.6		*	
Pulse-NMR T23 (ms)	29.0	3.4	27.9	3.4	31.1	2.1			
Pulse-NMR P21 (%) (NMR-cooking loss)	22.1	5.3	23.2	6.0	11.5	5.6		***	***
Pulse-NMR P22 (%)	18.1	6.0	19.5	4.7	14.7	4.2			
Pulse-NMR P23 (%)	61.8	5.0	57.1	5.6	75.1	4.6	P=0.07	***	***

[1] Significant differences between meat quality groups: *= P≤0.05, **= P≤0.01, ***= P≤0.001.

In Table 3 the structural traits of the cooked meat can be seen for the three different meat-quality groups. It is interesting to note, when comparing Tables 1 and 3, that the sarcomere length of the cooked meat has shortened (5-13%) compared with raw meat for all the three groups. This is, however, most evident for the DFD meat, when cooked, has significantly shorter sarcomere lengths than normal and PSE meat. This was not the case for raw meat.

For normal and PSE meat, the T22-relaxation time diminished, when cooked to 68°C, whereas the opposite occurred for DFD meat. Therefore, the significant differences in T22 observed for the three meat qualities in the raw state have vanished to a certain extent in the cooked meat. The T23-relaxation time decreased by about 10ms on cooking from about 40ms to around 30ms, for all three meat qualities. This is probably due to the aggregation of proteins upon cooking to 68°C. Part of the proteins are aggregated in PSE meat, and that also lowers the T23-relaxation for this meat, when raw, compared to other qualities (Table 1). This difference is, however, no longer obvious when the meat is cooked.

The percentage of water having the longest relaxation time $T_2 > 1$s, designated P21 (NMR-cooking loss), in Table 3, is considered to be more or less free water. This water is, as pointed out by Tornberg and Larsson (1986), the amount that is liable to loss, whereas the actual loss measured depends on the meat structure and the type of mechanical handling of the meat. As for the other losses given in Table 2 the NMR-cooking loss is significantly less for the DFD meat. It is also interesting to observe that the percentage of water primarily held within the myofibrillar compartment (P23) is significantly higher and lower (P = 0.07) compared with normal meat, for DFD and PSE meat, respectively, when cooked.

Correlation between the sensory properties of the cooked meat and the measured attributes of both the raw and the cooked meat gave the following results. Meat flavor is best characterized by the parameters describing the degree of DFD formed in the meat, such as pH (r = -0.54**) and NMR-cooking loss (r = 0.61 ***). This is not surprising since one of the crucial factors for meat flavor is the formation of a crust on frying, which is more inhibited the higher the pH of the meat, i.e. the more DFD meat that occurs. Also the T22 relaxation time of the meat both in the raw and the cooked state is of relevance to the meat flavor (r = 0.57*** and r = 0.53**, respectively) (Figure 2). As reasoned above this relaxation time reflects the width of the gaps between the fiber bundles, which of course is another indicator of the degree of DFD. It could also be an interesting phenomenon *per se*, indicating that flavor is more easily released, when there are larger water pores in the meat.

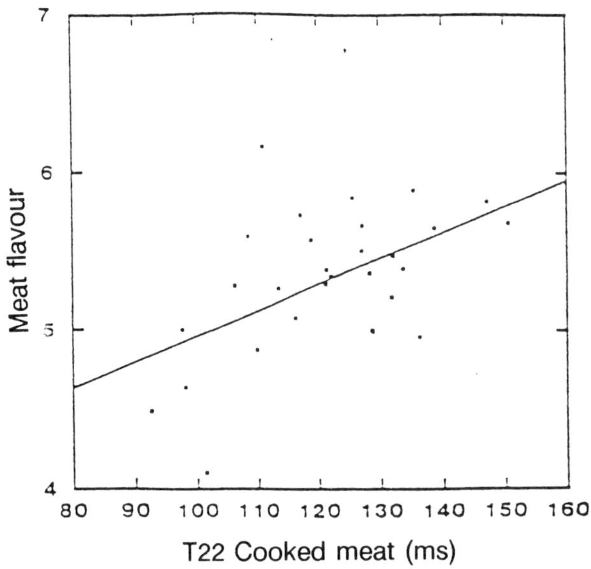

Figure 2. Meat flavor as a function of the T22-relaxation time (ms) in porcine LD muscle cooked to 68°C (r=0.53**).

Very few significant relationships between tenderness and all the parameters investigated in this study were found. In fact the only one of any relevance is the one between tenderness and the sarcomere length of the raw meat, which can be seen in Figure 3 (r = 0.52**).

This finding is somewhat in opposition to the results obtained by Fjelkner-Modig and Tornberg (1986a), who found strong correlations (r = 0.97**) between tenderness and the shortest relaxation time T23 of the raw meat. However, this good correlation was only restricted to meat from the pure Hampshire breed of normal quality and when the meat was fried to an internal temperature of 68°C. The relationship given in Figure 3 suggests that the degree of contraction is inversely related to the tenderness of porcine LD fried to 68°C. This has also been observed by

Westervelt and Stouffer (1978), for pork, using an instrumental method to register tenderness.

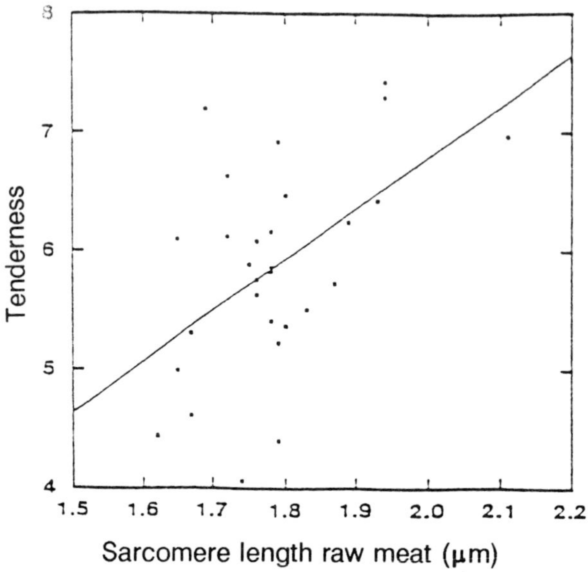

Figure 3. Tenderness as a function of the sarcomere length (µm) in raw porcine LD muscle (r = 0.52**).

The IMF experiment

The material was divided into 5 groups for the anterior and posterior pieces of LD muscle, respectively, depending on the estimated IMF content, as shown in Table 4.

Table 4. The material divided into five different IMF-groups

IMF-group	Last rib	Anterior piece		Posterior piece	
	IMF %	IMF % (mean)	nr of loins	IMF% (mean)	nr of loins
1	≤ 0.99	0.98	2	0.77	7
2	1.00- 1.49	1.30	10	1.18	7
3	1.50- 1.99	1.88	5	1.76	11
4	2.00 - 2.49	2.17	9	2.22	2
5	≥ 2.50	2.89	3	2.62	2

Figure 4 shows the results of the sensory evaluation for the anterior and posterior pieces of the LD for the five different IMF groups. The mean assessment of the tenderness of the pork loins was more or less independent of IMF group. The pork loins used in this investigation were of good quality, the mean values for tenderness lying in the upper half of the scale. But, as shown by the large standard deviation, wide variations were noted between assessors in the sensory evaluation (Figure 4). The anterior piece was significantly more tender than the posterior sample in IMF groups 2, 3 and 4. This does not depend on the variation in IMF content, which has often been argued to be the reason for variation in tenderness along the LD muscle.

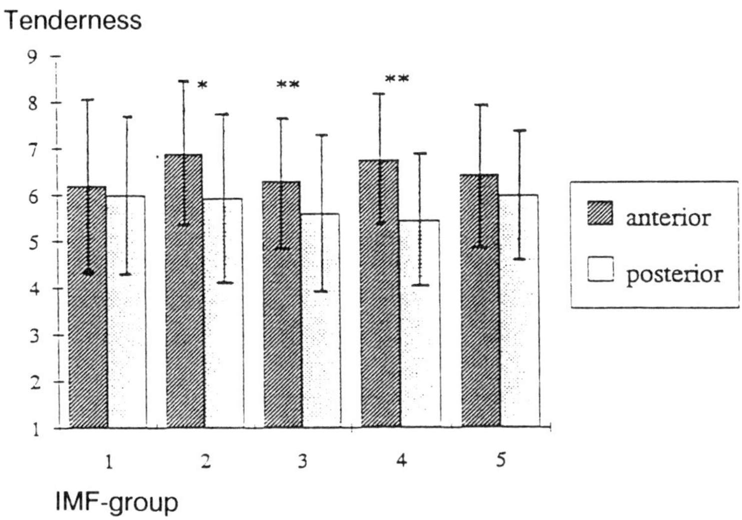

Figure 4. Mean values and standard deviations from the sensory evaluation of the anterior and posterior pieces of porcine LD. Significant differences between the anterior and posterior pieces are given in the figure, *** = $P \leq 0.001$; ** = $P \leq 0.01$; * = $P \leq 0.05$.

An analysis of variance was performed for the anterior and posterior pieces of LD muscle, respectively. The independent variables were normalized to vary between $^-1$ and $^+1$, in order to be able to compare the degree of influence of each variable on the sensory properties. The model used to describe the tenderness was:

$$\text{Tenderness} = \text{constant} + \text{assessor} + \text{sex} + \text{IMF} + \text{pH}_{24} + \text{protein} + \text{percentage lean} + \text{slaughter weight}$$

Most of the variation in tenderness was caused by the assessors. Slaughter weight or sex did not affect tenderness, neither in the anterior nor in the posterior piece. There was an obvious difference for the anterior and the posterior sample with regard to the models that best described tenderness. Tenderness for the posterior sample was only significantly affected by the assessors and the model explained just 45% of the total variation. The tenderness in the anterior piece, on the other hand, was affected significantly by pH_{24}, IMF and percentage lean. This is illustrated in Figure 5, where the degree of influence on tenderness for the normalized parameters can easily be compared. A notable result is that a higher IMF deteriorates tenderness, not as was previously believed that it improved it (Bejerholm and Barton-Gade, 1986). Of the total variation in tenderness, 55% was explained by this model.

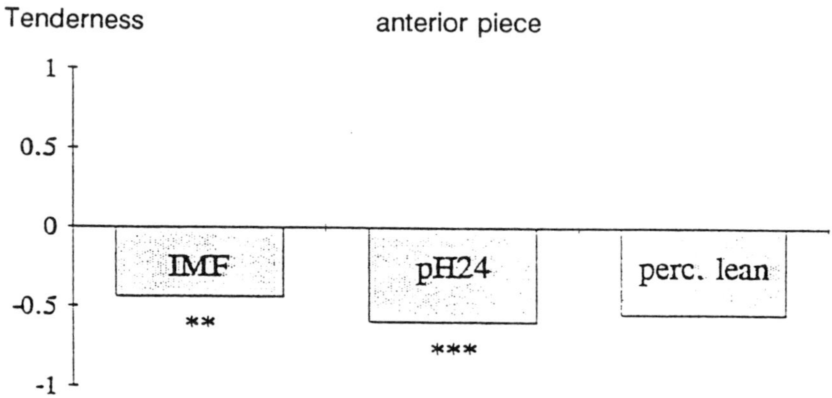

Figure 5. The normalized influence of IMF, pH_{24} and percentage lean on tenderness (mean (min-max) = 6.43 (3.50-9.00)) in the anterior sample of LD. Significant influence, ***= $P<0.001$; ** = $P<0.01$; * = $P<0.05$.

Conclusions

The water-distribution experiment

1. The water distribution in the meat was registered as a three-component T_2 relaxation using ^{1}H-pulse NMR measurements. The major fraction (70-94%) of the muscle water had a relaxation time (T23) between 30-50ms. Most of the remainder of the water relaxed in the range of 70-180ms (T22). The longest relaxation time (T21) was regarded to be the "expelled water".
2. Among the raw meat samples, the PSE meat had the fastest T23 and differed significantly in comparison with the other two meat qualities. The DFD meat had the fastest T22 and the PSE meat had a tendency to give the slowest T22.
3. The meat quality parameters characterizing the raw meat were best correlated with the T22-relaxation time of all the data obtained from the ^{1}H pulse NMR measurements. It was shown that the T22-relaxation time is dominated by the diffusion of "free" water to the exchange site of the hydration water of proteins, where the average diffusion distance is between fiber bundles.

4. There was no significant difference in the sarcomere length of the raw meat between the different meat quality groups. For the cooked meat, however, DFD had significantly shorter sarcomeres than normal and PSE meat. On cooking the sarcomeres shortened 5-13%.
5. Meat flavor was lower, and there was a tendency for the tenderness to be higher for DFD meat compared with normal and PSE meat. No significant differences in sensory properties were found between PSE and normal meat.
6. Cooking losses were substantially less for DFD meat than for normal and PSE meat, where the two latter qualities do not differ significantly.
7. On cooking, both the T22- and T23-relaxation times decreased by about 10ms, probably due to the contraction/aggregation of the meat proteins.
8. The percentage of water mainly held within the myofibrillar compartment (P23) is significantly higher and lower ($P = 0.07$), compared with normal meat, for DFD and PSE meat, respectively, when cooked.
9. Meat flavor is, in these experiments, primarily governed by the ultimate pH; the higher the pH, the less developed the flavour. This could be due to the greater degree of inhibition of a frying crust the higher the ultimate pH.
10. Tenderness was best correlated with the sarcomere length of the raw meat. This indicates the importance of having a meat with a low degree of contraction for obtaining tender pork.

The IMF experiment

1. The anterior piece of the LD was significantly more tender than the posterior piece of the same muscle. The difference did not depend on the variation in IMF, since the samples were compared at the same IMF content.

2. The good eating quality of the pork used in this investigation was maintained even at low IMF values, when cooked to 68°C.

Acknowledgements

The authors wish to thank the Farmers' Research Council Information and Development and the Swedish Council for Forestry and Agricultural Research for supporting this research in part. They also appreciate the skilful technical assistance of Ms. Susanne Lexhammar, Ms. Pia Olsson and Ms. Lisbeth Persson.

References

Barton-Gade, P. and Bejerholm, C. 1985. Eating quality in pork. Manuscript No. 698 E.

Bejerholm, C. and Barton-Gade, P.A. 1986. Effect of intramuscular fat level on eating quality of pig meat. Manuscript No. 720 E.

Borisova, M.A. and Oreshkin, E.F. 1992. On the water condition in pork meat. Meat Sci. 31, 257.

Buchter, L. and Zeuthen, P. 1971. The effect of ageing on the organoleptic qualities of PSE and normal pork loins. Proc. 2nd Intern. Symp. Condition and Meat Quality of pigs, Wageningen, The Netherlands, p. 247.

Cloke, J.D., Davis, E.A., Gordon, J., Hsieh, S.-Z., Grider, G., Addis, P.B. and McGrath, C.J. 1981. Scanning and Transmission Electron Microscopy of Normal and PSE Porcine Muscle. Scanning Electron Microscopy/III SEM. Inc. A.M.F., O'Hare, IL 60666, USA. p. 435.

Cross, H.R., West, R.L. and Dutson, T.R. 1981. Comparison of methods for measuring sarcomere length in beef semitendinosus muscle. Meat Sci. 5, 261.

DeVol, D.L., McKeith, F.K., Bechtel, P.J., Novakofski, J., Shanks, R.D. and Carr, T.R. 1988. Variation in composition and palatability traits and relationships between muscle characteristics and palatability in a random sample of pork carcasses. J. Anim. Sci. 66, 385.

Dransfield, E., Nute, G.R., Mottram, D.S., Rowan, T.G. and Lawrence, T.J.L. 1985. Pork quality from pigs fed on low glycosinate rapeseed meal: Influence of level in the diet, sex and ultimate pH. J. Sci. Food Agric. 36, 546.

Einstein, A. 1956. Investigations on the Theory of Brownian Movement. Meuthen, Dover.

Fjelkner-Modig, S. and Ruderus, H. 1983. The influence of exhaustion and electrical stimulation on the meat quality of young bulls: Part 2 - Physical and Sensory properties. Meat Sci. 8, 203.

Fjelkner-Modig, S. and Tornberg, E. 1986a. Intramuscular lipids in *M. longissimus* dorsi from pork, as related to breed and sensory properties. J. Food Qual. 9, 143.

Fjelkner-Modig, S. and Tornberg, E. 1986b. Water distribution in Porcine *M. longissimus dorsi* in relation to Sensory roperties. Meat Sci. 17, 213.

Hamm, R. 1975. Water-holding Capacity of Meat. In: Cole, P.J.A. and Lawrie, R.A. (eds), Meat, Butterworths, London. p. 321.

Honikel, K.O. and Kim, Ch.-J. 1986. Causes of the development of PSE pork. Fleischwirtschaft. 66, 349.

Kemp, J.D., Montgomery, R.E. and Fox, J.D. 1976. Chemical, palatability and cooking characteristics of normal and low quality pork loins as affected by freezer storage. J. Food Sci. 41, 1.

Kotter, L. 1955. Neue Wege bei der histologischen Wurstuntersuchung. Berl. Münch. Tierärtzl. Wschr. 68, 360.

Larsson, G. and Tornberg, E. 1988. An Attempt to Relate Meat Quality of pork *(M. longissimus dorsi)* to Meat Structure. Proc. 34th International Congress of Meat Science and Technology, Brisbane. p. 588.

Lillford, P.J., Clark, A.H. and Jones, D.V. 1980. Distribution of Water in Heterogenious Food and Modelsystems. In: Rowland, S.P. (ed.), Water in Polymers, ACS Symposium series No. 127, p. 177.

Lundström, K., Bjärstorp, G. and Malmfors, G. 1984. Comparison of different methods for evaluation of pig meat quality. Proc. Biophysical PSE-muscle analysis, Vienna, Austria. p. 311.

Meiboom, S. and Gill, D. 1958. Modified Spin-Echo Method for measuring Nuclear Relaxation times. Rev. Sci. Instrum. 29, 688.

Nilsson, R. 1968. Swedish Meat Research Inst., Anal. Instr. 10.

Nilsson, R. 1969. Swedish Meat Research Inst., 6.

NMKL 1974. Fedt, Bestemmelse i köd og ködvarer efter SBR. No. 88.

Offer, G. and Knight, P. 1988. The Structurel Basis of Water-Holding in Meat. Part 1: General Principles and Water Uptake in Meat Processing. In: Lawrie, R. (ed.), Developments in Meat Science - 4, Elsevier Applied Science, London. p. 63.

Renou, J.P., Kopp, J., Gatellier, Ph., Monin, G. and Kozak-Reiss, G. 1989. NMR-relaxation of water protons in normal and malignant hypothermia-susceptible pig muscle. Meat Sci. 26, 101.

Renou, J.P., Monin, G. and Sellier, P. 1985. Nuclear magnetic resonance measurements on pork of various qualities. Meat Sci. 15, 225.

Rhodes, D.N. 1970. Meat quality: Influence of fatness of pigs on the eating quality of pork. J. Sci. Food Agric. 21, 572.

Skelley, G.C., Handlin, D.L. and Bonette, T.E. 1973. Pork acceptability and its relationship to carcass quality. J. Anim. Sci. 36, 488.

Suess, I. and Honikel, K.O. 1987. Über die Ursachen der Zunahme von Saftverlust und Festigkeit beim Erhitzen von Fleisch. Mitteilungsblatt BAFF. 96, 7464.

Tornberg, E. and Larsson, G. 1986. Changes in water distribution of beef muscle during cooking - as measured by pulse NMR. Proc. 32nd European Meeting of Meat Research Workers, Ghent. p. 437.

Tornberg, E. and Nerbrink, O. 1984. Swelling of whole meat and myofibrils - as measured by pulse NMR. 30th European Meeting of Meat Research Workers, Bristol. No. 3-10, p. 112.

Voyle, C.A. 1971. Sarcomere length and meat quality. 17th European Meeting of Meat Research Workers, Bristol. p. 95.

Weber, R. 1973. Technical Report No. 7, Geneve: Technicon International Division SA.

Westervelt, R.G. and Stouffer, J.R. 1978. Relationships among spiral cord severing, electrical stimulation and post-mortem quality characteristics of the porcine carcass. J. Anim. Sci. 46, 1206.

Wismer-Pedersen, J. 1971. Chemistry of Animal Tissues. Part 5. Water. In: Price, J.F. and Schweigert, B.S. (eds), The Science of Meat and Meat Products. 2nd edn. W.H. Freeman and Co., San Fransisco. p. 177.

Discussion

*Kallweit - I would like to know whether you homogenized the samples, and how you got them into the tube.

*Tornberg - I just put the samples as whole meat into the tube.

*Kallweit - And, how much did you fix them, were there air bubbles?

*Tornberg - No, the air bubbles don't matter. And, if you homogenize, the sample is destroyed, and you can't measure the widths of the gaps between fiber bundles, and that's what I'm looking for.

*Essén-Gustavsson - In your first water experiment, were there a lot of different breeds used?

*Tornberg - Yes, three pure breeds, and a crossbreed.

*Essén-Gustavsson - How would that correlate? The muscle fiber areas may differ from the beginning, and the metabolic profiles within the fiber. So, if the diffusion, the capillarization, and so on is different from the beginning, it may influence your results.

*Tornberg - Yes, but that would give rise to variation, which we wanted to look for, so I am happy for that variation. But we have made as analysis of variance, looking at the development of PSE, and DFD, as fixed variables, and the breeds as fixed variables. The breeds never had any significant influence, compared with normal, PSE, and DFD. So, there was no problem.

*Essén-Gustavsson - About the intramuscular fat, I just want to mention again, that IMF is IMF, it is all kinds of fat. So, still, the question remains to be answered, are there some other fat stores that make a difference? It sounds like you say that IMF is not important for tenderness, which you show nicely here. But, the question remains, then what about the fat, located, or stored within each fiber, does that have an influence or not? I just wanted to raise that question.

*Tornberg - The type of fat that I have discussed here was mostly situated between the fiber bundles, at the perimysium. And, some is between the fibers. But, the fat within each fibers I haven't measured. But it's so small compared with this amount. I think the fat with inside the fibers, if it controls metabolism, and in that way, the rigor, then it could influence the tenderness, because the rigor influences the tenderness.

*Essén-Gustavsson - The fat within the fiber is not small, it's 25% of all intramuscular fat. We have just measured the triglyceride content within each muscle fiber, so it does contribute. I just want to raise the question, but, we should not forget that there are different kinds of lipids we are talking about.

*Moller - I think we have to be a bit careful when we discuss IMF. When we discuss all these correlations, I think we must realize that these involve interactions. And, sometimes, in some material, we can find a correlation, and another time, in another material, we cannot. That's the same as the correlation between sarcomere length and tenderness. It could be high in one case, and in another case, low. I think, to discuss those relationships, we have to be very much aware of whether or not there could be some interactions in the sample material overshadowing the specific correlations under study.

*Tornberg - The variables that I've studied in this experiment explained 54% of the variance. But, in the analysis of variance, we made some crosses, and we didn't get any high explanation. So, with regard to those variables studied in this investigation, I don't think we have any cross effects.

*Moller - Some results from the Danish Meat Research Institute have shown that if there is PSE, there could be a high correlation with IMF, if you have contracted muscle.

*Tornberg - Yes, I agree, there could be an effect. But, this was normal meat. It was a starting point. I think we should go on and study the influence of IMF on PSE meat also.

*Smulders - Only one addition to what Dr. Moller has just said about the relationship between sarcomere length and tenderness. It might be a good suggestion to identify fast- and slow-glycolizing muscle. He has shown in his studies that in slow-glycolizing muscles, the relationship may be much higher than in fast-glycolizing muscles. We have reproduced that finding in beef. On another topic, I was a little bit surprised to find that DFD has less flavor than normal meat because of "formation of the crust." Could you expand on that, because I could easily come up with another explanation, for example, that there was less lactate, and therefore the flavor was less accentuated .

*Tornberg - Exactly, the Maillard reaction is much less, because you have less glucose.

*Smulders - There's less lactate, and therefore, there is less accentuation of the normal meaty flavor.

*Tornberg - It could be that too, yes.

*Jorgenson - I'm not quite certain I understand what you meant about the T2, and the diffusion distance. There's clearly a linear relationship, but, what is the biological background for the diffusion distance?

*Tornberg - Well, I said, "why do we got a T2 relaxation that is much shorter than in free water?" It's due to the protons in the free water, exchanging with the hydration water adjacent to a protein molecule, consisting of 2 or 3 layers of water. So what you really measure is the diffusion properties of the protons in water.

*Honikel - Can you explain why DFD meat, on cooking, shortens more than PSE, or normal meat? It's an interesting observation.

*Tornberg - No.

*Puolanne - Maybe there is no real rigor?

*Honikel - That's what I would expect, because DFD muscles can be extended quite easily, but in normal and PSE muscle, much more force is required.

*Puolanne - If the muscle is in the carcass until the onset of rigor, it is not able to shorten freely. If it is excised pre-rigor, it shortens, and the extent of shortening is related to the temperature of the muscle. Consequently, there will be a difference between a cold boned muscle and a hot boned muscle. So, I do not expect any large differences in the sarcomere lengths of different muscles, if they go into rigor in the carcass, but the difference may come up during cooking. Could you comment on that?

*Tornberg - Yes, but the muscles used in this investigation were not excised pre-rigor but post-rigor. I cannot see how cooling itself influences the shortening during rigor.

*Puolanne - Yes, but they weren't in the carcass when you cooked them.

Relationships Between Halothane Sensitivity, Carcass Quality and Meat Quality in Belgian Slaughter Pigs

S. De Smet[1], H. Pauwels[2], W. Eeckhout[3], D. Demeyer[1], I. Vervaeke[1], S. De Bie[4], G. Van De Voorde[1] and M. Casteels[5]

[1]Research Centre for Nutrition, Animal Production and Meat Technology, Proefhoevestraat 10, 9090 Melle;
[2]Ministry of Agriculture, Scheldewindeke;
[3]National Institute for Animal Nutrition, Melle-Gontrode;
[4]Lokerse Vleesveiling, Lokeren;
[5]Study Centre for Pig Production, Melle-Gontrode

Introduction

It is now well established that the halothane gene in pigs has a great influence on both carcass and meat-quality traits. Depending on their reaction to halothane anaesthesia, pigs are classified as halothane-positive (H^+) or halothane-negative (H^-). The effect of these phenotypes on traits of economical importance has been studied extensively (reviewed by Webb *et al.*, 1982; Monin, 1989). As no easy and quick method of distinguishing the three underlying genotypes was available until recently, fewer studies compared homozygous reactors (nn) with heterozygous (Nn) and homozygous (NN) non-reactors (Eikelenboom *et al.*, 1980; Lundström *et al.*, 1989).

Halothane-sensitive pigs are generally assumed to yield more lean when compared with halothane-resistant pigs, but meat quality (rate of pH decline, color, water-holding capacity) is inferior, and PSE frequency is higher. Whereas the halothane gene is recessive for stress susceptibility, it appears additive for lean content and certain meat quality criteria (Simpson and Webb, 1989; Sellier, 1987). However, it is not always clear to what extent heterozygotes differ in meat quantity and quality from homozygous non-reactors. One important question is whether the negative (genetic) correlation between meat quantity and meat quality, generally encountered in populations with a high incidence of halothane susceptibility, remains in the absence of the halothane gene. Furthermore, relationships between different measurements of meat quality (e.g. color and water-holding capacity) may also be influenced by the halothane phenotype (Eikelenboom and Nanni Costa, 1988).

The incidence of the halothane-positive reaction is widely different between countries and breeds (Kallweit, 1985). In some countries, selection has successfully reduced the frequency of H^+ pigs. On the other hand, no selection has taken place in some areas, and stress susceptibility in some heavy muscled Landrace breeds, especially the Belgian Landrace breed, remains very high (Lampo, 1984). However, increasing interest exists in Belgium for the development of maternal NN lines. By mating these females to H+ males to produce heterozygous slaughter pigs, it should be possible to exploit the beneficial effects of the halothane gene. Acceptable carcass quality may then be obtained without the adverse effects of stress susceptibility on meat quality and stress deaths. In addition, H^- sow lines show higher reproductive performance (Webb *et al.*, 1985).

Ultimate meat quality is determined both by genetic factors and slaughter conditions. Pre-slaughter handling and methods of slaughter may greatly affect the development of PSE (Warriss, 1987; Tarrant, 1989). Although many recommendations can be made at present in order to reduce the amount of stress imposed upon pigs in the pre-slaughter period and to limit PSE frequency, it remains unclear which factors are most important in influencing meat quality. Also, there may be an interaction between slaughter conditions and genotype. Meat quality of H^- pigs may vary widely according to different pre-slaughter handling conditions, whereas H^+ pigs may yield meat of inferior quality, irrespective of pre-slaughter management (Barton-Gade, 1984).

In this paper, some results are reported on the relation between carcass quality and several meat quality criteria in Belgian Landrace pigs and Pietrain x Belgian Landrace crosses, as influenced by halothane status. In addition, two abattoirs were compared for their effect on meat quality of very similar pigs.

Material and methods

Animals and slaughtering

A total of 949 fattening pigs, obtained from three commercial farms, were slaughtered over a 2-year period, on 34 occasions. The mean number of pigs slaughtered per day was 28 (range 12-42). Animals were either halothane-sensitive (H^+; 39.9%) or halothane-resistant (H^-; 60.1%). They belonged to the Belgian Landrace breed (BL; 49.0%) or were cross-breds (CR; 51.0%) of mainly Pietrain x BL and some German Landrace x BL. Gilts and barrows were almost equally represented, 54.2% and 45.8% respectively. Overall mean ± SD carcass weight was 86.8 ± 8.0 kg.

The vast majority (89.7%) of pigs was slaughtered in the same commercial abattoir (SLH-1). On eight occasions, half of the pigs (10.3% of total) from one farm were simultaneously slaughtered at the Institute's abattoir (SLH-2). Transportation times varied between 20 and 60 min. Lairage time was kept fairly constant, at approximately 2h (between 1 and 3h). On the other hand, time of feed withdrawal before slaughter was rather variable (between < 1 and 17h). SLH-1 differs from SLH-2 in several aspects. In SLH-1 pigs are driven via a restrainer to the stunning equipment and electric goads are sometimes used, while automatic high-voltage stunning, followed by immediate sticking and bleeding out in lying position is applied. In contrast, in SLH-2 manual low-voltage stunning is applied using two tongs. In SLH-2, pigs are stuck only after shackling, which increases time between stunning and commencement of bleeding.

The genotype of the parents of 827 pigs could be derived from their own reaction to the halothane test, and from the incidence of reaction among their progeny. The proportion of homozygous and heterozygous H- pigs could thus be estimated. The expected frequency of NN and Nn pigs among the non-reactors was 16.4% and 83.6% respectively. Three genotypic groups of pigs could thus be distinguished: a first group consisting of only H^+ pigs (group nn); a second group consisting of exclusively heterozygous H^- pigs, descending from a H^+ boar or a H^+ sow (group Nn) and a third group of H^- pigs consisting of homo- and heterozygotes, descending from two H^- parents (group (Nn+NN)). The expected frequency of Nn and NN in the latter group was 0.516 and 0.484, respectively.

Carcass and meat-quality measurements

Objective carcass grading was only performed in SLH-1 using a SKGII device (Casteels, 1988). Four physical carcass traits (angle of ham, loin fat thickness, width of ham and width of loin) were measured to estimate lean content and conformation (low score for the latter means good conformation).

At 40 min post-mortem, pH was measured in the *M. longissimus dorsi* (LD) (13th rib) and in the *M. semimembranosus* (SM) on the right carcass side (pH_1). Carcasses were transported for cutting after overnight conventional chilling. At about 22h post-mortem, just before cutting, pH was measured again (pH_2). Internal light reflectance (fibre optic probe, FOP) and conductivity (Tecpro pork quality meter, PQM) measurements were performed in the same muscles. Two slices 2-3cm thick of the LD around the 13th rib were removed and transported to the laboratory for further determinations. Meat color was assessed using a surface reflectance meter (GÖFO) and a Hunter Labscan apparatus. Transmission value, an indicator of protein denaturation, was determined according to Hart (1962). Drip loss was determined as the percentage weight loss of a sample (*ca.* 60g) after 48h in a plastic bag at 4°C. Cooking loss (%) was assessed after heating for 50 min at 75°C (Honikel, 1987). The filter-paper press method of Grau and Hamm (1952) was also applied, but expressible fluid was expressed as the difference between total area and meat area.

Results and discussion

Halothane genotype and meat quality

As expected, halothane sensitivity was accompanied by better carcass quality and worse overall meat quality (Table 1). H^+ pigs were characterized by a higher lean content and a better conformation as compared to H^- pigs. On the other hand, much lower pH_1 values and higher FOP and PQM values indicated impaired meat quality in H^+ pigs. No differences in ultimate pH were found. Lower values with GÖFO instrument and higher lightness (L_*) values reflect increased paleness of the meat from halothane-susceptible pigs. Drip loss, cooking loss and expressible fluid of LD samples were all highest in H^+ pigs. These results confirm the findings of several other studies (see reviews of Webb *et al.*, 1982 and Monin, 1989).

As group (Nn+NN) consists of only 48% homozygous H^- pigs, one may expect the real difference between NN and Nn pigs to be higher than the difference between group Nn and group (Nn+NN) in this study. So, it is impossible to determine the exact position of heterozygotes compared with both homozygotes. Regarding SKGII parameters, heterozygotes appear to be intermediate between both homozygotes for carcass quality. Lean content and conformation score were significantly different between all three genotypic groups. Mean values of the four SKGII parameters for group Nn were all between the means for both other groups, but differences were not always significant. Taking into account the ±50% homozygous character of group (Nn+NN), it appears that heterozygotes are closer to nn than NN pigs for all carcass traits in this study, except for the ham angle.

In general, heterozygotes also appear to be more or less intermediate for meat-quality traits. Again, differences between the three genotypic groups were not always significant and the pattern was not consistent. For some criteria, the mean of group Nn was closest to the mean of group nn, while for other parameters it was closest to the mean of group (Nn+NN). For all measurements made on the carcass before cutting, except ultimate pH in SM, a significant difference between group Nn and group (Nn+NN) was found. Heterozygotes appear to be rather close to NN for pH_1 and more intermediate for FOP and PQM. Mean ultimate pH of group (Nn+NN) was slightly higher than mean values of group nn and Nn, which did not differ. On the other hand, meat lightness expressed as GÖFO units and L_* values were equal in both H^- groups, despite clear differences in internal reflectance. The latter may be more affected by fluid losses than the former. Significantly lower drip loss and cooking loss was also found for group Nn, compared with group (Nn+NN), but no difference in expressible fluid was apparent between these groups. Mean transmission value was significantly different between all three genotypic groups. The data thus suggest that relationships between factors responsible for impaired meat quality, i.e. pH and protein denaturation, and final meat-quality criteria, i.e. color and water holding capacity, may differ according to halothane genotype. A similar finding was reported for FOP and drip loss by Eikelenboom and Nanni Costa (1988). Sellier (1987, 1990) reviewed within-population comparisons between halothane genotypes and concluded that the halothane-sensitivity gene is incompletely recessive for certain meat quality criteria related to PSE.

Table 1. Mean values ± SD of carcass traits and meat quality criteria for the three halothane genotypic groups[1]

Genotypic group Number of pigs[2]	100% nn 256-379	100%Nn 215-334	52%Nn/48% NN 121-172
Lean (%)	61.7 ± 3.1^a	60.9 ± 2.9^b	58.4 ± 2.6^c
Conformation	1.74 ± 0.57^a	2.03 ± 0.43^b	2.41 ± 0.36^c
Loin fat (mm)	13.3 ± 4.6^a	13.9 ± 4.8^a	17.6 ± 4.5^b
Ham width (mm)	206 ± 12^a	203 ± 10^b	195 ± 9^c
Loin width (mm)	112 ± 8^a	111 ± 7^{ab}	110 ± 6^b
Ham angle	36 ± 11^a	45 ± 7^b	46 ± 6^b
PH_1 LD	5.69 ± 0.33^a	6.07 ± 0.30^b	6.16 ± 0.30^c
PH_1 SM	6.00 ± 0.33^a	6.25 ± 0.28^b	6.32 ± 0.29^b
PH_2 LD	5.56 ± 0.15^a	5.56 ± 0.15^a	5.61 ± 0.15^b
PH_2 SM	5.82 ± 0.26^a	5.83 ± 0.26^{ab}	5.89 ± 0.26^b
FOP LD	54 ± 17^a	42 ± 13^b	37 ± 12^c
FOP SM	45 ± 18^a	39 ± 18^b	32 ± 14^c
PQM LD	10.5 ± 2.4^a	8.5 ± 2.7^b	6.7 ± 2.8^c
PQM SM	9.4 ± 2.9^a	9.1 ± 3.1^a	7.4 ± 2.8^b
GÖFO (μA)	58.2 ± 11.1^a	68.3 ± 9.0^b	69.7 ± 8.4^b
L_*	59.3 ± 4.3^a	55.6 ± 3.8^b	55.8 ± 3.6^b
a_*	8.5 ± 1.7^a	8.6 ± 1.6^a	7.9 ± 1.7^b
b_*	15.5 ± 1.4^a	14.7 ± 1.3^b	14.4 ± 1.3^b
Transmission (%)	25 ± 21^a	12 ± 12^b	9 ± 9^c
Drip loss (%)	6.9 ± 2.5^a	6.2 ± 2.5^b	5.3 ± 2.6^c
Cooking loss (%)	36.0 ± 2.0^a	35.6 ± 1.9^a	34.5 ± 2.3^b
Press fluid (cm^2)	4.9 ± 0.9^a	4.4 ± 0.9^b	4.3 ± 0.7^b

[1] Means with different superscripts are significantly different at $P<0.05$

[2] Number of pigs varies among variables due to missing values.

Relation between lean content and meat quality

The relatively large standard deviations of meat quality and quantity traits suggest that there is considerable overlap between the three genotypes. This urged us to look closer at the relationship between lean content and several meat-quality criteria in each genotypic group. Therefore, pigs were divided in four lean classes: <57.5%; 57.5-60%; 60-62.5%; >62.5%. Means of meat-quality criteria were calculated for each genotypic group and lean class. H^- pigs are of course relatively more numerous in the lowest lean classes and *vice versa* for H^+ pigs. However, all means are based on at least 25 observations, except for group (Nn+NN) in the class >62.5% that was based on only six animals.

Mean pH_1 (LD) sharply declines as lean content increases in halothane-sensitive pigs, whilst there is only a small reduction for both H^- animal groups (Fig. 1). Furthermore, mean pH_1 (LD) of H^- pigs remains, even in the highest lean content class, at a relatively high level, well above the value of 5.8, generally considered as a threshold value for the incidence of PSE meat (Bendall and Swatland, 1988). So, a slight depression in pH of Nn or NN pigs with higher lean content, does not necessarily mean that one has to worry about an increased risk of PSE. The same applies to internal reflectance measured in LD at 22h post-mortem (Fig. 2). As lean content increases, mean value of FOP increases regularly for H^+ pigs, but not for H^- pigs. The distinction between H^+ and H^- pigs is even greater for GÖFO reflectance (Fig. 3). Meat lightness remains very constant with increasing lean content in group Nn and (Nn+NN) and is very similar in both genotypic groups. In group nn, however, meat becomes paler as lean content increases. Drip loss, however, shows a different pattern. In all three genotypic groups, there is an overall increase in drip loss as lean content increases, although the pattern is irregular. Mean value for drip loss of the exclusively heterozygote group is rather close to the mean of group nn for the lean classes <57.5 and 57.5-60.0, while it is closer to group (Nn+NN) for the lean class 60.0-62.5 and intermediate for the leanest class. As poor water-holding capacity is probably the most important meat-quality defect, compared with paleness and soft structure, this finding deserves attention. These figures, however, should be interpreted carefully, due to the different number of pigs in each group and the low number of (Nn+NN) pigs in the leanest class. Thus, increased lean content in H^+ pigs coincided with impaired meat quality for the four criteria studied. Little or no deterioration was observed for pH_1, FOP and GÖFO in H^- pigs, but it seems that drip loss is affected more. One may expect, however, an advantage of NN to Nn pigs for drip loss.

Anyway, these figures illustrate again the prevailing influence of the halothane status on meat quality, irrespective of carcass quality.

The more pronounced inverse relationship between lean content and certain meat quality criteria in H^+ pigs compared to H^- pigs is also clear from simple linear relations (Table 2).

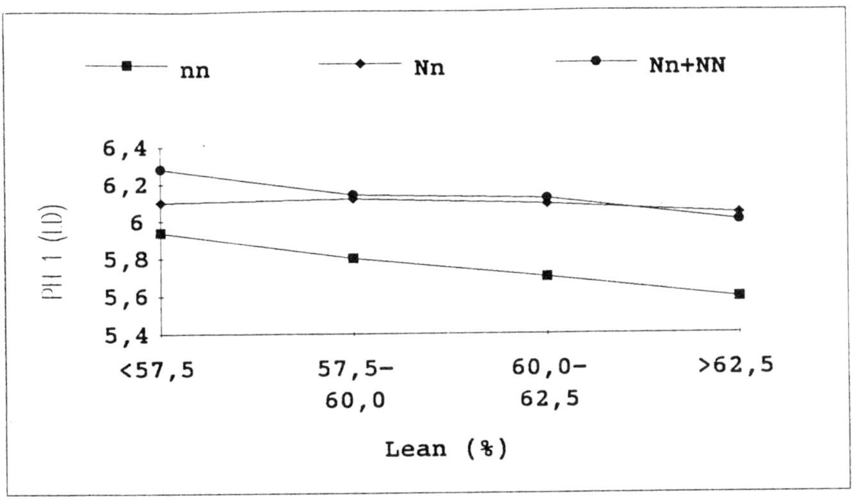

Figure 1. Relation between mean pH_1 (LD) and lean content (%) for three halothane genotypic groups (nn= homozygous reactors; Nn= heterozygous reactors; NN= homozygous non-reactors).

Correlations and regression coefficients (the latter not shown) between lean content and pH_1, FOP and GÖFO for all pigs and for H^+ pigs separately are very similar, and much higher than those for H^- pigs. As could also be seen in Figure 3, no linear relationship exists between lean content and GÖFO in H^- pigs. On the contrary, no such discrepancy is apparent between both populations for the relationship between lean content and drip loss. Regarding correlations among meat-quality criteria, FOP and GÖFO are better related to pH_1 in H^+ pigs than in H^- pigs, whilst this is again not true for drip loss. Correlations among FOP, GÖFO and drip loss are not significantly different for all pigs and H^+ or H^- pigs separately.

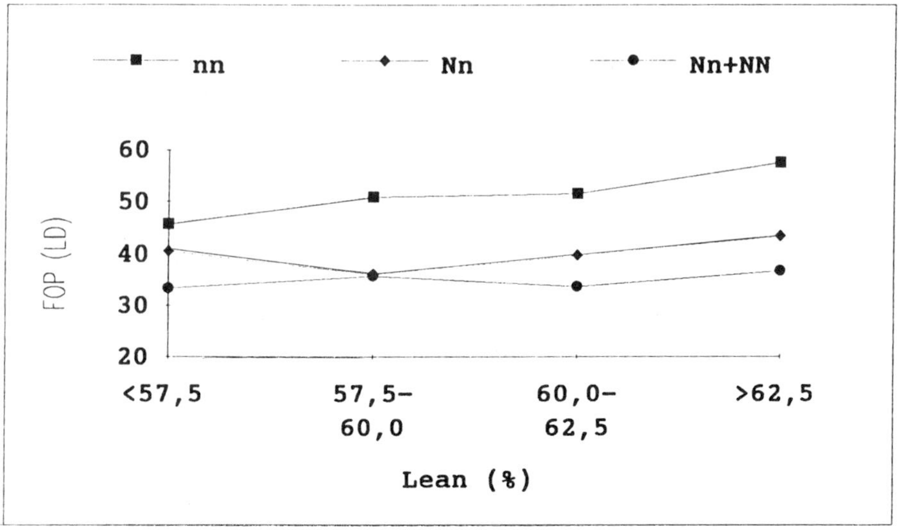

Figure 2. Relation between mean FOP (LD) and lean content (%) for three halothane genotypic groups (see Fig. 1).

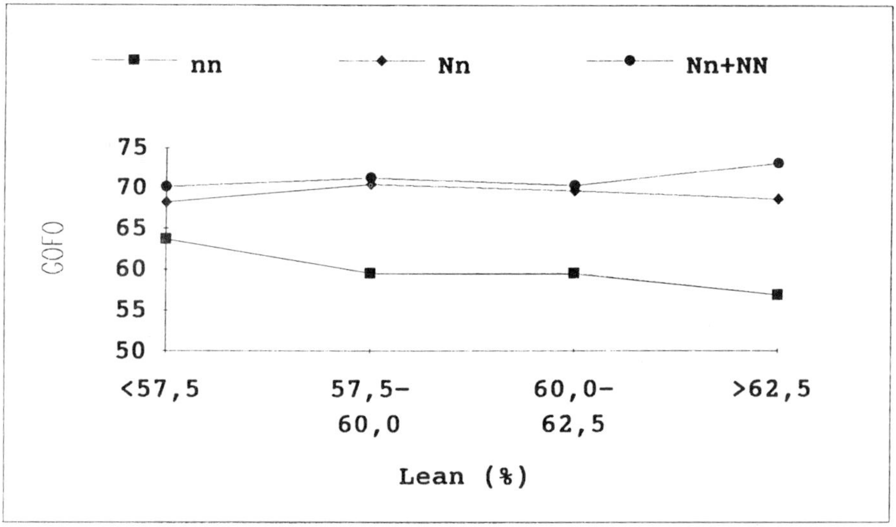

Figure 3. Relation between mean GÖFO and lean content (%) for three halothane genotypic groups (see Fig. 1).

In a review, Sellier (1988) concluded that the genetic antagonism between meat quality and meat quantity tends to be stronger when the incidence of halothane sensitivity is relatively high in the population under consideration, and/or the meat-quantity trait is related to muscle development, and/or the meat-quality trait is a PSE indicator. Indeed, Sönnichsen *et al.* (1984) found a fairly strong negative correlation between LD area and GÖFO in Pietrain and Belgian Landrace. Cole *et al.* (1988) found a more pronounced inverse relationship between meat quantity and quality in Belgian Landrace compared to Large White and French Landrace. In a study on Dutch slaughter pigs, essentially free of halothane sensitivity, Merks *et al.* (1989) concluded that selection for meat quality traits was possible, without losing substantial genetic progress for lean percentage in the carcass, due to correlations close to zero of most meat-quality characteristics with lean percentage. Malmfors *et al.* (1980) found only a small positive genetic correlation between light reflectance and carcass lean percentage. Although no genetic correlations were calculated in our study, the results may give some indications, since Cameron (1990) showed that the direction of responses in meat quality traits to selection on carcass traits was consistent with the responses predicted from phenotypic correlations. Our results on heavy-muscled Belgian slaughter pigs suggest that further selection towards increased lean will decrease pH_1 (LD) and GÖFO, and increase drip loss in H^+ pigs. On the other hand, a smaller decrease in pH_1 (LD) and no change in GÖFO value may be expected to selection for more lean in H^- pigs, but drip loss, although at a lower level, may be affected to the same extent as in H^+ pigs. The genetic opposition between meat quality (PSE) and meat quantity is thus largely determined by the incidence of the halothane gene in the population under investigation, which is not surprising since halothane sensitivity strongly affects both muscular development and PSE frequency.

Table 2. Correlation coefficients between and among some carcass and meat-quality traits for all pigs (higher line) and for H^+ / H^- pigs separately (lower line)[1]

	pH_1 LD	FOP LD	GÖFO	Drip loss
Lean meat	-0.366	0.295	-0.231	0.240
	-0.380/-0.172	0.259/0.148	-0.235/-0.003^{ns}	0.174/0.204
PH_1 LD		-0.588	0.683	-0.454
		-0.538/-0.419	0.657/0.507	-0.348/-0.470
FOP LD			-0.780	0.543
			-0.746/-0.711	0.512/0.527
GÖFO				-0.611
				-0.560/-0.635

[1] All correlations are significant at $P<0.01$, except when ns is mentioned.

Non-genetic factors affecting meat quality

Besides the important influence of genetic factors, it is well known that stress during pre-slaughter handling and method of slaughter can affect meat quality. In Table 3, mean values of pH_1 (LD), FOP (LD), GÖFO and drip loss for H^- and H^+ pigs slaughtered in two different slaughterhouses (SLH-1 and SLH-2) are shown. Pigs originated from the same farm. On eight occasions, a total of 188 pigs were transported with the same truck, and approximately one half of the pigs were unloaded in SLH-2 after 30 min of transport and the other half in SLH-1 after a further 30 min of transport. Resting time was approximately 2.5h and 2h in SLH-2 and SLH-1 respectively, so that total time delay between loading on the farm and start of slaughtering was always equal in both slaughterhouses. It is unreasonable to suggest that these small differences in transport and lairage time might have affected meat quality (Warriss *et al.*, 1990). In SLH-1, immediate pre-slaughter stress was probably somewhat higher than in SLH-2, due to the use of electric goads and a restrainer (Troeger, 1991). In SLH-2, however, less-efficient stunning, a more than five-fold longer time delay between stunning and sticking and bleeding out in hanging position were expected to have a negative impact on meat quality (Woltersdorf and Troeger, 1987). The latter stress factors apparently prevailed, since meat quality of both H^- and H^+ pigs was inferior in SLH-2 compared with SLH-1. Differences between SLH-1 and SLH-2 in H^- pigs were very significant for all four criteria. Differences between both slaughterhouses were somewhat smaller in H^+ pigs, but nevertheless significant, except for pH_1 (LD). According to Warriss (1987), the interaction between genotype and handling seems particularly important. Stress-susceptible pigs may respond less to changes in pre-slaughter handling, since the trauma of slaughter itself is sufficient to initiate a rapid rate of acidification and hence the development of PSE. In our study, pre-slaughter handling up to the stunning phase was fairly similar in both slaughterhouses, but differences in the method of stunning and bleeding were apparently responsible for the seriously impaired meat quality in SLH-2.

Although differences were slightly smaller in H^+ pigs compared to H^- pigs, no clear interaction between genotype and environment was apparent. Considering the ever increasing importance of good meat quality and the even larger impact that differences in fasting, transport and lairage may have upon it, it is clear from the foregoing that efforts should be made to follow slaughter recommendations whenever possible.

Conclusions

In this study on heavy-muscled Belgian Landrace and Pietrain x Belgian Landrace slaughter pigs, halothane status had a preponderant influence on both carcass and meat-quality traits. Although no full distinction was possible between heterozygous and homozygous non-reactors, it is suggested that heterozygotes are intermediate for lean content and conformation. For meat-quality traits, heterozygotes appear to be intermediate (FOP, PQM, transmission, water-holding capacity) or closer to NN pigs (pH_1, color). As lean content increased in H^+ pigs, pH_1 (LD) and GÖFO values decreased, and FOP (LD) and drip-loss values increased, all demonstrating impaired meat quality. This inverse relationship between meat quantity and quality was weaker for pH_1 (LD) and FOP (LD), non-existing for GÖFO and similar for drip loss in H^- pigs compared to H^+ pigs. The genetic opposition between meat quantity and quality thus seems to depend on the incidence of halothane sensitivity and may differ among meat-quality traits. Besides genetic factors, it was demonstrated that the method of slaughter could have a large impact on meat quality of both H^+ and H^- pigs.

Table 3. Mean values ± SD of some meat quality traits for H^- and H^+ pigs slaughtered in two slaughterhouses

		SLH-1	SLH-2
Number of H^-/H^+ pigs[1]		45 / 45	55 / 43
pH_1 LD	H^-	6.13 ± 0.21	5.91 ± 0.22*** [2]
	H^+	5.78 ± 0.36	5.69 ± 0.22ns
FOP LD	H^-	39.5 ± 8.0	50.7 ± 14.3***
	H^+	51.2 ± 15.3	63.9 ± 19.3**
GÖFO	H^-	67.8 ± 7.2	62.2 ± 10.4**
	H^+	58.0 ± 10.9	53.3 ± 9.8 *
Drip loss	H^-	5.6 ± 2.4	8.1 ± 2.7 ***
	H^+	7.0 ± 2.3	9.1 ± 2.8 ***

[1] Number of missing values ≤ 7
[2] Significance of difference between SLH-1 and SLH-2: *= P<0.05; **= P<0.01; ***= P<0.001; ns= not significant

Acknowledgements

The farmers who kindly supplied the pigs are gratefully acknowledged. This research was supported by a grant from IWONL, Brussels.

References

Barton-Gade, P. 1984. Influence of halothane genotype on meat quality in pigs subjected to various pre-slaughter treatments. Proc. European Meeting of Meat Research Workers 30, 8-9.

Bendall, J.R. and Swatland, H.J. 1988. A review of the relationships of pH with physical aspects of pork quality. Meat Sci. 24, 85-126.

Cameron, N.D. 1990. Genetic and phenotypic parameters for carcass traits, meat- and eating-quality traits in pigs. Livest. Prod. Sci. 26, 119-35.

Casteels, M. 1988. Evaluatie van het S.K.G.II-apparat in de slachtlÿn van vier Belgische exportslachtuizen. Rijksstation voor Veevoeding, Publikatie n° 717, 46pp.

Cole, G.; Le Hénaff, G. and Sellier, P. 1988. Paramètres génétiques de quelques caractères de qualité de la viande dans les races porcines Large White, Landrace Francais et Landrace Belge. Journées Rech. Porcine en France 20, 249-54.

Eikelenboom, G. and Nanni Costa, L. 1988. Fibre optic probe measurements in landrace pigs of different halothane phenotypes. Meat Sci. 23, 9-19.

Eikelenboom, G., Minkema, D., van Eldik, P. and Sybesma, W. 1980. Performance of Dutch landrace pigs with different genotypes for the halothane-induced malignant hyperthermia syndrome. Livest. Prod. Sci. 7, 317-24.

Grau, R. and Hamm, R. 1952. Eine einfache Methode zur Bestimmung der Wasserbindung im Fleisch. Fleischwirtschaft 4, 295.

Hart, P.C. 1962. Fysisch-chemische kenmerken van gedegenereerd vlees bij varkens. Tijdschr. Diergeneesk. 87, 156-57.

Honikel, K.O. 1987. How to measure the water-holding capacity of meat? Recommendation of standardized methods. In: Tarrant, P.V., Eikelenboom, G., Monin, G. (eds), Evaluation and Control of Meat Quality in Pigs. Martinus Nijhoff Publishers, Dordrecht, The Netherlands, pp. 129-42.

Kallweit, E. 1985. Selection for stress resistance in pigs in various West-European countries. In: Ludvigsen, P.V.J.B. (ed.), Stress Susceptibility and Meat Quality in Pigs. EAAP Publication, no. 33, 60-67.

Lampo, PH. 1984. Stress susceptibility in the Belgian Landrace pig - frequency and possible implications for breeding. In: Results of Pig Research, IWONL, Brussels, pp. 119-126.

Lundström, K., Essén-Gustavsson, B., Rundgren, M., Edfors-Lilja, I. and Malmfors, G. 1989. Effect of halothane genotype on muscle metabolism at slaughter and its relationship with meat quality: a within-litter comparison. Meat Sci. 25, 251-63.

Malmfors, B., Eriksson, J. and Lundström, K. 1980. Effects of including meat quality in a selection index for pigs. Acta Agric. Scand. 30, 405-17.

Merks, J.W.M., van der Wal, P.G., Buiting, G.A.J. and Bolink, A.H. 1989. The genetic variation for pork quality within herdbook and breeding companies in the Netherlands. 40th Annual Meeting of the European Association for Animal Production, Dublin, Ireland.

Monin, G. 1989. Genetic effects on technological qualities of pig meat. 40th Annual Meeting of the European Association for Animal Production, Dublin, Ireland.

Sellier, P. 1987. Cross-breeding and meat quality in pigs. In: Tarrant, P.V., Eikelenboom, G., Monin, G. (eds), Evaluation and Control of Meat Quality in Pigs. Martinus Nijhoff Publishers, Dordrecht, The Netherlands, pp. 329-42.

Sellier, P. 1988. Aspects génétiques des qualités technologiques et organoleptiques de la viande chez le porc. Journées Rech. Porcine en France 20, 227-42.

Sellier, P. 1990. Genetic control of meat quality. 41st Annual meeting of the European Association for Animal Production, Toulouse, France.

Simpson, S.P. and Webb, A.J. 1989. Growth and carcass performance of British Landrace pigs heterozygous at the halothane locus. Anim. Prod. 49, 503-9.

Sönnichsen, M.-L., Claus, J. and Kalm, E. 1984. Parameterschätzung und Indexkonstruktion für die Populationen Deutsche Landrasse B und Pietrain in Schleswig-Holstein. 2. Schätzung von genetischen und phänotypischen Korrelationen. Züchtungskunde 56, 249-61.

Tarrant, P.V. 1989. The effects of handling, transport, slaughter and chilling on meat quality and yield in pigs - a review. Ir. J. Food Sci. Technol. 13, 79-107.

Troeger, K. 1991. Slaughtering: animal protection and meat quality. Fleischwirtschaft 71, 298-302.

Warriss, P.D. 1987. The effect of time and conditions of transport on pig meat quality. In: Tarrant, P.V., Eikelenboom, G., Monin, G. (eds), Evaluation and Control of Meat Quality in Pigs. Martinus Nijhoff Publishers, Dordrecht, The Netherlands, pp. 245-264.

Warriss, P.D., Brown, S.N., Bevis, E.A, and Kestin, S.C. 1990. The influence of pre-slaughter transport and lairage on meat quality in pigs of two genotypes. Anim. Prod. 50, 165-72.

Webb, A.J., Southwood, O.I., Simpson, S.P. and Carden, A.E. 1985. Genetics of porcine stress syndrome. In: Ludvigsen, P.V.J.B. (ed.), Stress Susceptibility and Meat Quality in Pigs. EAAP. Publication, no. 33, 9-30.

Webb, A.J., Carden, A.E., Smith, C. and Imlah, P. 1982. Porcine stress syndrome in pig breeding. Proc. 2nd World Congr. Genet. Appl. Livest. Prod., Madrid, Vol, 5, 588-608.

Woltersdorf, W. and Troeger, K. 1987. Schlachttechnik zur Verringerung des PSE-Anteils beim Schwein. Fleischwirtschaft 67, 724-30.

Discussion

Discussion after De Smet's presentation:

*Lingaas - I see you have this interesting inverse relationship between meat quantity, and meat quality. So, now you are in a sort of cost-benefit situation. What are you going to do? Are you going to get rid of the halothane gene or not?

*De Smet - We are certainly not going to get completely free of the halothane gene in the near future. Some farmers, but not all, are now selecting maternal lines completely free of the gene. They will then cross these saws with halothane-postitve boars to obtain heterozygous fattening pigs. Until now, meat quality has received very little attention in our country, everyone has always been paid for carcass quality.

*Lingaas - So, I understand then that the economic value of meat quantity is greater than the meat quality losses.

*De Smet - Yes, until now, the economic value of meat quality in our country has been very low. The main reasons why some farmers are now selecting halothane-negative maternal lines are higher fertility and less stress deaths among negative fattening pigs. It has little to do with meat quality, but both aspects are closely linked of course.

*Lundström - How many halothane-positive pigs were in the low lean group? There can't be many, I suppose, when you have the lean percent below 57.5%.

*De Smet - There were at least 25 pigs in each group, except for the halothane-negative mixed group in the high lean class. The number of halothane-positive pigs in the low lean class must have been around 30.

*Smulders - I would like to complement you on a most interesting paper, showing, once again, that the slaughter effects can be even more important than genetic effects. In your Table 3, you state "halothane-positive, and halothane-negative." I was wondering what exactly the halothane-negative is. That is the mixed population of heterozygotes, and homozygous negative animals, right?

*De Smet - Yes, right. But, the estimated frequency of homozygous-negative animals in this mixed population is low.

Paleness, Softness and Exudation in Pork - Review

H.J. Swatland

Department of Animal & Poultry Science, University of Guelph, Guelph, Ontario N1G 2W1, Canada

Introduction

In earlier presentations we followed a sequence of events starting with muscle formation in the embryo and concluding with lactate production, rigor development and loss of membrane capacitance after slaughter. Here we will consider how post-mortem metabolism affects the physical properties of pork and how pH is a dominant factor in the determination of pale, soft, exudative (PSE) pork. However, we must not forget that the overall sensory properties of pork are affected by many other factors that are beyond the scope of this presentation. A few that come to mind are connective tissue content, intramuscular adipose tissue, sarcomere length, fiber-type composition, and degree of post-mortem conditioning.

The early and classic studies on this topic were reviewed by Bendall and Swatland (1988), with a compilation of meat-quality data and a bibliography. This presentation covers recent research with the objective of presenting ideas on the cellular basis of post-mortem metabolism. Of the three aspects of the PSE condition, exudation is the best understood (although many problems remain), there has been recent progress on paleness, but softness still is relatively unknown.

Exudate

Data on pork exudation are quite variable because of the multiplicity of breeds, slaughter methods, refrigeration and analytical methods. Figure 1 shows Bendall's compilation of data collected at Bristol, similar trends, but somewhat different values, have been measured in Ontario.

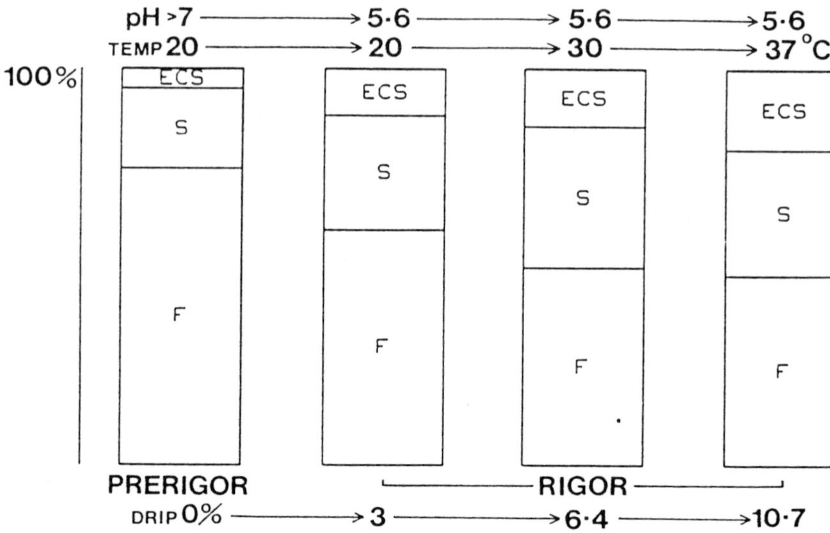

Figure 1. Relative magnitudes of the fluid compartments formed by the extracellular space (ECS), sarcoplasmic space (S), and fibrillar space (F) in pork at different pH and temperature (Bendall and Swatland, 1988).

The efflux of fluid from the fibrillar space when the pH is lowered is a reversible effect caused by the decrease in negative electrostatic repulsion between filaments. It is difficult to avoid tissue shrinkage with fixation and embedding procedures for transmission electron microscopy (TEM), so that interfilament spacing is given most accurately by low-angle X-ray diffraction (Fig. 2). The problem with both methods is that samples are generally maintained in some type of fluid medium while being processed, thus, the pH of this medium determines the interfilament spacing. The pH of the medium may be matched to that of the carcass to demonstrate what the filament spacing might be in the carcass (Fig. 3), but the logic of such experiments is one step removed from reality. Ideally, we would like to know how different histochemical fiber types behave, and how interfilament spacing is affected by the state of the cytoskeleton - particularly desmin, but these are not easy topics to investigate.

Another problem that requires further investigation is the shape of the relationship between pH and filament spacing. In pork samples at Guelph, filament spacing does not decrease until the pH drops below 6.4, then it decreases linearly and predictably down to a spacing of 40nm at pH 5.2 (Fig. 4). These results are consistent with data published for other species, but have been criticized by other meat scientists.

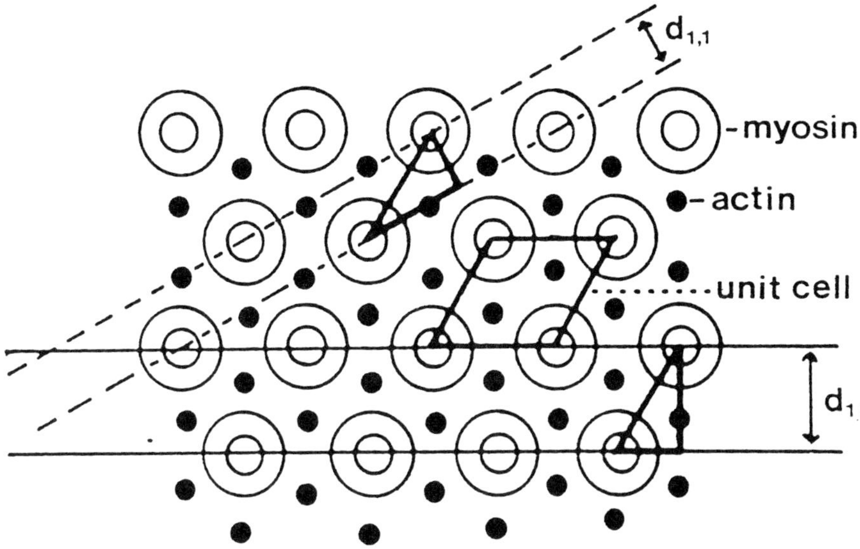

Figure 2. Low-angle X-ray diffraction for the measurement of interfilament spacing in pork (Diesbourg *et al.*, 1988).

A similar pattern was reported by Penny (1969) who showed that water-holding capacity does not change from pH 7 down to 6.2, but then decreases as the pH is lowered below this point. Most of the data on drip losses show the opposite pattern, however, with losses increasing as the pH is decreased down to pH 6.1, then not changing much at lower pH. In other words, studies on drip loss give the opposite pattern to what one might expect from filament spacing and centrifugation. Further research is clearly needed: either we have missed a major factor (such as the state of the cytoskeleton) or downstream factors (such as the trapping of fluid by membranes) are more important than previously realized.

The efflux of fluid from the fibrillar space into the sarcoplasmic space may be observed by TEM, using percentage calculations in an attempt to minimize the bias caused by fixation and embedding (Fig. 5).

The fluid volume may be measured by light microscopy further downstream in the intercellular space. However, special care is needed to extract the meat samples because the fluid between muscle fibers readily escapes as the sample is dissected and trimmed. Essentially, the more care that is taken, the higher the values for interfiber space. After freezing the samples rapidly in liquid nitrogen, it is a risk to do anything more than sectioning them in the frozen state because fixation, dehydration in alcohol, embedding or staining may introduce artifacts. Thus, the best method of observation is to use a technique such as differential interference contrast microscopy that requires nothing other than mounting dry sections in a medium. Given the opportunity for experimental error, interfiber area is reasonably well correlated with water-holding capacity (Fig. 6).

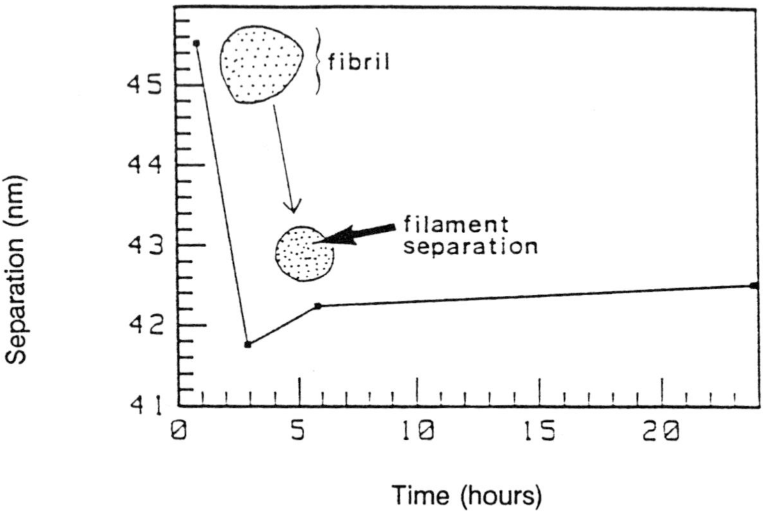

Figure 3. Decrease in interfilament seperation in porcine *longissimus dorsi* post-mortem (Diesbourg *et al.*, 1988).

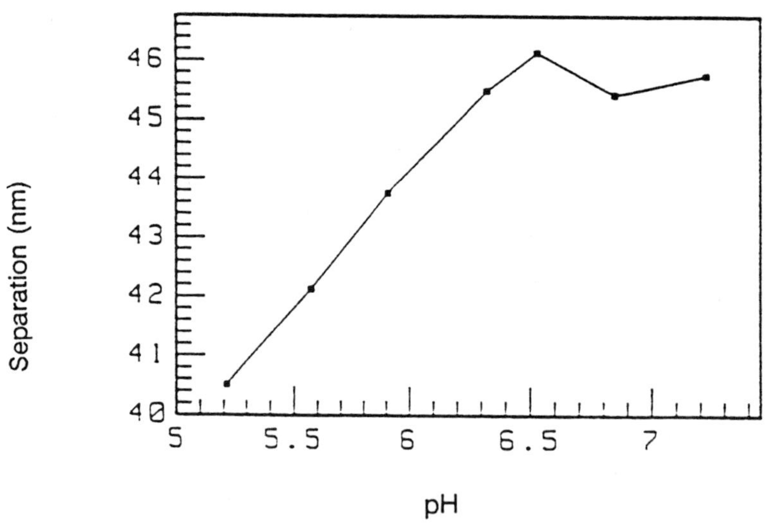

Figure 4. Effect of pH on interfilament spacing measured by X-ray diffraction in porcine *longissimus dorsi* muscle (Irving *et al.*, 1990).

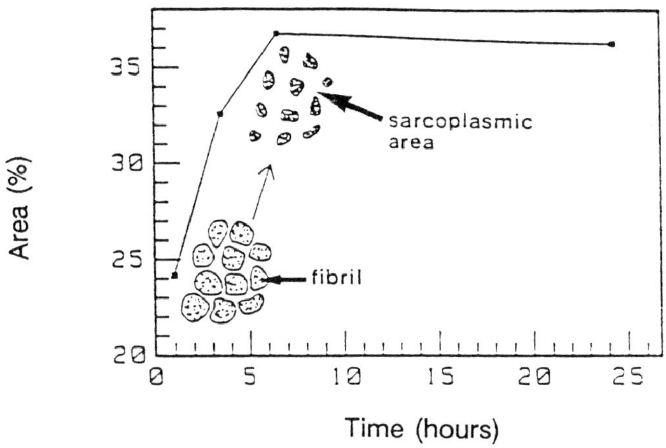

Figure 5. Increase in intermyofibrillar area in porcine *longissimus dorsi* muscle post-mortem (Swatland and Belfry, 1985).

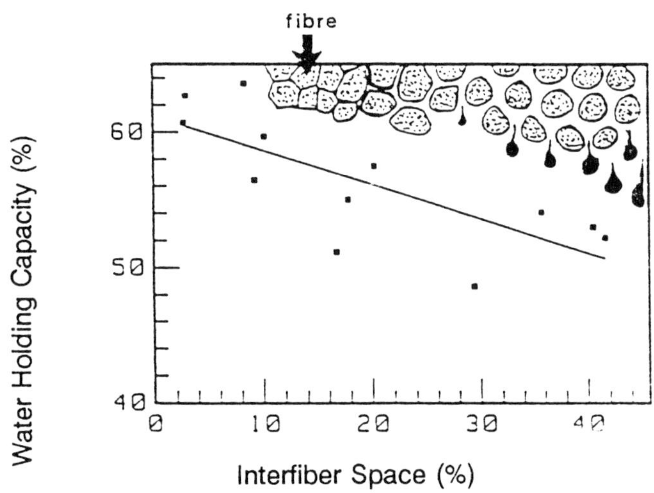

Figure 6. Relationship between water-holding capacity and interfiber spacing in porcine *longissimus dorsi* muscle (Swatland *et al.*, 1989).

Paleness

When light reaches the surface of pork, some of it is reflected from the wet surface (specular reflectance) with no change in wavelength. Of the light that enters the meat, most of it is strongly scattered so that its original direction is lost. A fraction of this scattered light makes a U-turn, after being scattered at a myriad of points, and returns from whence it came.

Some eventually escapes from the meat surface and is visible to the observer - the more light that is returned, the more pale the meat appears to be. While passing through the meat, some wavelengths are absorbed (much of the green light is absorbed by myoglobin) so that the pork may appear to be pink to the observer. If a high degree of scattering shortens the light path through the meat, the opportunity for this selective absorbance is decreased, and the meat is less pink than normal (as well as being more pale than normal). In combination, these two factors may cause the almost white appearance of severe PSE pork - when the observer must look closely to find the boundary between muscle and fat.

In severe PSE, Bendall (1973) concluded that increased light scattering originated from denaturation of sarcoplasmic proteins, and there is no reason to doubt this. However, protein denaturation cannot be responsible for the well-known relationship between paleness and pH that is observed in all types of fresh meat across a wide pH range, from pH 7 to pH 5, as from dark-cutting meat to PSE meat. If individual muscles fibers are isolated under the light microscope, their transmittance is decreased at a low pH (Fig. 7). Scattering tends to be inversely proportional to the fourth power of wavelength under certain conditions, and this probably accounts for the skewed spectrum, which is not unlike the reflectance spectrum of meat that has been exhaustively washed to remove its myoglobin. There are several possibilities to explain this effect. Offer *et al.* (1989) suggest that the increased separation of myofibrils (as in Fig. 5) increases the chances of scattering at the surfaces of myofibrils.

Alternatively, or in addition, it is well known that myofibrils are birefringent, that is, that they have two refractive indices. The A (anisotropic) and I (isotropic) bands of the sarcomere were named from this property. When light passes through myofibrils, it takes two routes (the ordinary and extraordinary ray paths) whose relative lengths may be measured by ellipsometry. The path difference of the two rays is strongly affected by pH (Fig. 8). Thus, changes of the refractive index within the fibril might also be a cause of pH-related increases in light scattering. The relative magnitudes of these two mechanisms, scattering at the myofibrillar surface *versus* scattering caused by the angles of passing through the myofibril, have yet to be measured against each other. But at least some progress is being made in explaining pork paleness. There are also other possibilities, and the field is wide open for further investigation.

The internal reflectance (interactance) spectrum obtained via optical fibers directly from the surface or interior of the meat is relatively flat, but may transformed by the cubic power of wavelength to resemble a conventional surface reflectance spectrum measured at the meat surface (Fig. 9). The curvilinear shape of the conventional surface reflectance spectrum is caused by the selective absorbance of myoglobin superimposed on a skewed, but otherwise relatively flat spectrum (as in Fig. 7) created by the meat trapping light at low wavelengths.

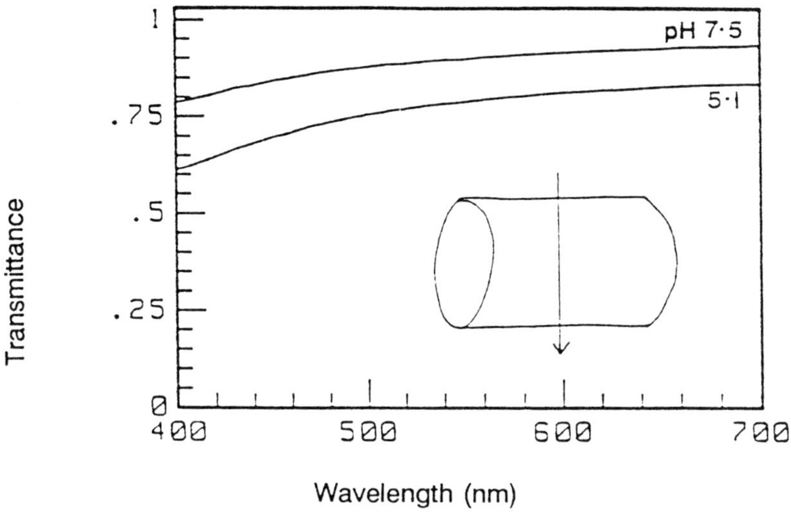

Figure 7. Effect of pH on the transmittance perpendicularly across a muscle fiber (Swatland, 1990a).

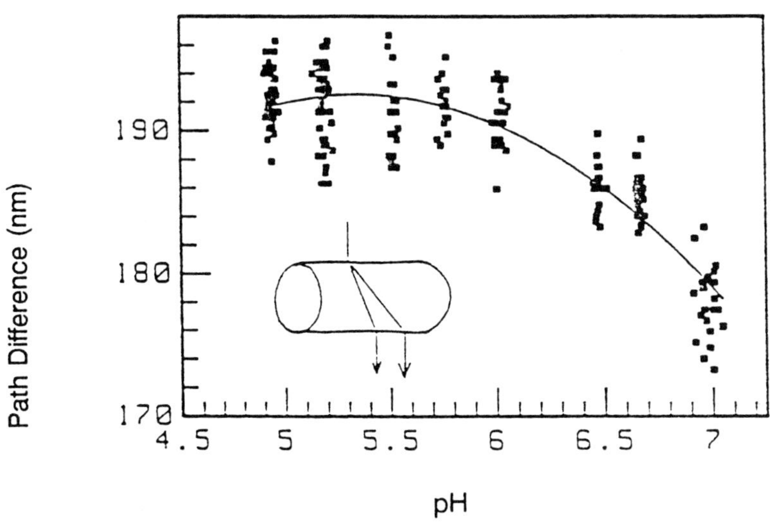

Figure 8. Effect of pH on the path difference of bifringence through a muscle fiber (Swatland, 1989a; 1990b).

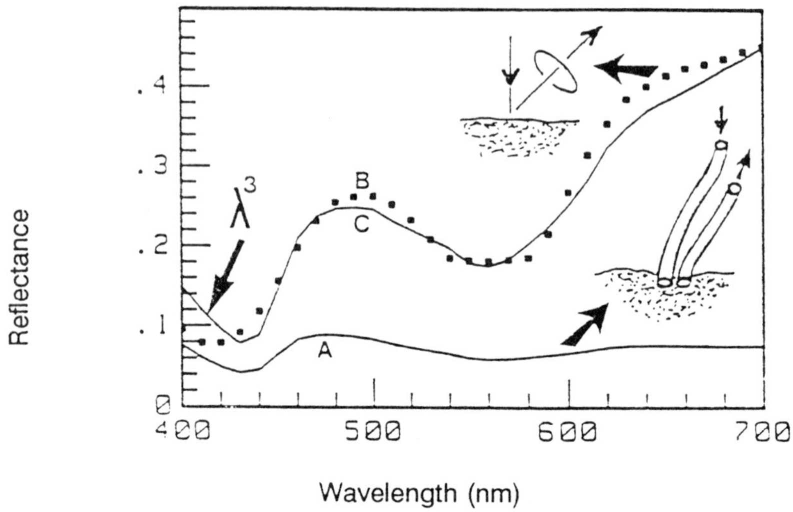

Figure 9. Reflectance spectra of pork. Spectrum A is an internal reflectance (interactance) spectrum obtained directly from the meat via optical fibers. Spectrum B (dotted) is a conventional surface reflectance spectrum obtained at 45° from the meat surface. Spectrum C is the internal reflectance spectrum transformed by the third power of wavelength (Swatland, 1989b).

Softness

Pork softness is the least understood aspect of pork quality. It seems reasonable to suppose that the collapse of the filament lattice and the resulting expansion of sarcoplasmic and intercellular fluid compartments might decrease the firmness or turgidity of meat, much as the osmotic condition of plant cells determines the turgidity or softness of vegetables and fruits. However, the finger-pressure forces involved in the subjective testing of pork are very weak and are coupled with an extremely sophisticated human response in determining the resistance of the pork, the extent to which the pork is deformed, and the extent to which it returns to its original cut surface condition. Thus, the human observer may use additional information, such as whether a finger indentation in the pork disappears or remains as a fluid-filled puddle.

Something a lot more sensitive than a Warner-Bratzler shear press is needed in attempts to measure pork softness objectively. Repeated and relatively non-destructive testing may be of more use than a single destructive test, and it may be an advantage to use some type of electromechanical system that will automatically release when a certain force is reached, although this introduces a variety of additional problems in the interpretation of the data (Fig. 10).

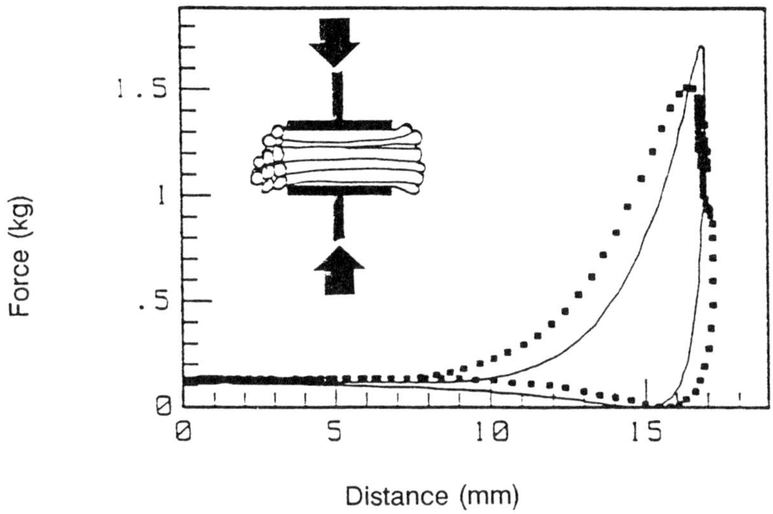

Figure 10. Non-destructive testing of softness in a sample of porcine *longissimus dorsi* muscle, with (dotted line), and without (solid line), activation of maximum force cut-off (Swatland, 1992).

Softness may involve microstructural changes, in addition to changes in fluid distribution, because laser diffraction measurements of sarcomere length may be correlated with X-ray diffraction measurements of filament spacing (Fig. 11). Post-mortem denaturation of heavy meromyosin subfragment 1 may occur in PSE pork (Stabursvik *et al.*, 1984). This, in turn, may cause the mass of subfragment 1 to be more widely distributed in the filament lattice when measured by X-ray diffraction. This loss of rigor bonding may allow sarcomeres to lengthen when samples are taken for examination, and there is evidence that PSE pork may have longer sarcomeres than normal (Eikelenboom and Nanni Costa, 1988; Irving *et al.*, 1989). However, if samples are taken very carefully with a minimum of imposed tension, the sarcomeres do not lengthen and the relationship is lost. Thus, some of the softness of PSE pork may originate from the lengthening of sarcomeres when the meat is handled.

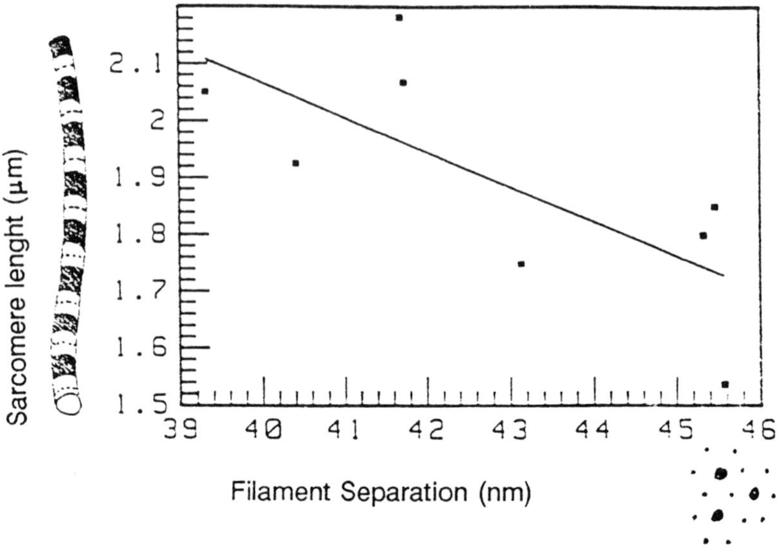

Figure 11. A relationship between sarcomere length measured by laser diffraction and filament spacing measured by X-ray length diffraction (Irving *et al.*, 1989).

PSE - interrelationships and detection

Although pH appears to be the primary factor that determines the severity of the PSE condition, there are many subtleties in the relationship. The most obvious is the interaction of temperature and pH, and it is normally thought that a low pH at a high body temperature is more damaging than the same low pH attained when the carcass has cooled. This raises the question of whether paleness, softness and exudate are always linearly related. There are many factors to be taken into account, such as the possible non-linearity of human subjective responses, and there is no time to consider them here. However, in methods for PSE detection, it is important to remember that paleness, softness and exudate may be partly independent in some circumstances.

Direct electronic measurement of pork quality, either by capacitance or conductivity, is very interesting because it is relatively simple and amenable to commercial implementation. Meat is electrically anisotropic and the direction in which it is measured must be standardized. Precise temperature correction is required for anything that involves resistance (such as impedance, conductivity, etc.), and both capacitance and resistance tend to vary with electrode design (Fig. 12).

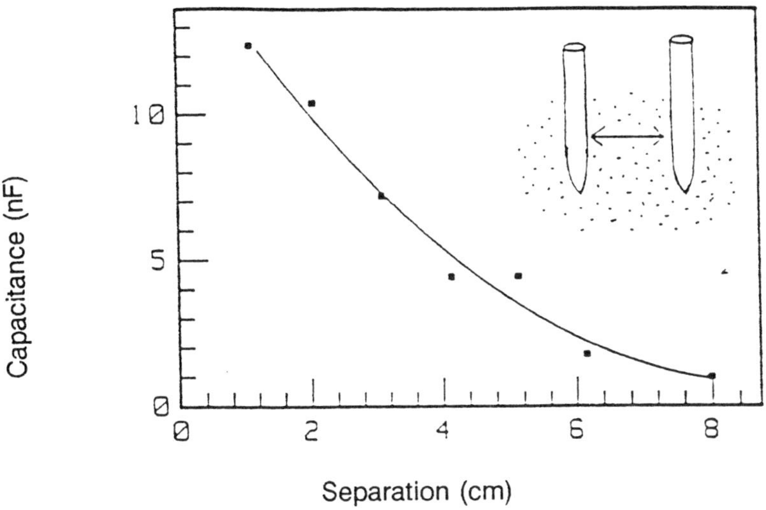

Figure 12. Effect of interelectrode distance on pork capacitance (Swatland, 1981).

Clear separation may be obtained between resistivity and capacitance decline rates in normal and PSE pigs under experimental conditions (Fig. 13), but the results of electronic measurement may be no better than pH under commercial conditions, although easier to make and to standardize.

Direct measurements of paleness made via optical fibers have much in their favor, because they are based directly on a parameter of commercial importance. The principle of MacDougall's fibre-optic probe may be pushed much further than monochromatic measurements, and diode-array spectrophotometry may enable the measurement of fat translucency and muscle myoglobin, as well as paleness. More recently, spectrophotometry has been combined with goniophotometry to develop a correlation matrix that optimizes the prediction of any predictable aspect of the pH condition (Fig. 14).

This method combines the principle of goniophotometry, which was introduced by Birth *et al.* (1978) as a laser method for the detection of PSE, with the principle of spectrophotometry. Figure 14 is for the prediction of Japanese pork color standards (JPCS; where PSE = 1, and dark pork = 6), and wavelength-dependent aspects of scattering in pork are detectable. For example, violet light at 400nm is strongly scattered in pork. When the pork is PSE, more light than normal at 400nm is passed through the meat by scattering, and there is a negative correlation of intensity with JPCS. At the other end of the visible spectrum, red light at 700nm is subjected to less scattering than violet light and tends to be transmitted more readily through the meat. When the pork is dark and scattering is low, the transmittance intensity of red light is positively correlated with JPCS. The data shown in Fig. 14 are very useful in designing optical probes because they show the best combinations of angles for the optical fibers and wavelengths for the spectrometer. Optimum conditions for the prediction of paleness may differ from those for the prediction of exudate and softness, so that it may be possible to improve the prediction of these parameters.

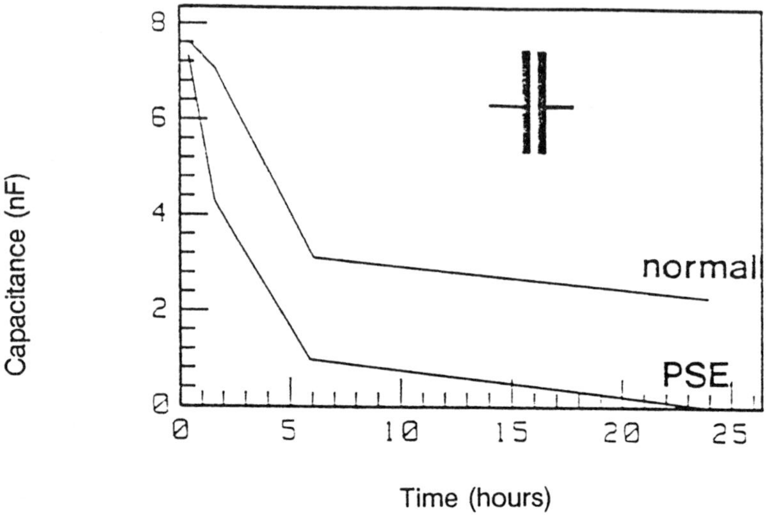

Figure 13. Post-mortem decline in capacitance in normal and PSE pork (Swatland, 1980).

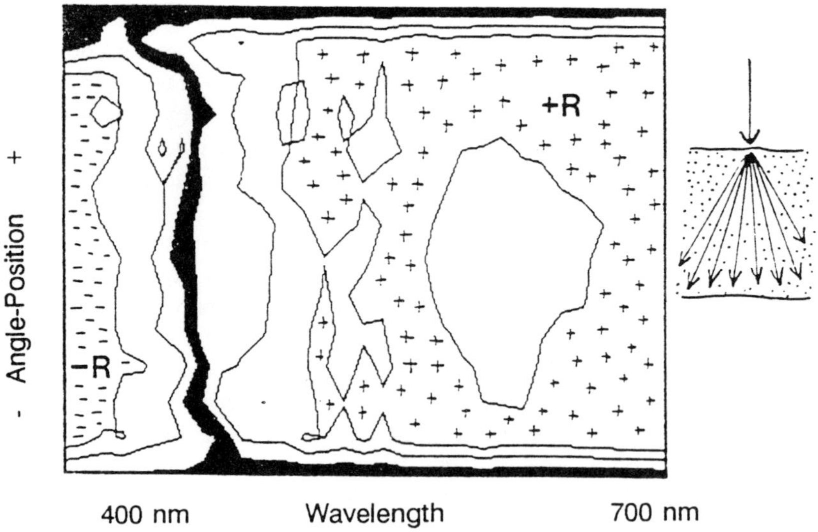

Figure 14. Countour map of a correlation matrix for the prediction of paleness from optical measurements taken at different angles, pathlengths and wavelengths, through a set of samples of porcine *longissimus dorsi* (Swatland and Irie, 1991). Countour interval of r = 0.2, and the zero-correlation zone is solid black.

References

Bendall, J.R. 1973. Postmortem changes in muscle. In: Bourne, G.H. (ed.), The Structure and Function of Muscle, vol. 2, part 2, 2nd edn. Academic Press, New York, pp. 244-309.

Bendall, J.R. and Swatland, H.J. 1988. A review of the relationships of pH with physical aspects of pork quality. Meat Sci. 24, 85-126.

Birth G.S., Davis, C.E., and Townsend W.E. 1978. The scatter coefficient as a measure of pork quality. J. Anim. Sci. 46, 639-45.

Diesbourg, L., Swatland, H.J. and Millman, B.M. 1988. X-ray diffraction measurements of post-mortem changes in the myofilament lattice of pork. J. Anim. Sci. 66, 1048-54.

Eikelenboom, G. and Nanni Costa, L. 1988. Fibre optic probe measurements in Landrace pigs of different halothane phenotypes. Meat Sci. 23, 9-19.

Irving, T.C., Swatland, H.J. and Millman, B.M. 1989. X-ray diffraction measurements of myofilament lattice spacing and optical measurements of reflectance and sarcomere length in commercial pork loins. J. Anim. Sci. 67, 152-6.

Irving, T.C., Swatland, H.J. and Millman, B.M. 1990. Effect of pH on myofilament spacing in pork measured by x-ray diffraction. Can. Inst. Food Sci. Technol. 1(23), 79-81.

Offer, G., Knight, P., Jeacocke, R., Almond, R., Cousins, T., Elsey, J., Parsons, N., Sharp, A., Starr, R. and Purslow, P. 1989. The structural basis of the waterholding, appearance and toughness of meat and meat products. Food Microstruct. 8, 151-70.

Penny, I.F. 1969. Protein denaturation and water-holding capacity in pork muscle. J. Food Technol. 4, 269-73.

Stabursvik, E., Fretheim, K. and Froystein, T. 1984. Myosin denaturation in pale, soft, and exudative (PSE) porcine muscle tissue as studied by differential scanning calorimetry. J. Sci. Food Agric. 35, 240-4.

Swatland, H.J. 1980. Post-mortem changes in electrical capacitance and resistivity of pork. J. Anim. Sci. 51, 1108-12.

Swatland, H.J. 1981. Electrical capacitance measurements on intact carcasses. J. Anim. Sci. 53, 666-9.

Swatland, H.J. 1989a. Birefringence of beef and pork muscle fibers measured by scanning and ellipsometry with a computer-assisted polarizing microscope. J. Comput. Assist. Microsc. 1, 249-62.

Swatland, H.J. 1989b. Relationships between surface and internal reflectance in pork. Can. Inst. Food Sci. Technol. 1. 22, 165-9.

Swatland, H.J. 1990a. Effect of acidity on the optical transmittance of bovine myofibrillar proteins: how meat traps light at low wavelengths. J. Anim. Sci. 68, 1284-90.

Swatland, H.J. 1990b. Effect of pH on the birefringence of skeletal muscle fibers measured with a polarizing microscope. Trans. Amer. Microsc. Soc. 109, 361-7.

Swatland, H.J. 1992. Computer-assisted apparatus for measuring physical properties of meat during compression. Comput. Electron. Agric. 7, 149-62.

Swatland, H.J. and Belfry, S. 1985. Post mortem changes in the shape and size of myofibrils from skeletal muscle of pigs. Mikroskopie 42, 26-34.

Swatland, H.J. and Irie, M. 1991. Wavelength-position matrices in the fiber-optic analysis of meat. J. Comput. Assist. Microsc. 3, 149-55.

Swatland, H.J., Irving, T.C. and Millman, B.M. 1989. Fluid distribution in pork, measured by x-ray diffraction, interference microscopy and centrifugation compared to paleness measured by fiber optics. J. Anim. Sci. 67, 1465-70.

Poster Abstracts

Allele specific PCR in the diagnosis of porcine malignant hyperthermia*

M. Chaput[1], V. Claes[1], M-F. Thielemans[2], C. Bodart[2], R. Kettmann[1] and A. Burny[1]

[1]UER Molecular Biology and Animal Physiology- Faculty of Agronomy, B-5030 Gembloux, Belgium
[2]Animal Husbandry Research Station, Agronomy Research Center, B-5030 Gembloux, Belgium

Malignant hyperthermia (MH) in pigs is an inherited skeletal muscle disorder triggered by anesthetics (i.e. halothane), and stress. This disorder has an autosomal recessive inheritance. It is responsible for major economic losses in the swine industry, resulting from sudden animal death, and pale, soft, exudative meat.

The biochemical defect in porcine MH, appears to be an abnormal regulation of the Ca^{2+}-release channel (or ryanodine receptor) of skeletal muscle sarcoplasmic reticulum, leading to an increased myoplasmic Ca^{2+} concentration. Fujii *et al.* (1991), have identified a single point mutation in the porcine gene for the ryanodine receptor. The tight correlation between this mutation and the disorder, as well as its presence in various breeds suggests that the mutation could be responsible for porcine MH.

Our sequence information and previous results have enabled us to develop a molecular test for genotyping Belgian Landrace pigs at the MH locus, using the allele specific polymerase chain reaction (PCR) methodology. The aim of our study was to determine whether the same C to T mutation was also detectable in our breeds, and to evaluate the feasibility of a large-scale molecular testing of pigs.

The presence of each haplotype was separately checked, using 2 PCR reactions that allow amplification of a 1kb fragment from ryanodine receptor gene. Both reactions include the same forward primer while the reverse primers were allele-specific. Those last primers had exactly the same sequence except for the last 3' nucleotide that was the mutation site. PCR conditions were chosen so that amplification occurred only when the last 3' nucleotide of the reverse primer was correctly paired.

170 animals were tested by halothane challenge and by the PCR technique. As in other breeds, the C to T mutation was detectable in Belgian Landrace breed, and correlated with halothane sensitivity. The discrepancies between both techniques observed in 10% of the cases resulted certainly from the inaccuracy of halothane test. Indeed, the reliability of the molecular test was confirmed, in those 10% cases, by direct sequencing of the gene region around the mutation.

* This work was supported by grant N: 5426A from IRSIA RK is Research Directorat the FNRS.

Glycogen depletion in different fiber types of red and white pig muscle in response to a paired contest: Quantification of glycogen using computerized image analysis.

X. Fernandez, M.-C. Meunier-Salaün, L. Lefaucheur and P. Ecolan

INRA, Station de Recherches Porcines, 35590 L'Hermitage, France

The purposes of this study were: i) to study the plasma cortisol and catecholamines (results not available yet) and metabolic responses to a standardized stress (paired contest); and ii) to determine the resulting muscle glycogen depletion as a function of muscle and fiber types in pigs.

Paired contests (eight pairs of tested animals, and eight control animals, catheterized 5 days before the test) were arranged between unfamiliar pigs. Agonistic behavior was recorded for 30 min. Blood samples were collected at different times for enzymatic determination of plasma metabolites (lactate, glucose and free fatty acids (FFA)). Pigs were slaughtered 4h after the beginning of the test. Samples of *longissimus* (LD) and *semispinalis* (SS) muscles were taken during bleeding. Four serial cuts of both muscles were stained in red with azorubine (reference stain), for ATPase activity at pH=4.35, for SDH activity and using PAS coloration for photometric determination of glycogen. A computerized image analysis system (Lefaucheur *et al.*, Meat Sci., in press) allowed fibre classification as αW, αR or, βR and automatic recording of the average PAS staining luminance in individual fibers.

Plasma metabolite concentrations increased during the paired contest and reached maximum values at the end of the test (lactate, glucose) or still increased after the test (FFA). Image analysis revealed that glycogen was significantly depleted to a similar extent in LD and SS muscles (difference between control and tested pigs of *ca.* 5μmol/g). However, glycogen depletion was significant in fast-twitch fibers only (αW and αR fibers). Classification of fiber types according to glycolytic response was as follow: αR (*ca.* 10μmol/g) > αW (*ca.* 5μmol/g) > βR (no difference between tested and control pigs) for m. LD and αR = αW (*ca.* 10μmol/g) > βR (no difference between tested and control pigs) for SS. Overall, the stress-induced depletion of muscle glycogen was not sufficient to significantly increase ultimate pH (24h post-mortem). A close relationship was found between glycogen content (photometric determination) and the number of aggressive acts performed by the animal during paired contest, for SS but not for LD ($r=0.67$; $P<0.01$ *vs.* $r=0.09$, NS, respectively). This suggests that in the red muscle, *semispinalis*, muscular activity is an important factor for the extent of glycogen depletion.

The relationship between pH_1 value and early post-mortem metabolic activity in pork

M. Honkavaara

Finnish Meat Research Centre, Box 56, SF-13101 Hämeenlinna, Finland

The aim of this study was to determine the relationship between pH_1 values, and the temperature, rate of glycogen breakdown, and lactate formation in pork during the first hour after slaughter. Based on the pH_1 value, pigs were placed into one of three groups. The pigs in group 1 (G1) were classified as PSE ($pH_1<5.8$), group 2 (G2) pigs were considered normal ($5.8<pH_1<6.4$), and group 3 (G3) pigs were classified as slow ($pH_1>6.4$).

G1 pigs had the highest carcass temperature after scalding and before chilling, whereas these temperatures were lowest in G3 pigs. During the first hour after slaughter, the decline of muscle glycogen in G1 pigs was more rapid (3.1μmol/kg-s) than in G3 pigs (1.1μmol/kg-s, $P<0.05$). At the same time, muscle lactate concentrations of G1 pigs increased more rapidly (9.1μmol/kg-s) than the G3 pigs (5.2μmol/kg-s). Values recorded for the rate of glycogen breakdown, and lactate production of G2 pigs were intermediate between those values recorded for G1 and G3.

Introduction

It is well known that meat quality is greatly affected by post-mortem biochemical processes, such as glycolysis. The breakdown of muscle glycogen via anaerobic glycolysis will produce lactic acid, which causes a pH drop in pig muscle. If too much lactic acid is accumulated during the first hour after slaughter, there is an increased risk of developing PSE meat. The purpose of this study was to investigate how the carcass temperature, glycogen breakdown and lactate production in pork are affected by the early pH changes of the carcass during the first hour after slaughter.

Material and methods

The test animals consisted of 103 pigs, 52 of which were of known halothane genotype. The animals were from the following breeds: 39 Finnish Landrace, 8 Finnish Yorkshire and 56 Landrace x Yorkshire crosses.

Carcass temperature and muscle pH_1 (45 min post-mortem) were measured in the posterior part of the *longissimus dorsi* muscle. Carcasses were classified, according to pH value, into three groups: G1 (PSE, $pH_1 \leq 5.8$), G2 (normal, $5.8 < pH_1 \leq 6{,}4$) and G3 (slow, $pH_1 > 6.4$). This classification was based on the earlier results of Honkavaara and Tuominen (1982), and Honkavaara (1985).

The meat sample (0.1g) used for the of glycogen and lactate analyses was taken from the posterior part of LD muscle with a modified biopsy needle. Samples were taken before scalding (at exsanguination ≈0 min post-mortem), after scalding (10 min post-mortem), and before chilling (45 min post-mortem). Each sample was immediately placed into an ampule, and immersed in liquid nitrogen for 1 to 2 weeks of storage at -196°C.

The rates of glycogen breakdown and lactate formation were calculated by the following formulae:

μmol glycogen broken down per 1kg wet tissue in 1s

$$= \frac{\text{glycogen before scalding - glycogen before chilling}}{\text{time elapsed from exsanguination to chilling}} \quad \frac{[\mu\text{mol}]}{[\text{kg-s}]}$$

μmol lactate formed per 1kg wet tissue in 1s

$$= \frac{\text{lactate before chilling -lactate before scalding}}{\text{time elapsed from exsanguination to chilling}} \quad \frac{[\mu\text{mol}]}{[\text{kg-s}]}$$

Results

The percentage of PSE occurrences among halothane reactors (6 pigs), non-reactors (7), Hal-gene carriers (16), and non-carriers, and pigs of unknown halothane genotype (50) were 100.0, 14.3, 13.0, 12.5 and 12.0% respectively.

As shown in Figure 1, carcass temperature increased most rapidly in G1 (+7.1%, from 38.3 to 41.0°C) and slowest in G3 (+1.3%, from 38.6 to

39.1°C) during scalding. The rate of temperature decrease between scalding and chilling was fastest in G3 (-2.6%, from 39.1 to 38.1°C) and slowest in G2 (-1.0%, from 39.4 to 39.0°C).

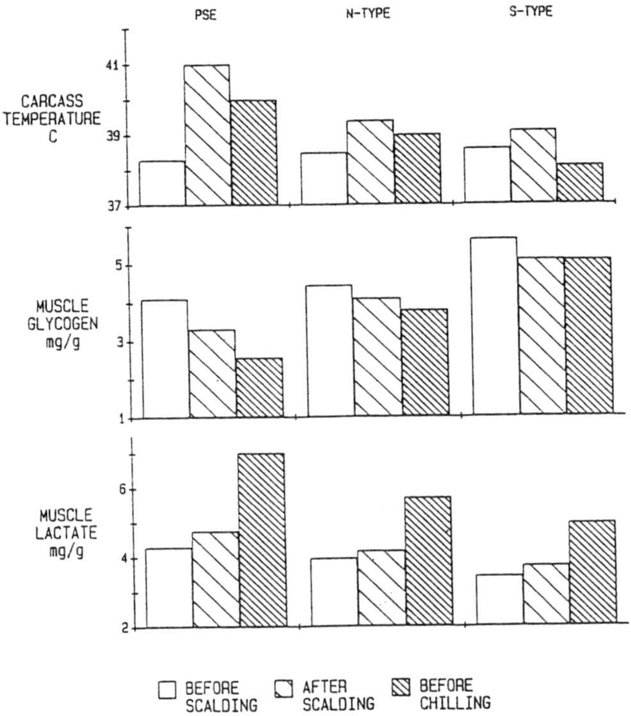

Figure 1. Early post-mortem changes in carcass temperature, muscle glycogen and lactate of three pig types.

In general, muscle glycogen content was highest in G3 and lowest in G1 ($P<0.05$). The glycogen level decreased most in G1 (-19.0%) and least in G2 (-7.9%) during scalding. After scalding, it decreased most in G1 (-23.0%) and least in G3 (0.2%) until chilling. As a result, the rate of glycogen breakdown was fastest in G1 (3.1µmol/kg-s) and slowest in G3 (1.1µmol/kg-s, $P>0.05$, Table 1).

Muscle lactate content was usually highest in G1 and lowest in G3 carcasses ($P<0.05$, Fig. 1). Lactic acid level increased most in G1 (+11.0%) and least in G2 (+5.0%) during scalding. After scalding, it increased most in G1 (+48.0%) and least in G3 (+33.0%) between scalding and chilling. Finally, the rate of lactate formation was higher in G1 (9.1µmol/kg-s), than in G2 (5.9 µmol/kg-sec, $P<0.05$), or G3 (5.2µmol/kg sec, $P<0.01$, Table 1).

Table 1. The rate of glycogen breakdown and lactate formation of three carcass groups during the first hour after slaughter

Type	µmol glycogen broken per 1kg of meat in 1s	µmol lactate formed per 1kg of meat in 1s
G1 (PSE)	3.1	9.1
G2 (Normal)	1.3	5.9
G3 (Slow)	1.1	5.2

Conclusions

Early post-mortem metabolic activity in pork is significantly affected by changes in pH shortly after slaughter. G1 (PSE type) pigs had the most prominent increase in carcass temperature during scalding. In addition they had the fastest rate of glycogen breakdown and lactate formation 45 min post-mortem. G3 (slow type) pigs had the lowest carcass temperature and the slowest rate of glycogen breakdown and lactate formation. Rate of glycogen breakdown and lactate formation in pigs with a moderate pH fall (G2) was, as expected, intermediate between G1 (PSE) and G3 (slow) pigs.

References

Honkavaara M., and R. Tuominen R. 1982. Effect of PSE Meat on Processing Properties of Cooked Sausage and Ham. Proc. 28th Eur. Meet. Meat Res. Workers, Madrid, Paper 4.20.

Honkavaara M. 1985. Influence of Halothane Positive Pigs on Technological Yeild of Cooked Ham. Proc. 31st Eur. Meet. Meat Res. Workers, Albena, paper 1.3.

Pig meat tenderness: repeatabilities and relationships

R. Hovenier[1], E. Kanis[1] and J.A.M. Verhoeven[2]

[1]Department of Animal Breeding, Wageningen Agricultural University, PO Box 338, 6700 AH Wageningen, The Netherlands

[2]Co-öperative Brabantse Vee- en Vleescentrale van de NCBBA, Boseind 10, 5281 RM Boxtel, The Netherlands

Repeatabilities of three tenderness measurements, and their phenotypic correlations were calculated. The measurements were: i) tenderness, as assessed by a taste panel; ii) Warner-Bratzler maximum shear force; and

iii) collagen value (collagen N/total N). Over the course of 8 weeks, 64 Duroc and Dutch-Yorkshire boars and gilts were slaughtered. Eleven slices were taken from the *M. longissimus dorsi* of each animal, from both carcass halves, at 24h post-mortem to measure several meat quality traits. Mean values per carcass half were corrected for week of slaughter, breed, sex, and their interactions. Variance and covariance components were calculated using REML procedures.

Repeatabilities were: 0.53 for taste-panel scores (12.4 observations per carcass half), 0.73 for shear force (4.0 obs.) and 0.43 for collagen value (1.5 obs.). Those repeatabilities depend on the number of observations per trait. In the case of only one observation, estimated repeatabilities would have been 0.08, 0.41 and 0.33 respectively.

Phenotypic correlations were: -0.50 between panel and shear force scores, 0.01 between panel and collagen value, and -0.10 between collagen value and shear force. If repeatabilities of the measurements are assume to be equal to one, correlations are -0.74, 0.02, and -0.18, respectively.

Assuming that taste panel scores give the true meat tenderness, it can be concluded that collagen value does not give significant information about pig meat tenderness. Maximum shear force can be used as an effective index trait to improve pig meat tenderness.

The content of vitamin E in porcine skeletal muscles

Marianne Jensen Waern

Department of Medicine and Surgery, College of Veterinary Medicine, Swedish University of Agricultural Sciences, 750 07 Uppsala, Sweden

The exposure of biological membranes to oxygen-derived radicals can result in progressive degeneration of membrane structures, reduction of cell activity and possibly cell death.

By acting as a free-radical scavenger, the fat-soluble vitamin E is the first line of defence against peroxidation of phospholipid membranes. Enzymes such as catalase, superoxide dismutase, xanthine oxidase and glutathione peroxidase also take part in the defence against oxidative damage and lipid peroxidation.

In both wild and domestic animals, vitamin E deficiency may result in skeletal and cardiac muscle degeneration, microangiopathy and yellow fat. Young animals are especially affected. In primates, the vitamin deficiency may cause hemolysis.

The present study was undertaken in order to obtain information about the vitamin E content in muscle bundles with differing fiber type composition. Pigs were fed on a barley based diet with graded levels of vitamin E for a 7-week period. Biopsy specimens from different skeletal muscles were collected under general anaesthesia. All pigs were then fed on a vitamin E deficient diet for another 7-week period, until slaughter.

The results show that skeletal muscle bundles with a higher proportion of oxidative type I fibers contain more vitamin E than those muscles with a lower oxidative capacity. This result can probably be explained by the higher fat content in type I fibers, which will mean a better storage capacity for lipid-soluble compounds.

The myocardium contained a higher concentration of vitamin E than the skeletal muscles which illustrates the importance of vitamin E in tissues highly exposed to oxygen.

In the unsupplemented group of pigs, two animals died of vitamin E deficiency. The histopathological examination showed lesions typical for vitamin E deficiency such as skeletal muscular degeneration and microangiopathy in the cardiac muscle. In addition, the histochemical examination of the skeletal muscle revealed an increased number of giant fibers (about 5%). This finding was interpreted as a pathological aberration of normal myofibers, due to vitamin E deficiency.

Finally, this study demonstrates that if pigs are fed on a diet supplemented with 45mg of vitamin E/kg, a pork chop (200g) will provide the human being with 10% of the daily vitamin E requirement.

Intramuscular fat content in pigs of various genetic origins, measured by NIR

E. Kallweit[1], K.-H. Hoppenbrock[2] and U. Baulain[1]

[1]Institut für Tierzucht und Tierverhalten (FAL) Mariensee, D-3057 Neustadt 1
[2]Lehr- u. Versuchsanstalt für Tier- u.Pflanzenproduktion Haus Düsse, D-4772 Bad Sassendorf

Introduction

Intramuscular fat content (IMF) is an important quality trait in pork. As the proportion of fat in the carcass has gradually been decreased by selection during the last decades, IMF has been reduced to a very low level. At present, it is discussed whether the IMF should be considered in the breeding goal. Therefore the phenotypic and genetic variation of this trait has to be evaluated as well as its relation with other economically important performance traits.

Materials and methods

Data were collected from seven different genetic groups of pigs that went in for a performance test at "Haus Düsse." They represented commercial and hybrid crosses. The number of animals per group varied between 46 and 50, the total number was 340. Slaughter and evaluation of carcass and meat quality was carried out according to the guidelines of the German progeny testing scheme. IMF was measured by the "near infrared spectroscopy" (NIR) utilizing transmission values (Infratec 1255 Food and Feed Analyser).

Results

High accuracy of the NIR-calibration is demonstrated by applying it to a validation sample. The lean content of almost 55% in average indicates a high level of muscularity. The average IMF in the *M. longissimus dorsi* (MLD), however, was only 1.1%. The results of meat-quality evaluation are in the normal range. The correlation analysis confirms some well-established relations. IMF is negatively correlated with percentage lean body mass and positively correlated with backfat thickness. The importance of IMF for meat quality is also confirmed by the positive correlation with pH_1 and the negative one with LF_{24} (conductivity). Between genetic groups, IMF varies from 0.95% to 1.43%. From breed comparisons, it is obvious that increasing IMF values, do not always mean increasing amounts of fat tissue in the carcass. Also, lean body mass and IMF are not closely related. Meat-quality traits generally are improved as IMF increases.

Conclusion

There are differences in IMF between the seven genetic groups. However, they all performed at a low level. Comparison of mean values between groups as well as correlation coefficients would allow IMF to increase without affecting other traits negatively. If the breeding goal preferred higher IMF levels than presented here, additional breeds would have to be introduced into the cross-breeding scheme.

Variation of color and pH in a cut of *M. longissimus dorsi* in pigs

R. Kangasniemi

Agricultural Research Centre, Dept. of Animal Production, Unit of Animal Breeding, SF-31600 Jokioinen, Finland

Muscle color and pH differences were studied in a cut of *M. longissimus dorsi* at the last rib. The purpose of this study was to determine how measurements should be taken to get a representative and repeatable measure from the traditional cut for progeny testing purposes. Altogether, 749 pigs from four progeny testing stations were available for these measurements. Color measurements were made from 5 sectors of a muscle surface, 60 min after the muscle was cut, with a Minolta Chromameter CR 200 (8mm measuring area) using a triplicate measuring mode and CIE L*a*b* color space. pH values were measured using a Knick Portamess 752 with an Ingold Xerolyte electrode.

The darkest L* values were obtained on the muscle sector close to the spine (L=51.0) and the lightest ones on the ventral sector (L=54.8), although values from the muscle center (L = 54.2), sector close to ribs (L = 53.7) and dorsal sector (L = 54.1) were similar to ventral values. Similarly a* values were the highest (6.4) and b* values the lowest (5.3) on the spine sector. The lowest a* values were found on the dorsal sector (5.5) and the highest b* values on a rib sector (6.0). Variation of L* values was the largest in rib sector (SD=3.82) and the smallest in ventral sector (SD=3.09).

To simplify the measurements, an overall measure was taken independently from those five measurements. The measuring head was moved over the most representative color area of the muscle between the three illuminations. Averages and standard deviations for these measurements were: L* 54.1, 3.42 (SD); a* 5.7, 1.02 and b* 5.7, 1.30, respectively.

The standard deviations for an L* difference between the middle sector and one of other four sectors varied from 1.7 to 2.2 showing large variation compared with the standard deviation of the trait itself (3.42). Similar relative magnitudes were obtained also for a* and b* values. For pH, the importance of measuring site was relatively smaller. The difference between the average of five measurements and the overall measurement gave much smaller standard deviations: L* 0.92, a* 0.30 and b* 0.39. It is concluded that a reasonable repeatability of color values can be obtained using a triplicate measuring mode of Minolta CR 200 meter, by moving the measuring head over the representative color area of a muscle surface between the three illuminations.

The use of principal components from light-reflectance profiles, together with other technological meat-quality traits, in a combined principal component analysis (PCA)

Karlsson, A., K. Lundström, R.E. Klont, A.-C. Enfält, and A. Andersson

Dept. of Food Science, Swedish University of Agricultural Sciences, Uppsala, Sweden

Principal component analysis (PCA) may be appropriate when a large number of measurements are available, to ascertain whether they can be replaced by fewer measurements without a significant loss of accuracy. PCA reduces a complete set of correlated variables to uncorrelated linear functions. The first principal component (PC) is the linear combination of all the variables that display the maximum variation among the objects, while the second PC represents the next largest variation. The extent of the variation is indicated for each PC by the size of the corresponding eigen value.

Data from probe measurements of internal reflectance in lean meat can include information about protein structure, intramuscular fat content, and pigment content, depending on the wavelength used. The Hennessey Grading Probe (GP2-Q) was used in the present study to measured internal reflectance at 24h post-mortem each 0.2mm, at wavelengths of 590 and 700nm, from 4cm in the lean part of *M. longissimus dorsi.* PCA was used to reduce 200 reflectance values to five PCs per wavelength. giving 10 variables instead of 400.

These five PCs accounted for 92% and 95% of the total variance of the profile data, at 590 and 700nm respectively, with 78% and 82% in the first PC. The information from the five PCs was used to predict the following technological meat quality traits: surface reflectance (EEL), pH_u, drip loss, filter paper wetness, shear force, and intramuscular fat content (IMF) in three experiments with pigs of differing genetic origin. The PCs from the profile data, combined with the meat-quality traits in a new PCA were very much alike, irrespective of genetic origin. PCs from both profile wavelengths behaved very similarly. Prediction of meat-quality traits by means of a Partial Least Squares Regression (PLSR) showed that EEL, and pH_u, were quite well predicted (r^2=0.5-0.8, depending on the material), mostly with information from the first profile PC. Drip losses were well predicted (r^2=0.3-0.4) and the information was shared between the first and second PC. IMF was predicted only to a minor extent (r^2=0.1-0.3), the second profile being more informative. Shear force value could not be predicted at all (no significant component).

The eradication of the halothane gene and simultaneous breeding for better carcass quality

Seppo Kuosmanen and Matti Puonti

The Finnish Animal Breeding Association, PO Box 40, 01301 Vantaa, Finland

In the 1970s, increasing numbers of stress-susceptible pigs caused more transport mortality of pigs, and increased the occurrence of PSE-meat in meat processing. In 1979, 12.4% of Landrace pigs were halothane-positive, but only 0.5% of Yorkshire pigs. Halothane-testing of artificial insemination (AI) boars was started and only hal-negatives were accepted for AI. In selection of breeding pigs, pale meat color and the a-allele in the H-blood grouping were avoided.

In 1982, it was decided to eradicate the halothane-gene from Finnish pig populations and halothane-tests were started in breeding units. In 1983, meat quality was added as part of the progeny test index and the price of breeding pigs became dependent on their halothane-status. To rid the population of the halothane-gene, progeny testing system was started in 1983. The number of progeny to be tested depended on the frequency of halothane-gene in that particular breed and the known halothane-status of mating partners.

In 1985, tests for different types of blood markers - system H of red cell blood groups, the serum postalbumin P02, and enzymes PHI, and PGD, were included in the halothane program. The genes controlling these markers are situated on the same chromosome, close to the halothane-gene. Use of these blood markers enabled control of halothane-test results, regardless of whether or not all possible combinations of haplotypes were tested in one particular test mating. It also enabled the selection of halothane-free animals from litters where one or both of the parents were carriers of halothane-gene.

Since 1988, only halothane-free animals are accepted into the elite class of the herd book. In 1992, all AI and farm boars, and over 95 % of sows in breeding farms are registered halothane-free. A retesting project was been started in 1992, to ensure that conclusions made earlier were correct. So far, all retested halothane-free boars have been correctly classified. Transport mortality of progeny test pigs decreased significantly due to the halothane program - from 0.8% in 1979, to 0.09% in 1989. The occurrence of PSE has also gone down.

Along with the halothane-program, the national breeding program has improved daily gain, feed conversion and carcass quality. The progress in lean-meat percentage was slightly less during years 1983-1985, when heavy culling was carried out, according to the halothane-status of pigs. After

1985, when the use of blood markers enabled selection of breeding pigs from parents that were carriers, the progress in selection for higher lean-meat percentage has continued as successfully as before the halothane program. The eradication of halothane gene did not affect the selection for better daily gain or feed conversion.

Fiber types in porcine *longissimus dorsi, gluteus medius* and *semimembranosus muscles**

Matassino[1], D., Barone[1], C.M.A., Girolami[2], A., Caiazzo[1], M. and Gigli [3],S.

[1]Dipartimento di Scienza della produzione animale - Università degli Studi di Napoli 'Federico 11' - 80055 Portici (Napoli), Italia
[2]Dipartimento di Scienze animali, vegetali e dell'ambiente - Università degli Studi del Molise, Italia
[3]Istituto Sperimentale per la Zootecnia di Roma, Sezione Operativa 'Produzione della carne'- Monterotondo Scalo (Roma), Italia

The study of the distribution of fiber types, together with the knowledge of the rate of pH change in muscle, can help to provide for an early diagnosis of some anomalies of pork, such as PSE.

The purpose of this study was to evaluate the distribution of three fiber types: fast glycolytic, (FG), fast oxidative glycolytic, (FOG), and slow oxidative, (SO), in three muscles of Goland and Large White x (Landrace x Large White) [LW x (L x LW)] pigs using the simultaneous technique (Solomon and Dunn, 1988, modified) for the detection of the three fibers in a single tissue section. The area and fiber type percentages were measured on muscle samples of *M. longissimus dorsi* (LD) and *M. gluteus medius* (GM) from Goland and LD and *Semimembranosus* (Sm) of LW x (L x LW) directly from the slide, using an image analyser: MOP-VIDEOPLAN (Zeiss).

The most important results are the following : i) the FG fiber area was significantly larger than FOG and SO ($P<0.05$-0.001), except in the LD of the Goland type; ii) the FOG fiber tends to be smaller than SO in LD of both genetic types. No difference was observed between muscles, except for the SO in the Goland (LD>GM, $P<0.05$) and for the FOG in the LW x (L x LW) (LD<Sm, $P<0.05$).

In both genetic types: i) the FG fibers were more numerous than FOG and SO, the last being the least numerous; and ii) the distribution of three fiber types was similar in all muscles examined.

In conclusion, the variability in meat quality might be largely explained by differences in muscle fiber types, and hence the muscle typing should be performed first and rapidly. The simultaneous staining provides useful information about the meat quality if employed in slaughter line jointly with the video imaging analysis (VIA) system and the pH value of the muscle.

* Part of this research has been granted by Ministero Agricoltura e Foreste.

Pig meat quality for dry ham production

V. Russo[1], L. Nanni Costa[1], D.P. Lo Fiego[2] and A. De Grossi[1]

[1]Istituto di Allevamenti Zootecnici, Agricultural Faculty-University of Bologna. Villa Levi, 42100 Reggio Emilia, Italy
[2]Istituto di Zootecnica, Agricultural Faculty-University of Bologna. Via S. Giacomo,
11-40126 Bologna, Italy

A fresh thigh with an excellent aptitude for salting and seasoning is required for the production of seasoned ham. Such aptitude may be defined as the capacity of ham to achieve a high technological performance, presenting low weight losses at the end of the process (seasoning loss), and high-quality standards as a finished product, i.e. ideal organoleptic characteristics. At present, experts subjectively evaluate the suitability of the fresh thigh for salting and seasoning. However, the processing industry requires objective evaluation methods in order to classify, and possibly submit the raw material to differentiated technological processing.

Four experiments, involving a total of 620 pigs, have been carried out to study the relationship between several parameters normally used in the evaluation of carcass and meat quality and the suitability of the ham for salting and seasoning.

Carcass lean meat, and lean cuts percentage appear positively correlated with the seasoning loss, showing coefficients ranging from 0.28 to 0.69.

In general the relationship between meat-quality traits (pH, color and water-holding capacity) and seasoning loss results were very poor. When the correlations were statistically significant ($P<0.05$), their coefficients were never higher than 0.36. Moreover, these measurements did not appear to be related to the organoleptic characteristics (saltiness and firmness) of the seasoned ham. Some process parameters (i.e. liquid loss, absorbed salt and weight loss) recorded during the dry salting step appeared to be more strictly related to the seasoning loss. Among these parameters, the liquid loss results are the most interesting predictor to evaluate the ham's suitability for seasoning, as it shows a high correlation coefficient (0.73) and its determination can be done just 9 days after slaughtering.

The fiber-type composition and capillary density in *M. l. dorsi* of different pig cross-breeds

Marita Ruusunen

University of Helsinki, Department of Food Technology/Meat Section, Viikki E-building, 00710 Helsinki, Finland

The purpose of this study was to compare the fiber-type composition and capillary density in *M. l. dorsi* of 38 Hampshire pigs (H), 52 Finnish Landrace (L) or Yorkshire (Y) x (Finnish Landrace x Finnish Yorkshire) (L or Y x (LxY)) and 52 Hampshire x (Finnish Landrace x Finnish Yorkshire) (H x (LxY)) pigs.

Samples were taken from *M. l. dorsi* between ribs 13, and 14. Fiber types were analysed with the myosin ATPase method after an acid pre-incubation at pH 4.6. Fiber types were identified as type I, IIA and IIB fibers (Brooke and Kaiser, 1970). The amylase-PAS staining was used to visualize capillaries (Andersen, 1975). Capillary density was calculated in capillaries/mm^2.

M. l. dorsi of the Hampshire breed (n = 38) contained 15.3% ± 4.6% type I fibers, 9.2% ± 3.6% type IIA fibers and 75.5% ± 4.0% type IIB fibers and of the L or Y x (LxY) cross-breed (n = 52) 11.5% ± 4.6%, 8.2% ± 3.3% and 80.3% ± 5.1% and of the H x (LxY) cross-breed (n = 52) 11.6% ± 4.0%, 9.3% ± 3.2% and 79.2% ± 4.3%, respectively. In *M. l. dorsi* of the Hampshire breed there were significantly ($P<0.05$) more type I fibers and significantly ($P<0.05$) fewer type IIB fibers than in *M. l. dorsi* of either the L or Y x (LxY) or the H x (LxY) cross-breeds.

In *M. l. dorsi* of the Hampshire breed, there were significantly ($P<0.05$) more capillaries (256 ± 55 capillaries/mm^2) than in *M. l. dorsi* of the L or Y x (LxY) (187 ± 49 capillaries/mm^2) or the H x (LxY) (223 ± 52 capillaries/mm^2) cross-breeds. Also the difference in capillary density in *M. l. dorsi* between the H x (LxY) and the L or Y x (LxY) cross-breed were significant ($P<0.05$).

Conclusions

1. The properties studied were highly variable between the individual pigs of each treatment group.
2. The fiber type composition in *M. l. dorsi* of the H x (LxY) cross-breed more closely resembled the fiber type composition of the L or Y x (LxY) cross-breed than the Hampshire breed.
3. In the *M. l. dorsi* of H x (LxY) cross-breed the capillary density was greater than the density found in the *M. l. dorsi* of the L or Y x (LxY) cross-breed.

Proteolysis in pig muscles of different ultimate pH

Valérie M.H.[1] , Frans J.M. Smulders[1] and Gijs Eikelenboom[2]

[1]Department of the science of Food of Animal Origin, Faculty of Veterinary Medicine, University of Utrecht, The Netherlands
[2]Research Institute for Animal Production IVO-DLO 'Schoonoord', Zeist, The Netherlands

At 13h post-mortem, 20 pig carcasses (10 castrated males, 10 females) were selected on the basis of pH of the *M longissimus lumborum*, values ranging from 5.4 to 6.3. Meat quality of this muscle was assessed after 1, 2 and 7 days of vacuum storage at 4°C. Warner Bratzler shear forces, sarcomere lengths, and muscle proteolysis parameters (30kD and troponin T (TN.T) bands measured by densitometry) were assessed.

After 1 week of storage, shear force had decreased significantly ($P<0.05$) by 0.8kg/cm^2, 46% of which was achieved two days after slaughter. Ultimate pH and shear force were significantly, but not linearly correlated, supporting the observations of Bendall and Swatland (1988). Although the significance of r increased with time of storage, (Day 2: $r=-0.45$, Day 7: $r=-0.71$), r values between pH_{ult} and the 30kD/actin and TN.T/actin ratios of the pooled population were negligible ($P>0.05$).

Females were consistently tougher than males ($P<0.05$). This finding did not coincide with a significant difference in ultimate pH between the sexes, as reported by Shorthose *et al.* (1984), but rather with a distinctly lower 30kD/actin and higher TN.T/actin ratios as compared to the males. Disappearance of the TN.T on the SDS PAGE gel reflected the decrease in shear force, as has been reported for bovine muscle (e.g. Ouali, 1990). This was not always accompanied by appearance of a 30kD band, particularly in females.

References

Bendall, J.R. and Swatland H.J. 1988. A review of the relationship of pH with physical aspects of pork quality. Meat Sci. 24, 85-126.

Ouali, A. 1990. Meat tenderization: Possible causes and mechanisms. A review. J. Muscle Foods. 1, 129-65.

Shorthose W.R. *et al.* 1984 Some factors affecting the thoughness of pork. Proc. Eur. Meet. Meat Res. Work, Bristol, 186-7.

Recommendations and Conclusions

This workshop was organized into three main sections: porcine genetics, physiology, and pork quality. In addition to the plenary sessions, a team workshop was held for each main topic, as well as a final general meeting. The participants of this seminar have agreed on the following recommendations and conclusions, which were formulated by the chairpersons of each workshop.

Genetics

Dr. Asko Mäki-Tanila, chairman
Ms. Anne Kettunen, secretary

1. Animal breeders would like to see more precisely defined components of meat quality, and the economic values associated with these components.

2. In some countries, the PSE/DFD meat problem has been reduced by effective selection against the unfavourable recessive allele at the HAL locus. The recent results suggest strongly that malignant hyperthermia, and susceptibility to halothane exposure, are associated with the nucleotide 1843 change from C to T and the respective amino acid 615 change from Arg to Cys of the ryanodine receptor (or Ca^{2+}-release channel) gene. Future research should clarify the extent to which heterozygotes are expressing poor meat quality, and how the well-established association of leanness with the HAL_n allele is physiologically explained. There is an apparent need for the quantification of genetic variation in meat quality traits in halothane-free pigs, as many populations have negligible frequencies for the allele.

3. The importance of intramuscular fat (IMF) in determining the meat quality needs a thorough (re)assessment. Taste panels and objective measurements of IMF should be used to determine the optimum IMF content. There should also be studies to quantify the correlation between IMF and meat tenderness, taste, and juiciness. Apparently the fat composition is mostly affected by diet.

4. Animal breeders are very much aware of the need for improvement of meat quality, but aside from the lack of common agreement on selection objectives, there should be a continuous development of inexpensive routine measurement technology for quality traits. This is true for both, PSE/DFD meat and IMF, and also for heme-protein or pigment content, which was recently found to have an important role in meat quality. Principal component analysis or neural networks are now used to determine the importance and causal relationships between different measurements. For comparative studies, standardization of quality meters is required, and correction factors (e.g. for day of slaughter), need careful consideration. If extensive use of on-line measuring and electronic animal identification is possible, animal breeders would be able to screen very large numbers of animals and select in a very efficient way.

5. Molecular genetics provides animal breeders with entirely new tools for improvement. The European PiGMaP program has made a successful start in this area. The project aims at the development of a map with genetic markers at 20cM spacing. RFLP, VNTRs and possibly RAPDs are used as markers. There is a principal difference in the approach and outcome of the research done between, or within breeds. Most of the inter-breed work would probably not satisfy the needs of intra-breed selection programs, but the initial map can provide a useful framework for refining the map within breeds. DNA probes can now be inexpensively be used, e.g. PCR, for detecting genetic defects or major genes. In assessing the association between DNA markers and quantitative trait loci (QTL), all the economically important traits should be simultaneously considered. A multi-disciplinary approach is needed to produce a satisfactory outcome.

6. There is a need for more theoretical work on the optimum design and size of the studies aimed at the detection of major genes for production traits. The established major genes have been discovered merely by serendipity. The efficient use of molecular diagnostics in breeding programs and breed crosses needs more research. The genome mapping project produces large masses of data. Effective handling of this material can be accomplished with database routines.

Physiology

Dr. Birgitta Essén-Gustavsson, chairperson
M. Sci. Tiina Suuronen, secretary

1. There is a need for interface between fundamental and applied scientific aspects of animal and meat science.

2. A primary requirement is to examine the inter-relationships between genetic, environmental, physiological and biochemical mechanisms that affect animal performance, and carcass and meat quality.

These have implications for
- Husbandary and management of animals, on farms, and during the slaughter process
- Regulation of muscle growth and development
- Muscle metabolism, composition and deposition of fat
- Normal function of muscle
- The progression of muscle to meat
- The function *in vitro* and *in vivo* of the PSS gene
- Behavioural physiology

3. Further research is needed:
- To further define the relationship between the PSS/RyR gene effects observed *in vivo*
- To establish if the mutation in the RyR gene is the lesion responsible for PSS, or are other candidates still possible
- To define the mechanism whereby the PSS mutation enables physiological stressors to trigger the PSS response *in vivo*.

4. A standard methodology is required for histochemical fiber types in meat animals to improve communication and research applications. A special workshop should be proposed to consider the following methods and techniques:
- ATPase fiber types
- SDH, NADH and other aerobic indicators
- Capillary morphometry
- Microscope spectrophotometry
- Computer mapping of enzymes
- Statistical cluster analysis
- Correlations with biochemistry of muscle
- Correlations with commercial traits
- PAS and glycogenolysis
- Lipid stains and intramuscular fat
- Development of new methods for specific application in meat science.

5. It was concluded that OECD should continue to support research on meat quality for the following reasons:
- Meat quality is important in commercial competition, particularly when productivity reaches a plateau
- Meat quality is strongly related to animal welfare
- Meat quality is important for the survival of animal industries and the socio-economic structure of the rural environment

- Meat quality provides an important way for the commercial importance of basic scientific research to be demonstrated.

Quality

Dr. Eero Puolanne, chairman
Lic. Sci. Marita Ruusunen, secretary

1. Fresh pork quality *versus* eating quality
Pork quality is a product of pig breeding, production management, pre-slaughter handling, slaughter hygiene, preparation methods and technology, and product distribution. The quality traits of fresh and cooked pork are not identical. For economic reasons, fresh pork quality is of more concern at the industrial and retail level than processed pork. The quality of cooked pork, however, must also be evaluated terms of its organoleptic properties. More research is needed to establish the causal relationships between the quality traits of raw pork and cooked pork, i.e. eating quality.

2. Integrated quality systems
Efforts must be directed towards quality optimization by organizing integrated quality systems. Such systems should include breeding, feeding, handling and slaughter practices, processing methods, and, quality control systems. When organizing these integrated quality systems, the overall economy, including environmental costs, must be taken into the consideration. Special care must also be taken to ensure that animal welfare will not be compromised.

3. Need for standardized methods
The lack of standard methods of quality assessment makes it difficult to utilize various studies effectively for comparative, quantitative research. It was decided that a survey on meat quality methods should be established, as follows:
3.1. Four to six senior experts on various fields of meat-quality research will be nominated to define the project objectives, and to plan future work. These senior experts met for the first time during the 38th International Congress of Meat Science and Technology in Clermont Ferrand, France, in August, 1992, to discuss the choice of the methods to be reviewed, and how the reviews and development should be performed. The senior experts will meet in 1993, somewhere in Central Europe, for a 2-day meeting to discuss the review of the methods, and also to plan a workshop on that theme. An application will be made to the OECD to provide funding for the senior expert meeting.

3.2. On the basis of the work of the senior experts, a workshop will be organized. The invited speakers should each review the methods of the field given to them as a topic.

The main purpose is not to recommend a single method for each particular measurement, but merely to discuss the limitations of the most commonly used methods, although a reference method would be desirable. Other aspects of importance would be the applicability of the methods e.g., in different muscles *versus* meat in general, methods for meat products, routine methods *versus* research methods, organoleptic *versus* instrumental methods, etc. The workshop should not be limited only to pork, it should include discussions about the methods applicable to all species.

4. Lean percentage of pork

There was a recommendation to continue the efforts to increase the leanness of the pigs in general. However, pork quality, defined e.g. in terms of color, taste, tenderness, and juiciness, must simultaneously be taken into account. It was also mentioned that in some breeds, or in some pig populations, a level has been reached on which further increases in leanness may adversely affect the organoleptic quality of pork. The physiological capacity and welfare of the pigs should be studied when the eventual further increase of leanness is sought.

5. Intramuscular fat (IMF) level

It was very clear that the participants did not agree about the optimum level of IMF. There are social and cultural differences of opinion, as well as muscle to muscle variation, etc. The relationship between IMF and eating quality needs further study. IMF must be related to the overall amount of fat in a carcass. The correlations of IMF and metabolic factors in muscle should also be studied with more scrutiny.

6. The nutritional value of pork

Consumers seem to have incorrect, or incomplete, information about the nutritive value of pork, and especially pork fat. This may influence the consumer expectations of pork and may lead to the production of excessively lean pigs. This may further lead to decreased overall eating quality and, eventually, to decreased animal welfare.

7. Environmental aspects

The scientific community should pay more attention on the environmental effects of pork (meat) production, from feed production and farming, to the preparation and distribution of the final pork products.

General conclusion

The organization of multi-disciplinary meetings, such as this one, was highly recommended, so that applied and fundamental scientists can have a fruitful exchange of ideas. The topics for these meetings should be fairly specific, so that all aspects of a particular topic can be covered.

List of Participants

Aakula Kari, Mr.
Finnish Animal Breeding Co-operative
PL 40, 01310 Vantaa, Finland

Barone Carmela, Dr.
University of Naples
Dipartimento di Scienza della Produzione Animale
Via Universita, 100, 80055 Portici (NA), Italy

Chaput Mariella, Ms.
UER Molecular Biology and Animal Physiology
Faculty of Agronomy
B-5030 Gembloux, Belgium

Demeyer Daniel, Prof.
Department of Nutrition and Meat Science
University of Gent
Proefhoevestraat 10, B-9090 Melle, Belgium

Ellis Steven, Mr.
University of Connecticut
Storrs, USA

Enfält Ann-Charlotte, Mrs.
Department of Food Science
Swedish University of Agriculture
Box 7051, S- 750 076 Uppsala, Sweden

Essen-Gustavsson Birgitta, Dr.
Swedish University
of Agricultural Sciences
Faculty of Veterinary Medicine
75007 Uppsala, Sweden

Fernandez Xavier, Ingenieur de Recherches
INRA - Station de Recherches Porcines
Domaine de la Prise, Saint Gilles
35590 L'Hermitage, France

Frøysten Terje, Spes. Vetr.
Norway Meat
PO Box 60 Refstad
0513 Oslo, Norway

Haley Chris, Dr.
AFRC/Institut of Animal Physiology and Genetics Research
Roslin, Midlothan, EH25 9PS, United Kingdom

Harri Mikko, Prof.
University of Kuopio
PL 1627, 70211 KUOPIO, Finland

Hassinen Sinikka, Mrs.
Finnish Meat Research Centre c/o Itikka,
L 117, 60101 Seinäjoki, Finland

Henckel Poul, Mr.
National Institute of Animal Science
Department of Research in Pigs and Horses
P.O.Box 39, DK-8830 Tjele, Denmark

Honikel Karl-Otto, Prof.
Federal Centre for Meat Research
E.-C. Baumann-Strasse 20, D-8650 Kulmbach, Germany

Honkavaara Markku, Dr.
Finnish Meat Research Centre
Box 56, SF-13101 Hämeenlinna, Finland

Jensen Waern Marianne, Dr. Vet. Med.
Kirurgi Medicin
Box 7018, SLU, 75007 Uppsala, Sverige

Jørgensen Preben F., Dr.
Den KGL. Veteriner- og Landbohojskole
Bülowsvej 13
DK-1870 Frederiksberg C, Denmark

Kallweit Erhard, Prof.
Institut für Tierzucht und Tierverhaltung (FAL),
Mariensee, 3057 Neustadt 1, Germany

Kangasniemi Reima, Mr.
Department of Animal Breeding
Agricultural Researcd Centre
31600 Jokioinen, Finland

Karlsson Anders, Mr.
Dept. of Food Science
Swedish University of Agricultural Sciences
Box 7051, S- 750 07 Uppsala, Sweden

Kettunen Anne, Ms.
University of Helsinki
Department of Animal Husbandry
Viikki 21, 00710 Helsinki, Finland

Klont Ronald, Mr.
Research Institute for Animal Production
Driebergseweg 10 D
PO Box 501, NL-3700 Am Zeist, The Netherlands

Kuosmanen Seppo, Mr.
Finnish Animal
Breeding Co-operative
PL 40, 01301 Vantaa, Finland

Lingaas Frode, Mr.
Norwegian Pig Breedes' Association
Dept. of Animal Genetics
Norwegian College of Veterinary Medicine
PO Box 8146 Dep. N-0033 Oslo, Norway

Lister David, Dr.
CAB International
Wallingford, Oxon OX10 8DE, United Kingdom

Lundström Kerstin, Dr.
Department of Animal Breeding and Genetics
Swedish University of Agricultural Sciences
Box 7023, S-750 07 Uppsala, Sweden

Mickelson James R., Dr.
University of Minnesota
Department of Veterinary PathoBiology
295, Animal Sci./Vet. Med.
St. Paul, Mn 55108, U.S.A.

Møller Anders J., Dr.
Department of Dairy and Food Science, Sect. Meat Technol.
The Royal Veterinary and Agriculture University
Howitzvej 11, DK-2000 Fredriksberg, Denmark

Mäki-Tanila Asko, Prof.
Department of Animal Breeding
Agricultural Researcd Centre
31600 Jokioinen, Finland

Ojala Matti, Prof.
University of Helsinki
Department of Animal Husbandry
Viikki 21, 00710 Helsinki, Finland

Pajari Mikko, Mr.
Finnish Meat Research Centre
Box 56, SF-13101 Hämeenlinna, Finland

Puolanne Eero, Prof.
University of Helsinki
Dept. Food Technology
Meat Section
E-building, Viikki
SF-00710 Helsinki, Finland

Raichon Camille, Mr.
OECD
2, Rue Andr-Pawscal
75775 Paris Cedex 16, France

Rock Edmond, Dr.
INRA/Station de Recherches sur la Viande
Theix
63122 Ceyrat, France

Russo Vincenzo, Prof.
Istituto di Allevamenti Zootecnici
dell'Universita di Bologna
Villa Levi - Coviolo
42100 Reggio Emilia, Italy

Ruusunen Marita, Mrs.
University of Helsinki
Dept. of Food Technology
Meat Section
E-building, Viikki
SF-00710 Helsinki, Finland

von Seth Gertrud, Mrs.
Swedish Meat Research Institute,
Box 504, S-244 24 Kävlinge, Sweden

Sevon-Aimonen Marja-Liisa, Mrs.
University of Helsinki
Department of Animal Husbandry
Viikki 21, 00710 Helsinki, Finland

De Smet Stefaan, Dr.
Department of Nutrition and Meat Science
Proefhoevestraat 10
B-9090 Melle, Belgium

Smulders Frans J.M., Dr.
Dept. of the Science of Food of Animal Origin,
University of Utrecht, PO Box 80.174,
3508 TD Utrecht, The Netherlands

Storlien Hans, Mr.
Norwegian Pig Breeders' Association
Norsvinsenteret
2300 Hamar, Norway

Suuronen Tiina, Ms.
University of Helsinki
Dept. Food Technology
Meat Section
E-building, Viikki
SF-00710 Helsinki, Finland

Swatland Howard J., Prof.
Department of Animal and Poultry Science
University of Guelph
Guelph, Ontario N1G 2W1, Canada

Tarrant Vivion, Dr.
Director, The National Food Centre
Dunsinea, Castleknock, Dublin 15, Ireland

Tornberg Eva, Prof.
Swedish Meat Research Institute
PO Box 504, S-244 24 Kävlinge, Sweden

van der Wal Piet G., Dr.
Research Institute for Animal Production
"Schoonord" (IVO-DLO)
Driebergseweg 10 D, PO Box 501, NL-3700 AM Zeist
The Netherlands

Warkup Chris, Mr.
Meat Quality Unit
Meat and Livestock Commission
PO Box 44, Winterhill House,
Snowdon Drive, Milton Keynes MK6 1AX, United Kingdom

Vögeli Peter, Dr.
ETH-Institut für Nutztierwissenschaften
Tannenstrasse 1
ETH-Zentrum
8092 Zürich, Switzerland

Index